Construction Safety Management Systems

Construction Safety Management Systems

Edited by Steve Rowlinson

Spon Press
Taylor & Francis Group

LONDON AND NEW YORK

First published 2004 by Spon Press
11 New Fetter Lane, London EC4P 4EE

Simultaneously published in the USA and Canada
by Taylor & Francis Inc.,
29 West 35th Street, New York, NY 10001

Spon Press is an imprint of the Taylor & Francis Group

© 2004 Spon Press

Typeset by the author
Printed and bound in Great Britain by Antony Rowe Ltd, Chippenham, Wiltshire

British Library Cataloguing in Publication Data
A catalogue record for this book is available from the British Library

Library of Congress Cataloging in Publication Data
Construction safety management systems / edited by Steve Rowlinson. -
1st ed.
    p.   cm.
ISBN 0-415-30063-0 (alk. paper)
1. Building - Safety measures. 2. Industrial safety - Auditing.
I. Rowlinson, Stephen M.
TH443.C684 2003
690'.22 - dc21   2003012226

ISBN 0-415-30063-0

Contents

Preface

In May 2002 the Department of Real Estate and Construction at The University of Hong Kong hosted an international conference of CIB Working Commission 099 - Implementation of Safety and Health on Construction Sites. At this conference leading researchers from around the world presented their current research and exchanged ideas and opinions on best practice in safety and health management. The most important of the papers presented at the conference have been expanded and updated in order to create this book which presents best practice in health and safety management on construction sites.

The conference took place in Hong Kong, a city which has always been at the forefront of construction technology and management: the second Hong Kong and Shanghai Bank Headquarters utilized state of the art construction management methods in 1933. Hong Kong is currently in a period of transition from British colonial rule and is the second system in the 'One country, two systems' paradigm that has promised '50 years no change'. However life goes on, technologies change and, in this dynamic environment, roads, tunnels, houses, offices, in fact all of our infrastructure, has to be built at a rapid pace. The people of Hong Kong have adapted rapidly to this change and attitudes and expectations have also changed. Ten years ago, Hong Kong had a very poor construction safety record. Hence, changes in how we manage health and safety have also taken place and there has been a dramatic improvement in health and safety performance on construction sites in recent years.

The research and practice presented in this book is of particular importance as it draws together experiences from around the world, offering alternative perspectives on very similar problems, and the authors are to be congratulated on their combination of academic rigour and practical application, so essential in a field of study such as this. A number of important issues are raised, particularly the appropriateness of methodologies and ideologies when applied to safety management systems around the world. This is one of the themes currently developing in research worldwide. Another theme of great importance is the development of appropriate theory to underpin meaningful research in the safety field and it is hoped that this book will make a significant contribution in this respect.

It is our sincere hope that we have reached a turning point in the study of safety and health management systems and that a realistic agenda with a solid academic underpinning is being developed which will guide both practitioners and the working commission in the coming years.

In conclusion, it is important to bring to the readers' attention that a great deal of time and effort is expended in producing a book such as this and I would like to express my thanks to all of the authors who have contributed to the work and to the research, technical and administrative staff at The University of Hong Kong who have ensured that this book has come to press. In particular Kannex Chu, Joe Lee, John Leung, Donald Ng and C.M. Tang require a special mention.

Professor Steve Rowlinson
Department of Real Estate
and Construction
The University of Hong Kong

Hong Kong
2003

CHAPTER 1

Overview of Construction Site Safety Issues

Steve Rowlinson

INTRODUCTION

The aim of this chapter is to introduce the construction industry, its safety and health issues and to discuss the nature of the industry by way of background and to highlight how it differs from other industries as the construction industry is rather different from the majority of industries and has its own organisational and economic characteristics. This chapter will focus on the particular nature of management and organisation in the construction industry and introduce the nature of the work undertaken and the persons employed in industry. It draws on the work presented by the various authors and the research of the editor.

Beyond this basic introduction, the chapter lays out the structure of this book and identifies the book's six main themes, namely: international comparisons as a scene setter; safety management issues and the implementation of these systems; health issues and their importance (and relative neglect to date); the role of training and education in underpinning safety management systems; safety technology and its role *vis a vis* safety management systems; and accident analysis as a means of improving performance and providing feedback to the safety management system. Some basic statistics are presented and the key issues of prescriptive *versus* performance based safety and health legislation and the unique nature of the construction industry are discussed.

Despite the attention given to construction sites injuries in many countries, the statistics continue to be alarming. For instance, fatal accidental injury rates in the United Kingdom and Japan are reported to be four times higher in the construction industry.

When compared to the manufacturing industry (Bentil, 1990). Construction is often classified as a high-risk industry because it has historically been plagued with much higher and unacceptable injury rates when compared to other industries. In the United States, the incidence rate of accidents in the construction industry is reported to be twice that of the industrial average. According to the National Safety Council, there are an estimated 2,200 deaths and 220,000 disabling injuries each year (NSC, 1987). The extent of the problem in selected countries worldwide is indicated in Figures 1.1 and 1.2 below.

Hinze discusses what it is that makes construction work dangerous and he disagrees with the notion that construction work is inherently dangerous and

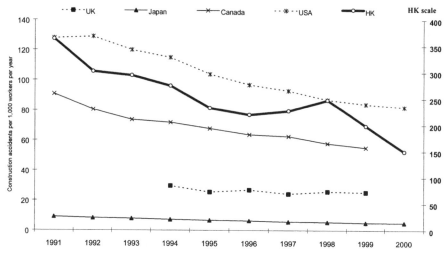

injuries are more likely to occur in this industry as he believes such a fatalistic view is inappropriate. Rather, Hinze indicates that if a proactive approach is taken

Figure 1.1   Accidents per 1,000 Construction Workers per year

(Sources: Japan Statistical Yearbook, Ministry of Public Management, Home Affairs, Posts and Telecommunications; Japan International Center for Occupational Safety and Health (JICOSH); Fatal Injuries to Civilian Workers in the United States, 1980-1995; United States Bureau of Labor Statistics Data, U.S. Department of Labor; Health and Safety Statistics, HSE publication; ILO Yearbook of Labour Statistics 2001, International Labour Office, Geneva; Occupational Safety & Health Branch of the Labour Department, Hong Kong Special Administrative Region.)

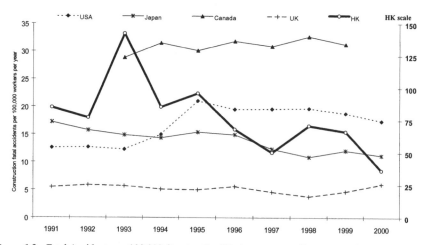

Figure 1.2   Fatal Accidents per 100,000 Construction Workers per year (Sources: as above)

to construction safety then it is possible to improve the safety record of the industry. He goes on to argue that although it has been possible to reduce accident rates in the United States by a considerable margin he believes there is still much room for improvement (Hinze, 1997:5-7).

It is obvious from the figures shown here that construction is an industry with a poor safety record worldwide. However, it is possible to improve this safety record by taking due credence of and paying appropriate attention to a series of operations and management systems that can be implemented. These systems are applicable worldwide but must be applied within the cultural context of each country, and indeed construction site, if they are to be fully effective. Later in this book it will be argued that many safety initiatives cannot be wholly successful without a high degree of top-level management commitment and without an appropriate safety infrastructure being in place.

The issue of construction site safety has engaged both practitioners and researchers for a long time. Hinze (1981) investigated the relationship between the safety performance of individual workers and individual worker attitudes. Hinze and Raboud (1988) explored several factors that apparently influence safety performance on Canadian high-rise building projects. In some studies, the usefulness of behavioral techniques to improve safety performance in the difficult construction setting was examined. The study by Mattila and Hyodynmaa (1988) revealed that when goals were posted and feedback was given, the safety index was significantly higher than when no feedback was given. Fellner and Sulzer-Azaroff (1984) analysed the industrial safety practices through posted feedback. In a study carried out on Honduras construction sites, Jaselskis and Suazo (1994) demonstrated a substantial lack of awareness or importance for safety at all levels of the construction industry. In addition, Laufer and Ledbetter (1986) assessed various safety measures. Other researchers examined costs of construction accidents to employers (Leopold and Leonard, 1987, Levitt and Samelson, 1993). With regards to construction site safety in Hong Kong, Lingard and Rowlinson (1994) investigated the theoretical background to commitment at the group and organisational level and presented a site-level research model which is illustrative of the possible effects that these have on performance. A more recent study by Tam, Fung and Chan (2001) explored the attitude change in people after the implementation of the new safety management system in Hong Kong.

It has been argued that the construction industry is essentially very different from manufacturing industries and so it is impossible for the techniques and systems used in those industries to be effective in the construction industry. The argument has some weight and will be further discussed in this book. However, this argument does not lead to the conclusion that construction works cannot be made safe. This would be an inappropriate conclusion but, for such improvements to be made, it would be necessary to adopt innovative and specifically focused measures for the construction industry. The following section will discuss the nature of construction works in order that the problem of construction site safety can be addressed in an appropriate context.

1.1   INTRODUCTION TO CONSTRUCTION SITE ORGANISATION

1.1.1   Civil and building works

Construction works have traditionally been divided into two distinct types: civil engineering works and building works. The former includes construction of roads, bridges, dams and the like and these are in most instances horizontal construction projects. They involve the use of large items of plant and equipment. Building works tend to focus upon vertical construction, such as offices, apartments, factories and involve mainly production of facilities in which people or equipment will be housed. Building works tend to be less plant intensive and more labour intensive than civil engineering works, in general. As a consequence, both types of construction work face particular problems that are peculiar to the building or civil engineering sectors. However, some of the problems faced are much more generic in nature and apply to both types of construction. Maintenance works form a significant part of the construction industry workload and contribute significantly to accident rates. The breakdown of work between building and civil engineering construction is roughly similar worldwide but that for each country the exact breakdown varies from year to year. During periods of development a country will normally have a higher percentage of civil works. Specific figures are not included in this book but it is generally held that civil engineering works are somewhat safer than building works. That does not mean that the accident record for civil engineering works is acceptable but it is purely a reflection of the nature of the works being undertaken.

1.1.2   Subcontracting systems

Most works undertaken in either building or civil engineering are undertaken on a subcontract basis. The building industry probably uses rather more subcontracting than does the civil engineering industry. Subcontracting is the system by which work is allocated to a main contractor that does not construct the works itself but employs subcontract organisations to produce the finished product. Typical subcontracting firms would specialise in areas such as concreting works, bricklaying, falsework and formwork erection and foundation construction. The organisation actually undertaking the construction works is rather small compared with the total size of the project. Herein lies one of the problems that besets the construction industry. When the size of the organisation undertaking construction work falls below a critical mass then the resources and facilities to enable safe construction are not readily available. This is a structural problem that occurs in most national construction industries and poses management difficulties in terms of construction site safety, productivity and quality management. Small firms, as are many subcontractors, are unable to adequately train and educate workers and this phenomenon has led to a problem of reduction in apprenticeships which has led to a shortage of skilled, trained labour. This leads to the consequence of poor performance in terms of productivity and quality and, inevitably, site safety. This is not an issue that can be dealt with by individual companies on their own very readily. If one company takes the lead in moving towards directly employed labour it is likely this company will be undercut by competitors making use of subcontractors. However, it is possible to address this issue through concepts such as partnering and this particular solution is discussed later in the book.

1.1.3 Construction contracting

Most construction contracts are awarded through competitive tenders, with the lowest bidder taking the contract. This system has been identified as the cause of a vicious circle of cost cutting and claims generation. This has impacted on project performance in terms of time, cost and quality but it also has an effect on safety. The first item that suffers in the competitive bidding system is often the safety budget. However, it is unlikely that the competitive tendering system will be quickly abandoned in the construction industry. Although concepts such as partnering have taken a strong hold in many industries, they still do not account for a large proportion of the construction works being undertaken. As a consequence, it is necessary to look at the nature of the construction contracting company and how these companies organise their construction sites.

Most construction companies operate with an organisation structure similar to that shown in Figure 1.3. In such a company the head office functions and the site functions operate fairly independently. As a consequence there is a great deal of decentralisation and much decision-making is made at the construction site level. Hence, in terms of safety there is an issue here. No matter how well organised the head office safety management system might be this system has to be implemented at the site level by independent groups or teams. The construction site team is normally headed by a project manager or site agent and this person is normally assisted by professionals such as engineers and quantity surveyors. However, the majority of the construction work is, of course, undertaken by construction workers who are supervised by foremen and gangers. These are the people who have most contact with the workers: the workers are of course those who are at most risk. Hence, the chain of command and the ability to influence workers is exceptionally long and the trail back to the head office is unlikely to be an easy one to follow. Hence, in a rather different manner to a factory, most people on the construction site operate autonomously and vigilance and monitoring is vitally important.

It is important to realise that the construction site is a highly autonomous organisation. As such it is very difficult for the head office to control what goes on the site on a day-to-day basis. In fact, on large construction sites, it is often difficult for the site agent to control what goes on a day-to-day or even hour-to-hour basis. Consequently, construction workers are expected to operate with a high degree of independence and initiative. In such situations, the non-standard work procedures are often undertaken. Thus, the potential for accidents and mishaps is large and this is undoubtedly a characteristic of the industry as it operates today. Given this very different type of organisation it is important that specific, tailored safety and health programmes are prepared for the construction industry. It is not possible to transplant systems that work in factories and offices on to the construction site.

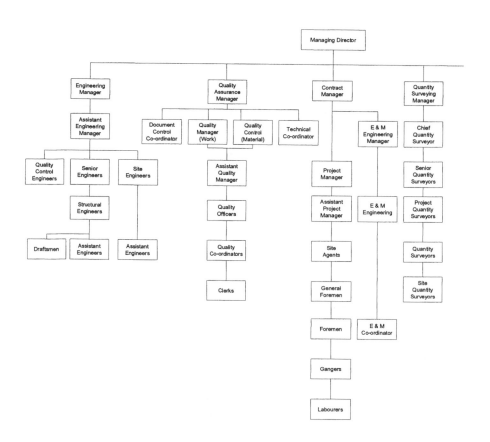

Figure 1.3 Typical organization structure of a construction company

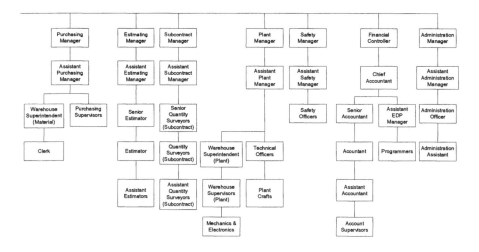

Figure 1.3 Typical organization structure of a construction company (cont'd)

1.1.4 Temporary multi-organisations

Construction project teams are generally temporary multi-organisations. By this it is meant that a group of organisations come together to form a project team, more correctly described as a coalition, for the purpose of the completion of the construction project. This team employs both design professionals and construction professionals, as well as financial controllers such as quantity surveyors. The nature of this temporary multi-organisation, TMO, is well described in the paper by Cherns and Bryant, 1984. The significance of the TMO is that it is an organisation that has no long track record of working harmoniously together. Hence, every time a team is set-up a new series of relationships, procedures and methods have to be established. The different organisations have to be integrated and the ways of operating have to be adjusted in order that they can work effectively together. This process takes time and, sometimes, is impossible to achieve due to structural or attitudinal problems within organisations or members of organisations. Hence, the industry is highly differentiated in terms of organisation and requires a high level of integration to ensure that all operational systems, including safety management systems, are implemented effectively.

1.1.5   Organic organisations

The construction project team could be described as an organic organisation. Organic organisations tend to have no firmly set procedures or rules of conduct laid down and are organisations which can react rapidly to changes in the environment. This is important in the construction industry where often it is not possible to plan more than a few days ahead due to changing circumstances, design changes and weather disturbances. As a consequence the organisation needs a great deal of integrative and co-ordinative effort in order to function properly. Mechanistic organisations, the opposite of organic organisations, do not need such devices as their procedures and rules are clearly set down. The paper by Wilson (1989) discusses this aspect of the construction industry in detail.

1.1.6   Temporary workers

The construction industry is a highly labour intensive industry. Many different trades and crafts are represented on the construction site. Most of the workers who work on the sites could be classed as temporary workers in that they are not employed directly by the main contractor and they are often recruited by the subcontractor on a job-by-job basis, sometimes only on a weekly, or even daily, basis. This is again a reflection of the subcontracting practices of the industry and these have serious implications for training. Few subcontractors are willing to train construction site workers as they are small enterprises and have no organisational slack to be able to do this. As a consequence, many workers arrive on sites only partially or even poorly trained and stay on any one particular site for a period of only a few weeks before moving on to a new project. This also reflects the nature of the industry where certain trades may work intensively on a particular aspect of the project for a few weeks or months and then their particular task is complete and they move on to another project. It will be obvious that such employment practices have strong implications for construction health and site safety.

1.1.7 Employment and reward structures

The temporary employment nature of the workers is reflected in the payment and reward structures adopted. Most workers are paid on a daily or weekly basis and negotiate their own wage rates within a fairly narrow band of norms. Most workers in South East Asia will in fact work a number of hours overtime per day on a regular basis in order to both complete necessary works and to raise their daily take-home pay. This is not the case in, say, Australia, the Carribean or France where higher levels of unionisation and different cultural values dictate a very different scenario. However, in many countries, weekend working is common in the construction industry, as is late night working during periods of intense effort towards the close of a construction project. Such times also raise the opportunity to pay bonuses for rapid completion of work and such bonuses often mitigate against safe working practices.

1.1.8 Site management structure

Most construction sites adopt a management or organisational structure as shown in Figure 1.4 below. This figure indicates that the main contractor supplies only the senior levels of management and supervision, and that virtually all the levels below this are part of a multi-tier subcontracting system. Such a system allows greater flexibility in terms of employment practices but makes the control of quality, safety and productivity throughout the construction site an extremely difficult task.

1.1.9 Formal organisation

As far as the construction contractor goes Figure 1.3 indicates a typical organisation structure based around the head office of a major contractor. It will be obvious that there are a whole series of different activities which have to be undertaken in the construction company and it will also be obvious that the safety department plays a key role in the construction process. However, how this department is staffed and run and its line of authority are key issues in determining safety performance. This is an issue that will be addressed later in this book from the perspective of a number of different nations.

1.1.10 Autonomy

Most construction workers are expected to exercise a high degree of autonomy. Construction work is not work that can be easily supervised as there is no set working area each day to which a worker will return. Workers progress around the site as the building works progress and each worker is expected to use his own judgement to ensure work proceeds in new and continually changing surroundings.

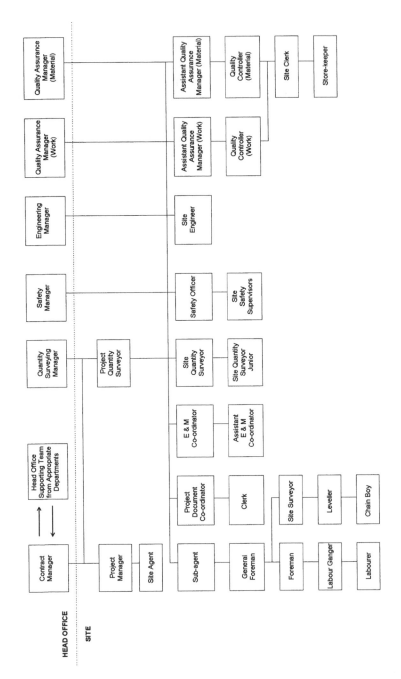

Figure 1.4 Overview of Construction Site Safety Issues

This exercise of autonomy leads to an attitude where workers believe that they know best and they conduct their daily work in a very independent manner. When properly trained and highly skilled and when provided with an appropriate safety infrastructure this is not a problem. However, when any of these prerequisites do not exist then the potential for accidents is high.

1.2   STRUCTURE OF THIS BOOK

This book is divided into six sections, namely:
1. International comparisons
2. Safety management issues
3. Health issues
4. Training and education
5. Safety technology
6. Accident analysis

The aim of this division is to take the reader logically through the process of reviewing and understanding the issues inherent in construction site safety and health worldwide. The approach adapted is top-down in that the issues move from the general: international comparisons, safety management systems to the specific training: technology and accident analysis.

Having provided a general introduction to the issues to be addressed in this book and the context within which they are addressed this chapter concludes with a review of the issues to be addressed in Section 1 of the book, International comparisons.

Peckitt, Glendon and Booth provide a discussion of safety culture as it is seen to operate in the UK and the Caribbean. The chapter deals with a series of related issues. In most societies, the construction of structures is integral to human activity providing, *inter alia*, places for shelter, business, religious ceremonies and learning. The construction industry produces the built environment, creates employment and generates wealth. Small businesses, specialists in one of numerous different construction related activities, dominate the industry resulting in a competitive, complex, dynamic and fragmented industry. The construction industry is commonly considered to be dangerous, difficult and dirty, and is one of the most hazardous land-based industrial activities, producing numerous serious accidents and cases of ill health to workers and members of the public.

The concept of safety culture is concerned with managing health and safety risks. The Advisory Committee on the Safety of Nuclear Installations (ACSNI) study group on human factors, provided one of the most quoted definitions of safety culture: *the product of individual and group values, attitudes, perceptions, competencies, and patterns of behaviour that determine the commitment to, and proficiency of, an organisation's health and safety management* (HSC, 1993). The authors consider what Geller, 1994, proposes as the concept of *Total Safety Culture* identifying
1. *personal* (knowledge, skills, abilities, motivation, personality);
2. *behavioural* (compliance, coaching, recognition, communication); and

Steve Rowlinson

3. *environmental* (equipment, tools, machines, housekeeping, environment, engineering)

factors as key aspects of safety culture. One of the most simple and general, but useful, definitions of safety culture is: *aspects of culture that affect safety* (Waring, 1992).

The chapter compares the industry in two different parts of the world in order to gain a better understanding of factors that significantly impact upon the safety culture of this industry. It describes some key issues, highlighting significant attitudes, behaviours and situations that were found to impact upon the safety culture of the construction industry in Britain and in seven anglophone Caribbean countries in the last decade of the twentieth century. It examines the following specific issues:

1. standards of construction site health and safety;
2. construction worker attitudes to health and safety related issues;
3. construction companies health and safety management practices;
4. societal factors that impact upon the safety culture of the construction industry, including legislation;

and so gives a broad but detailed overview of the health and safety problems in national construction industries.

Zeng and Deng discuss construction site safety in China and attempt to identify those factors which affect safety in the Chinese industry. By international standards, the construction industry in China performs very poorly in the area of safety. In 1999, 923 serious site accidents occurred in countryside construction works, in which 1,097 employees sustained fatal injuries and 299 serious injuries (*China Statistical Yearbook of Construction*, 2001). However, they report that until recently there was little research that attempts to examine construction site safety in the Chinese construction industry in international literature. This chapter explores the factors affecting construction site safety in China and identifies a number of factors which are important, including poor safety awareness of firm's top leaders, lack of training, poor safety awareness of project managers, reluctance to input resources for safety, reckless operation of plant and equipment and lack of certified skilled labour. To provide a more detailed view of the operation of a safety management system in China (and as something of an addendum to the previous chapter) Lu and Ji investigate construction safety supervision in Shenzhen, China's premier Special Economic Zone (SEZ), established in March 1979. They note that the local construction industry has made significant progress and the scale of construction has greatly expanded since its formation. As a result of such development, employment opportunities in the construction industry have flourished. Most of the construction workers were poorly educated and had little awareness of the safety hazards on site. Consequently, during the construction boom period of the 1990s, a high accident rate resulted. During the period 1982 to 1986, statistics indicated that 102 persons were killed, 213 seriously injured and 509 slightly injured in Shenzhen construction sites. In 1985, 41 persons were killed. During the following years, about 15 to 20 persons lost their lives annually working on construction sites. While the accident statistics in the industry have improved over the years, construction site safety continues to be a major issue on the industry agenda.

In the early period of Special Economic Zone (SEZ) establishment, supervision of construction safety was inadequately managed. From March 1979 to December 1983, there was only one part-time technical officer in charge of safety for the government. From January 1984 to March 1988, there was one full-time technical officer appointed to be in charge of construction quality and safety. Their duty was to manage the documentation and statistics, as well as to provide an annual report on the construction industry's safety developments. From 1988, a temporary Construction Safety Supervision Group was established to enforce the safety legislation on construction site safety. In 1989, *Standard of Construction Safety Inspection* was issued by the Construction Department of the Central Government and in the following year, a campaign was conducted to promote safety awareness in the industry. From February 1991 to October 1994, a series of documents and technical standards were issued by the municipal government. Together, these formed the foundation of construction safety supervision in Shenzhen. This chapter analyses the move towards better safety management during the course of rapid development of a construction industry and identifies the problems that must be addressed in such a situation.

Enshassi and Mayer produce an analysis of construction site safety in Palestine. They opine that construction sites are considered dangerous places. Many site injuries result from people falling from structures like roofs and scaffolds, or being hit by falling objects. Many others are caused by the misuse of mechanical plant and site transport, including hoists (Frayer, 1995). There is nearly always keen competition for new contracts and site personnel are often under pressure to work to tight time and cost constraints. It is hardly surprising that safety is often neglected. In Palestine over the years construction has been a major contributor to the national economy. Presently, however, its share of the GDP is about 5-7%. In the 1970s this percentage was roughly 7-10% (Koehn and Pan, 1996). The construction share of GDP in 1994 was about 16.8, in 1992 the percentage was about 13.0. Therefore, it is crucial to take safety into consideration in the construction industry in Palestine in order to minimise the number of injuries on construction sites. This chapter presents an analysis of statistical data on construction site injuries between 1999 to 2000 published by the Ministry of Labor.

1.3   REFERENCES

Bentil, K., 1990, The impact of construction related accidents on the cost and productivity of building construction projects. In *Proceedings of the CIB 90*, 6, Management of the Building Firm, Sydney.

Cherns, A.B. and Bryant, D.T., 1984, Studying the client's role in construction management. *Construction Management Economics*, 2, 177-184.

China Statistical Yearbook of Construction, 2001 (Beijing: China Statistical Press).

Fellner, D.J. and Sulzer-Azaroff, B., 1984, Increasing industrial safety practices and conditions through posted feedback. *Journal of Safety Research,* 15, pp. 7-21.

Frayer, B., 1995, *The Practice of Construction Management* (London: Collins).

Geller, E.S., 1994, Ten principles for achieving a total safety culture. *Professional Safety,* September, 18-24.

Hinze, J., 1981, Human aspects of construction safety. *Journal of the Construction Division,* ASCE, 107, pp. 61-72.

Hinze, J., 1997, *Construction Safety* (New Jersey: Prentice-Hall).

Hinze, J. and Raboud, P., 1988, Safety on large building construction projects. *Journal of Construction Engineering and Management,* ASCE, 114, pp. 286-293.

Health and Safety Commission, 1993, *ACSNI Human Factors Study Group Third Report, Organising for Safety* (London: HMSO).

Jaselskis, E.J. and Suazo, G.A.R., 1994, A survey of construction site safety in Honduras, *Construction Management and Economics,* 12, pp. 245-255.

Koehn, E. and Pan, C.S., 1996, Comparison of construction safety and risk between USA and Taiwan. In *Proceedings of CIB W65,* The Organization and Management of Construction, UK.

Laufer, A. and Ledbetter, W.B., 1986, Assessment of safety performance measures at construction sites. *Journal of Construction Engineering and Management,* 112, 4, pp. 530-542.

Leopold E. and Leonard, S., 1987, Costs of construction accidents to employers, *Journal of Occupational Accidents,* 8, pp. 273-294.

Levitt, R.E. and Samelson, N.M., 1993, *Construction Safety Management* (New York: John Wiley and Sons, Inc.).

Lingard, H. and Rowlinson, S., 1994, Construction site safety in Hong Kong. *Construction Management and Economics,* 12, pp. 501-510.

Mattila, M. and Hyodynmaa, M., 1988, Promoting job safety in building: an experiment on the behavior analysis approach. *Journal of Occupational Accidents,* 9, pp. 255-267.

Tam, C.M., Fung, I.W.H. and Chan, A.P.C., 2001, Study of attitude changes in people after the implementation of a new safety management system: The supervision plan. *Construction Management and Economics,* 19, 4, pp. 393-403.

National Safety Council, 1987, OSHA investigates Connecticut Apartment project collapse. *OSHA Up to Date,* XVI, 5, Chicago, Illinois.

Waring, A.E., 1992, Organisational culture, management and safety. *Paper presented at the British Academy of Management 6th Annual Conference* 14-16 September, Bradford University.

Wilson, H.A., 1989, Organizational behaviour and safety management in the construction industry. *Construction Management & Economics,* 7, 4, pp. 303-319.

Section 1

International Comparisons

CHAPTER 2

Societal Influences on Safety Culture in the Construction Industry

S.J. Peckitt, A.I. Glendon and R.T. Booth

INTRODUCTION

This chapter considers the influence of societal culture upon safety culture in the construction industry. It presents several models of construction accident aetiology to demonstrate that root causes of such accidents are embedded within the construction process and the societies within which construction activity takes place. We seek to demonstrate that the safety culture paradigm constitutes a holistic way of thinking about health and safety risk management to reveal underlying factors affecting safety performance in complex systems. It refers to research conducted by the first author in the Caribbean and the United Kingdom (UK) construction industry between 1992 and 1995. This work suggests that the comparatively highly developed societal systems of legislation, inspection and consultation in the UK have not resulted in lower rates of death and ill-health in its construction industry compared with the Caribbean (Peckitt, 2001; Peckitt, Glendon and Booth, 2002). It highlights the role of societal culture in influencing the health and safety culture of the construction industry.

2.1   THE CONSTRUCTION INDUSTRY

> *... building a wall involves physical and material factors, but it could not take place outside a system of meaning, institutionalised cultural knowledge, normative understandings and the capacity to conceptualise and use language to represent to oneself the task on which one was engaged and to build around it a collaborative and communicative 'world' of meanings in short, 'a culture'.* (Hall, 1997:232).

The construction sector is an important part of the economy in most countries, yet it is generally considered to be dangerous, dirty, hard and unreliable. Death rates within the construction industry around the world may indicate an inherently poor safety risk management culture. The construction industry produces 30% of fatal industrial accidents across the European Union (EU), yet employs only 10% of the working population. In the United States of America (USA) this sector accounts for 20% of fatal accidents and only 5% of employment (MacKenzie, Gibb and Bouchlaghem, 1999). Construction fatalities account for 30-40% of industrial fatal accidents in Japan, 50% in Ireland and 25% in the UK (Bomel, 2001). Fatal accident incidence rates (per 100,000) across the EU range from below ten in Britain and Germany to over 20 in France and Spain (Building, 2001). The fatal accident incidence rate in the South African construction industry was 53.5 in 1990 (Smallwood, 1996), and 87 in Hong Kong - ten times higher than the all-industry average (Tam and Chan, 1999).

In the British construction industry 114 workers were killed in the accounting year 2000/2001, giving an incidence rate of six, over five times the all industry average (HSE, 2002). High falls, vehicle movements and the collapse of structures are the leading causes of fatal accident injuries in the British construction industry. Occupational ill-health in the construction industry is a major issue that has received relatively little attention due, in part, to the difficulties of tracing causes of ill-health in a transient workforce. Deafness, dermatitis, occupational cancers and respiratory diseases plague the construction workforce. Almost 100,000 workers suffer from musco-skeletal injuries, while 750 construction workers die each year due to asbestos related diseases (HSE, 2002).

The construction industry comprises many organisations including property developers, architects, engineers, quantity surveyors, accountants, lawyers, management contractors, engineering contractors, civil engineering contractors, labour only subcontractors and specialist trades. The construction industry operates on international, national, regional and local scales, with participants ranging from large multinational organisations to single person operations. Construction projects vary from simple dwellings to complex structures such as nuclear power stations. Projects may involve just the client and the builder, others involve hundreds of suppliers, contractors and consultants. Prevalence of small firms, subcontracting, fragmentation, the one-off nature of many of its products, irregular employment, limited training and poor supervision are characteristic features of the construction industry.

The construction process involves hazardous activities such as working at height, manual handling, exposure to hazardous materials, demolition, frame erection, lifting operations, scaffolding and groundworks. Casual work, cash-in-hand payments, tax evasion, fraud and theft are commonplace. The industry is prone to 'boom and bust' cycles, under-production and over-capacity, intermittent work and climatic influences. Consequences of these negative characteristics of the construction industry include unnecessary financial and human loss. However, poor safety risk management does not have to be the norm in the construction industry (Ball, 1988; Dester and Blockley, 1995).

Casual work and poor working conditions are not inevitable consequences of building work but of the way that the actual physical process of building is organised and executed. (Ball, 1988:15)

2.2   CONSTRUCTION ACCIDENT CAUSATION MODELS

Knowledge of accident processes can help to identify root causes and appropriate remedial measures. Models of accident causation that are specific to the construction industry have been put forward by Hinze (1996), Leather (1987), Surabi, Duff and Peckitt (2001), and Whittington, Livingston and Lucas (1992). The focus of these models varies, ranging from the accident victim situated within a socio-technical system, to systems of work and worker interaction, to failures across the wider construction process.

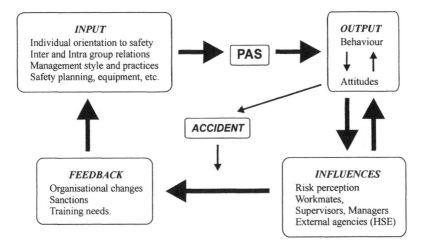

Figure 2.1   Leather's (1987) Potential Accident Subject (PAS) Model

The Potential Accident Subject (PAS) model (see Figure 2.1) is used by Leather (1987) to explain the construction accident causation process model. Internal and external inputs influence PAS behaviours and attitudes that may or may not result in an accident. The PAS is situated within a dynamic social system that includes co-workers, supervisors, managers and external agencies, all of whom can influence safety. Individual, organisational and job related variables interact within this input/output. On a construction site the potential accident subject (PAS) can be either a victim or contribute to an accident. Individuals' attitudes and behaviours are influenced by actions that are rewarded or punished, training received, instructions given, management systems and the impact of regulatory authorities. Rewards provided by bonus schemes for completing work quickly encourage workers to take shortcuts and ignore risks. The lack of site induction training can result in workers not being aware of specific site risks and safe systems of work. Prohibiting unsafe work activities by regulatory inspectors can both stop dangerous activities and result in changes to how a contractor manages specific site risks.

Hinze's (1996) Distractions theory of construction accident causation has a narrower, site based focus, which correlates productivity with risk, and predicts that workers under stress have a greater probability of being involved in an accident. Hinze's theory is based on the concept that where hazards exist workers will be distracted by them, increasing the probability of an injury and reducing productivity. As the focus shifts from the work to the hazard, productivity declines in proportion to the decline in the probability of an injury. However, safety and productivity are compatible when hazards are removed, thereby reducing distractions to workers. For example, providing a tower scaffold for work at height reduces the risk of falls and the potential distractions caused by trying to work safely from a less secure position - thereby increasing productivity compared with working from a ladder. Distractions can also take the form of personal life events, such as divorce, celebrations and illness, which can also distract workers from their work.

In their model of construction industry accident causation Whittington, Livingston and Lucas (1992) highlighted failures in project planning and execution. Their model illustrates how failures throughout planning and construction phase of a project can lead to accidents on site. The model identifies four key areas where failures can occur: company policy - e.g. procurement practices; project management - e.g., poor scheduling; site management - e.g. lack of segregation of pedestrians and vehicles; and by the individual - e.g. ignoring safety rules. Mediating influences on these failures include the efficacy of project risk management systems and the impact of regulatory authorities.

Ordraguez, Jaselskis and Russell (1996) highlighted the time dependent relationships between accidents and overall construction project performance. In their study of 17 major civil engineering projects, over budget projects were found to have increased accident incidence rates, particularly at the middle and end of the construction phase. Fast-track projects were found to have more accidents and that these occurred at an earlier stage of construction, compared with traditional projects that commence construction after most of the design is complete. More accidents occur when resources are expended at higher than average rates and when work rates increase. Projects that ran behind schedule or had accelerated cost expenditure had higher accident incidence rates, particularly during the early and middle portions of the construction phase.

Suraji, Duff and Peckitt (2001) proposed a constraint-response model, which highlights the complex and multi-causal nature of the accident process in the construction industry. Their model highlights the influence of both proximal factors, such as unsafe equipment, distal factors, such as client budget constraints and the links between them. Their model demonstrates how constraints that arise from the wider business environment can impact upon the health and safety of workers on site through a process of constraint and response by various players within the construction process. For example, failure by a quantity surveyor to include adequate allowances for safety requirements to reduce a tender price in order to get work can lead to workers being injured on site due to inadequate equipment provision. Adversarial relationships between client and contractors can cause communication problems that may result in key hazard information not being transmitted. This may result in workers being exposed to on-site hazards such as asbestos and contaminated ground.

These accident causation models provide insights into construction accident propagation. Construction accidents result from interactions within complex systems and need to be examined holistically in order to identify root causes and effective remedial measures. The safety culture paradigm can provide a holistic analysis of problems within the construction industry.

2.3   SAFETY CULTURE

The concept of safety culture has become part of predominant safety paradigms, being increasingly cited in discussions on the aetiology of accidents and managing high-risk processes. Cooper (1998) argued that an organisation's safety culture impacts not only upon accident rates, but also on work methods, absenteeism, quality, productivity, commitment, loyalty and satisfaction. Reason (1997) considers that few things are so important, yet so poorly understood as the concept of safety culture. Pidgeon (1998) highlights the paradox that organisational culture can act as both the precondition for safe operations and the cause of failures that can lead to accidents. Philosophical problems arise with the relationship between safety and risk and confusion is common regarding differences between safety climate and safety culture (Glendon, 2000; Glendon and Stanton, 2000; Guldenmund, 2000; Peckitt, 2001).

Functionalist approaches to safety management systems focus on the techniques and mechanics of managing safety. Rules, systems, policies and procedures may encourage safe actions, but they are only as affective as the consequences they predict and the extent to which they are implemented. As procedural systems expand and become increasingly restrictive, the drive to get the job done increases the likelihood of violations being committed (Reason, 1995). Disciplinary actions temporarily stop unsafe actions from being observed, but can foster resentment and negative attitudes. Work procedures lapse if neglected by management, or because operators are discouraged from following them by peer group pressures and/or the presence of production related bonuses (HSE, 1989).

Eckenfelder (1997) criticises traditional enforcement and behaviour based safety management programs for being myopic, a poor return on investment, not being self-sustaining, not placing enough responsibility on management, not promoting innovation, and failing to deal with root causes of accidents. The domination of the safety and risk literature by the functionalist, engineering biased perspectives limits it's applicability to socio-technical systems (Rochlin, 1999). Successful health and safety management requires both effective management systems and appreciation of human factors within an organisation (Waring and Glendon, 1998). The concept of safety culture emerged from the aftermath of

Chernobyl, as investigators tried to explain the human factors involved in this nuclear disaster.

Elitist paradigms view safety culture as organisational safety management best practice and consider that an organisation does not have a safety culture unless they can demonstrate excellent safety risk management performance. Relativists consider all organisations to possess a safety culture, which can be placed on a scale from good to bad. Waring and Glendon (1998) distinguish two safety culture paradigms: a top-down functionalist perspective that views safety culture as a structured system for managing risk imposed by management; and a bottom-up interpretative approach, which views safety culture as a complex phenomenon emerging from workplace social interactions. Both functionalist/top-down and interpretive/bottom-up processes are important influences on safety culture (International Atomic Energy Authority, 1998). The precursors of effective occupational health and safety risk management are both functional - e.g. involving formal management systems, and interpretative - e.g. involving social issues such as trust, blame, risk perception, learning, commitment and motivation.

The Advisory Committee on the Safety of Nuclear Installations (ACSNI) study group on human factors, produced one of the most quoted definitions of safety culture. They define safety culture as: *the product of individual and group values, attitudes, perceptions, competencies, and patterns of behaviour that determine the commitment to, and proficiency of, an organisation's health and safety management* (Health and Safety Commission, 1993:23). One of the simplest, general, but holistic definitions of safety culture is provided by Waring (1992), who defines it as aspects of culture that affect safety.

Health and safety management is primarily dependent upon the occupational risk-related attitudes and behaviours of directors, managers and workers, who are part of both the organisation and wider society. People's risk related perceptions and behaviours are primarily socially and culturally defined (HSC, 1993). Safety is therefore a dynamic social construct, formed in the interplay of human, environmental, economic and technological factors (Rochlin, 1999). Safety is linked with learned responses to risks present in an environment at a particular time (Douglas and Wildavsksy, 1982). Safety culture is influenced by factors that affect people's attitudes and behaviour in relation to safety, including race, religion, nation, community, group, history, technology and economics (Glendon and McKenna, 1995).

Every society has distinctive sets of values and priorities within its culture that guide its members' behaviour (Haralambos and Holborn, 1991; Ng, 1980). Expressions of culture include observable artefacts, patterns of behaviour, values and assumptions (Rousseau, 1990). Because culture is about human experience and interaction, it is complex and chaotic. Macro aspects of culture, located at societal level, include language, laws, rituals and worldviews, while micro aspects, - e.g. motivation, trust and violations, are located at the individual or group level. Cultures are continuums of variation making them difficult to delineate in precise terms (Alleyne, 1988). Referring to African, European or Caribbean culture does not mean that all people included by the term share a standard set of cultural traits. Different individuals within any culture possess commonly shared cultural features in different ways, to different degrees and in different mixes. Cultural analysis requires a tolerance for complexity that runs counter to a drive to simplify.

Safety culture is concerned with the attitudes, behaviours, systems and environmental factors that promote effective health and safety management. Safety risk management culture is a more inclusive term than safety culture, as the term *risk* implies health, safety and environmental hazards, while inclusion of *management* places the onus for action on all those with a degree of control over risk. Organisations can be described as having good/positive or poor/negative safety cultures depending upon the effectiveness of their safety risk management strategies.

2.4   SAFETY CULTURE TYPOLOGIES

Shillito (1995) identifies four categories of organisational safety culture – *rule book*, *engineered*, *procedural* and *behavioural*. Characteristic management attitudes in *rule book* cultures include regarding workers as lazy, careless, and in need of constant motivation and discipline. Workers should learn rules by heart because they cannot be expected to think for themselves. The *engineered* culture considers that engineers should eliminate all possible sources of harm. The *procedural* culture treats workers with more respect than the *rule book* culture, but still considers workers to be careless, forgetful, and in need of constant motivation (Shillito, 1995). Motivation is achieved by rephrasing rules into policies, objectives and targets. The *behavioural* culture considers that workers are motivated to give their best when they receive appropriate training, are empowered and accept responsibility for their work. The *behavioural* culture still contains policies, objectives, targets, procedures, manuals, records and audits, but sees these as tools to aid the performance of empowered teams.

Eckenfelder (1997) proposed a five-stage maturity grid of safety culture ranging from ignorance to perfection, based on ten key values. These key values include a concern for safety, integration of safety into the production process, recognition that safety management is a never ending process, putting the right people in the right places, using appropriate measures, and empowering staff. Characteristics of the ignorance stage of the maturity grid, include safety being: driven by regulations and costs of accidents; seen as a separate part of the management process; governed by quick fixes; managed by anybody; and based solely upon inspection and compliance. Safety is a burden, accident rates are the only measure of performance and safety comes last when allocating funds. Characteristics of a fully mature safety culture, include a sincere and continuous concern for employees and their safety, safety is seen as a profit centre not as a cost, and loss prevention measures are fully integrated into the business. The IAEA (1998) provided a simpler development scheme of safety culture identifying three different stages, namely: *Rule and Regulation; Organisational Objective,* and

Continuous Development. The Keil Centre also produced a progressive model of safety culture maturity with five stages of development from *emerging* to *continually improving* (HSE, 2000).

Within organisations sub-cultures form around different roles, functions and levels of power. Most organisations have separate executive, management and worker cultures (Schein, 1997; Williams, Dobson and Walters, 1989). Directors and senior managers may share common beliefs about strategic direction and appropriate behaviours, creating an executive culture. The management culture tends to focus upon managerial and resource issues, while worker culture is more likely to focus on production and rewards. These three cultures are often at odds and so work at cross-purposes. Operators know that rules they should follow do not always work in practice, engineers prefer technical and impersonal solutions to problems, while senior managers distanced from day-to-day operations, may view staff as a cost. Both management and executive cultures are primarily task-focused and consider workers to be either a cost or a source of error (Schein, 1997). To work effectively an organisation must ensure that these different cultures can mesh together and communicate effectively. Recent work highlights the domination of the engineering culture within construction organisations and the conflicts this can have with effective health and safety management on site (Meijer (1999); quoted in Hale, 2000).

2.5   SUMMARY

Some critics of the safety culture paradigm consider it to be a fad (Back and Woolfson, 1999), a 'catch-all' for human factors issues and a concept without substance (Cox and Flin, 1998). The lack of precise definitions of safety culture, the traditional functionalist and techno-centric orientation of the safety discipline and confusion between safety climate and safety culture has fuelled scepticism regarding the paradigm's utility. However, studies of high reliability organisations (HROs) - i.e. those with excellent health and safety management records, emphasise the importance of understanding how HROs work from a cultural perspective (Roberts, 1990, 1992; Roberts and Rousseau, 1989; Rochlin, 1993). Studies by Coote and Lee (1993), Guest, Peccei and Thomas (1994), Lockhart and Smith (1986), and Sayers (1994) demonstrate that low accident frequency organisations are characterised by positive management attitudes to safety and worker empowerment (Cox and Flin, 1998).

Positive safety cultures focus on the people aspects of an organisation (Eckenfelder, 1997). Erickson (1997) identified management systems, management concern and positive employee setting as crucial determinants of safety performance, asserting that positive employee setting is the best predictor of safety performance. Positive employee setting includes work environment, management actions and behaviours, communication, treatment of employees, innovative thinking, management feedback, employee commitment, employee morale, organisational fit of employees and ethics. Staff should adopt a questioning attitude; search for ways to improve safety; constantly be aware of what can go wrong; and feel personally accountable for safe operations. The most important indicator of a positive safety culture is the extent to which employees are involved in safety management (Cooper, 1998).

An organisation's culture comprises a socio-technical system involving interacting personalities in decision-making processes, planning and control procedures, technology and resources. The most successful safety cultures are those in which risk is viewed as omnipresent, dynamic and manageable. They are proactive in dealing with potential problems through competent, empowered, trusting employees. They develop a just culture based on good communications with clear distinctions between tolerable and culpable behaviour, rather than a 'no-blame' culture (Pidgeon, 1998; Reason, 1998). Management is committed to continuous improvement and spend time preparing and reviewing operating plans and procedures.

Externally, organisations are embedded in the wider socio-technical systems including political, legislative, economic and technological systems (Williams, Dobson and Walters, 1989). Culture varies both between and within organisations. Risk, feedback, technology, ownership, power, history, size, specialisation, environment and people all impact on organisational cultures. Figure 2.2 depicts the key positive influences on safety culture identified by the IAEA (1998).

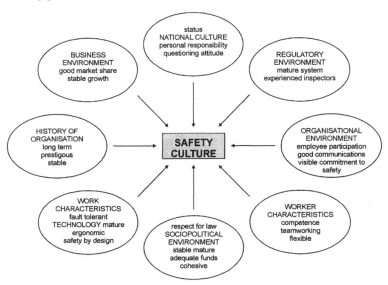

Figure 2.2 Positive Influences on Safety Culture (IAEA, 1998)

2.6   SAFETY CULTURE AND SOCIETY

As multinational corporate operations proliferate, it becomes increasingly important to be aware of the impact of societal culture upon organisational safety culture. From a social science perspective, safety culture is an important theoretical concept that explicitly examines some of the more elusive contributions to accident incubation addressing the wider social causes of accidents (Pidgeon and O'Leary, 1994). Fujita *et al.* (1993) and Rosen (1995) identify societal cultural differences as playing a significant role in determining the performance of nuclear power plants, influencing management styles, educational requirements, shift transfer procedures, adherence to written procedures, use of automation and standards of housekeeping.

Understanding the impact of societal culture upon safety culture allows antecedents of safety related attitudes and behaviours to be identified. Each culture has characteristics that impact both positively and negatively upon safety culture (IAEA, 1998). Risk perception, work group dynamics, and attitudes to work, religion, technology and time are factors that impact on safety culture (Meshkati, 1995). Goszczynska, Tyska and Slovic (1991) identify size of country, level of technological development and occupation as significant influences upon individual risk perception. In industrially developed nations natural events are labelled as disasters by an ideology that maintains that technology can control nature, while less technocentric societies consider the same disasters to be part of human existence (Adams, 1995).

In the industrial work environment the perception that safety competes with productivity often results in poor hazard management and deliberate risk taking to maximise productivity and minimise expenditure (Bird and Loftus, 1976; HSC, 1993). Risk takers often justify their actions by considering relevant risks to be benign because of their focus on the positive, rather than negative outcomes of risk taking behaviour (Cotgrove, 1982; Pederson, 1995). Adams (1995) described a *Risk Thermostat* model in which the propensity to take risks is a function of rewards, experience and perceived danger, and the balancing of behaviour produced by their interaction. Cultural filters are a key part of this model influencing risk-taking behaviour by dictating perception of hazards and rewards.

2.7   SOCIETAL CULTURAL DIMENSIONS

One of the major changes to orthodox psychological approaches to risk perception came from the grid/group cultural theory proposed by anthropologist Mary Douglas. In grid/group theory *group* refers to the extent to which an individual's choice is subject to group determination, while *grid* relates to the degree to which externally imposed prescriptions dictate individuals' lives (Thompson, Ellis and Wildavsky, 1990). Strong group boundaries coupled with minimal prescriptions produce social relations that are egalitarian. Where strong group boundaries and binding prescriptions exist, social relations are hierarchical. Where neither group incorporation nor prescribed roles exist, social relations are individualistic and boundaries are provisional and subject to negotiation. Binding prescriptions and exclusion from group membership, exemplify a fatalistic life orientation. Each world view perceives risk differently, for example individualist are associated with risk taking, while egalitarians are risk averse.

Hofstede (1980) initially identified four key cultural dimensions at a societal level: social inequality - including relationship with authority, community relationships, masculinity/femininity, and dealing with uncertainty - including control of aggression and emotional expression. Subsequently he added a fifth dimension to his analysis of differences among societal cultures - that of long-term versus short-term orientation to life (Bond and Hofstede, 1989; Hofstede, 1994; Hofstede and Bond, 1988). In a 65-nation study, the GLOBE project extended Hofstede's analysis to nine dimensions of culture - assertiveness, family collectivism, institutional collectivism, future orientation, gender egalitarianism, humane orientation, performance orientation, power distance, and uncertainty avoidance (House *et al.*, 2002).

Schein (1985) identified five basic assumptions around which cultural paradigms form - relationship to nature, truth and reality, human nature, human activity, and human relationships. Hampden-Turner and Trompenaars (1993), in a study of the seven most economically successful countries, identified seven cultural dimensions – universalism vs particularism, analysing vs integrating, individualism vs collectivism, inner-directedness vs outer-directedess, time sequential vs time synchronising, achieved status vs subscribed status, and equality vs hierarchy. Table 2.1 summarises the main dimensions identified by these analyses.

Cognitive processes, including attitudes, beliefs, values and perceptions, influence both human behaviour and situations. Key cognitive cultural factors include perceptions of nature, risk, time and uncertainty. Norms of behaviour, communication, discrimination, power, leadership and hierarchy are also key indicators of culture. Key situational factors influencing cultural bias include biology, climate, technology, industrial development, resources, history and competition. Hofstede's five key cultural dimensions – uncertainty avoidance, individualism, power distance, time and gender, are considered in the sub-sections below.

2.7.1   Uncertainty avoidance

Uncertainty avoidance describes the extent to which individuals within a culture feel threatened by uncertain or unknown situations (Hofstede, 1994). It is related to Durkheim's concept of anomie, as well as to stress and risk taking (Hofstede, 1980). Low uncertainty avoidance scores equate with a greater willingness to take risks. However, in his later work Hofstede warned that uncertainty avoidance should not be confused with risk avoidance because uncertainty is to risk as anxiety is to fear (Hofstede, 1994).

Low uncertainty avoidance orientation is associated with ambition, hope of success, high employee mobility and low satisfaction. McGregor's Theory X (=

workers are lazy) is supported and conflict and change are seen as inevitable. Broad outlines rather than detailed instructions are preferred, rules are viewed as flexible and there is little expression of emotion or anxiety. Many of the northwest European countries, including the UK show a bias towards low uncertainty avoidance (Hofstede, 1980). High uncertainty avoidance correlates with high anxiety, concern about the future, fear of failure and risk avoidance. High uncertainty avoidance cultures value large organisations, steep hierarchies, expert status, high levels of co-operation, teamwork, loyalty and emotional expression. Workers are considered, willing following McGregor's Theory Y, competition is not encouraged and detailed instructions are preferred and followed. Uncertainty avoidance orientation in a society is affected by the degree of individualism or collectivism within the dominant culture (Hofstede, 1994). Strong uncertainty avoidance combined with individualism tends to produce rule-based bureaucratic organizations.

Table 2.1   Cultural Dimensions

Bond and Hofstede (1989); Hofstede (1980, 1994); Hofstede and Bond (1988)	GLOBE Project (House *et al.*, 2002)	Schein (1985)	Douglas and Wildavsky (1982); Douglas (1992); Thompson, Ellis and Wildavsky (1990)	Hampden-Turner and Trompenaars (1993)
Time orientation; Uncertainty avoidance; Individualism; Power distance; Femininity/ Masculinity;	Future orientation; Uncertainty avoidance; Family collectivism; Institutional collectivism; Gender egalitarianism; Power distance; Humane orientation; Performance orientation; Assertiveness	Truth and reality; Human nature; Human relationships; Human activity;	Individualism Egalitarian Fatalist Hierarchist Hermit	Time Synchronous/ Segmented; Analysing/ Integrating; Universalism/ Particularism; Status; Hierarchy; Individualism; Inner directed/ Outer directed

2.7.2   Individualism

Individualism is linked to the worldview that people are responsible for their own destiny, and is placed by Hofstede (1980) at the opposite end of a bi-polar scale to collectivism.   The degree of bias towards individualism is of fundamental importance and is the most frequently identified dimension of cultural orientation (Hampden-Turner and Trompenaars, 1993).   Individualists are self-seeking, and are not bound by group incorporation nor by prescribed roles.   Hofstede (1980) found that relative wealth, cold climate, technological development and large organisations, are indicators of a society that places considerable emphasis on the individual.

Individualism is highly developed in very simple cultures - e.g. hunter-gatherer tribes, and very complex ones - e.g. the USA.   A bias towards individualism is noted in urban areas (Triandis, 1989).   Power, achievement, modernity, technology, wealth and well-developed legal systems are valued in individualistic cultures.   Workers feel emotionally detached from their employers and have high job mobility.   Social order is placed above community and the nuclear family dominates domestic arrangements.   Australia, Canada, Denmark, Italy, The Netherlands, New Zealand, UK and USA are most strongly oriented towards individualism (Hofstede, 1980).

Western views of self centre on the individual body, while African and Asian views of self centre on the collective group (Triandis, 1989).   Collectivist cultures conform to in-group norms, are respectful, nuturant and intimate. Hofstede (1994) considers that a majority of countries in the world are oriented towards collectivism, in which group interests prevail over individual interests. Community relations are strong, extended families are common and group decision-making is preferred (Hampden-Turner and Trompenaars, 1993). Organisations in collectivist cultures tend to be small, run as a family with strong emotional links, little job mobility, team working and communications are good, and workers spend many years with one organisation.

2.7.3   Hierarchy and power

Cross-cultural studies commonly focus on the nature of power relations in societies. Hofstede (1980) uses the term 'power distance' to describe the degree to which people can control the behaviours of others in hierarchies. Hampden-Turner and Trompenaars (1993) link hierarchy with inequality and status. Strong group boundaries and binding proscriptions characterise hierarchical ways of living. Hierarchies are central to the distribution and use of power in society.

In high power distance cultures managers are seen as father figures who have absolute power and deserve respect, personalities dominate power positions and status symbols are important (Hofstede, 1994). Large power distances inhibit the desire for equality and are seen as inevitable by those without power. High power distances prevail in countries with high degrees of inequality and correlate with low economic development, tropical climate and prevalence of the Roman Catholic faith. Governments tend to be autocratic, coercive, and unstable, with charismatic leaders. Employees accept inequality, fear to disagree with managers, and are reluctant to trust fellow workers.

Low power distances are found primarily among the middle and upper classes of wealthy, cold and temperate, Protestant countries. All northwest European countries and the USA show a positive bias to low power distances (Hofstede, 1994). Democratic governments, formal legal systems, technology and education correlate with low power distances. Independence, resistance to close supervision, mixed feelings about management and 'blame the system' attitudes are characteristic of low power distances. Organisations in low power distance countries, such as the UK and the USA tend to be mechanistic, structured hierarchies. Workers feel that they are a separate part of the organisation from management. Workers promoted to supervisory positions are commonly considered to change because power corrupts them.

Hofstede (1994) identified latitude, population size and wealth as key determinants of power distance. The degree to which climate promotes the use of technology for survival was the prime determining factor of power distance (Hofstede, 1980, 1994). Climate impacts upon lifestyle, concepts of time, architecture, pace of work, diet and appearance. Higher latitudes provide a less abundant natural environment than do tropical climates, promoting the need for people to interfere more with nature in order to exist and to learn to fend for themselves rather than depend on others (Hofstede, 1994). On the power distance dimension West Africa was ranked joint 10th with India. Africans and Asians often have difficulty understanding the impersonal machine-like bureaucracies of Western countries (Hofstede, 1994).

2.7.4   Gender

The masculinity/femininity dichotomy is based on the fundamental biological split within the human species. Masculinity is associated with materialism, achievement, independence, decisiveness, speed, assertiveness and living to work. Femininity is associated with humanitarian values, personal relationships, small and slow is good, intuition, sympathy, working to live and a sense of interdependence. Austria, Germany, Ireland, Italy, Japan, Mexico, the UK and Venezuela were among the top countries on Hofstede's masculinity index (Hofstede, 1980). Masculine cultures tend to try to resolve conflicts by fighting; feminine countries by compromise and negotiation.

2.7.5   Time

There are two ways in which cultures conceptualise time, which the Greeks perceived as two gods - *Kairos* and *Chronos*. Time can be thought of as linear and sequential, moving forward by increments (*Chronos*); or synchronous and circular, combining the past, present and the future (*Kairos*). Hampden-Turner and Trompenaars (1993) consider that France, Germany and Japan display bias to synchronous paradigms, while The Netherlands, Sweden, the UK and USA are oriented towards sequential paradigms. Sequential managers tend to do one thing at a time, regard time commitments very seriously, see time as a threat and tend to solve conflicts using a 'first come first served' strategy. Synchronous managers do many things at once, consider time commitments as desirable rather than absolute, and see time as a provider of opportunities. Synchronising managers value open communication, development of human resources and working in teams (Hampden-Turner and Trompenaars, 1993).

Americans and the British consider it rude and disrespectful to be kept waiting, however, it is not such an issue in many other countries - e.g. in Ethiopia the time required for a decision is directly proportional to its importance. In American and British culture time is viewed as sequential and therefore as a scarce and precious commodity. Common English and American proverbs attest to the importance of time - e.g. *time and tide wait for no man, time is money*, and *procrastination is the thief of time*. Without the cultural importance of the concept that time is money there would be no Taylorism and no time and motion studies (Hampden-Turner and Trompenaars, 1993). Negative aspects of this cultural value include an obsession with short-term performance, and an unwillingness to take time and invest in the long term. Americans and Northwest Europeans like to get to the point, not wanting to beat about the bush. Many other cultures find this approach rather crude and offensive and do not rush through their days on tight time-schedules (Hampden-Turner and Trompenaars, 1993).

In the so-called First World, people have experienced a speeding up of human behaviour through technology - e.g. consuming and travelling. Consequently, time becomes a cost due to the variety of activities that can be undertaken and their perceived opportunity cost - *time is money*! A consistent finding in industrial culture is the attitude that safety is an optional extra that costs time and money as opposed to being part of productive activity. People therefore often view safety measures as costly inconveniences that slow down the job (Andriessen, 1978; Bird and Loftus, 1976; HSC, 1993; Levine *et al.*, 1976).

2.7.6   The impact of societal cultural orientation

Preferred management styles have their roots in, and in turn, influence, national culture. The concepts of US authors Deming regarding quality circles, and Scanlon on self-managed groups, form the basis of Total Quality Management (TQM). Organisations in the USA generally rejected these ideas in favour of Taylorism in the first half of the 20th Century. Only after Japanese corporations prospered did US and European organisations consider employing TQM principles (Hampden-Turner and Trompenaars, 1993). The success of TQM in Japan is due, to a large degree, to its synergy with Japanese culture. Japanese culture is known for being collectivist, in contrast with the strongly individualist American culture (Hofstede, 1980). Japanese commentators consider that Taylorism and scientific management will be the downfall of American industry because they ignore the ideas of the majority of workers within an organisation (Hampden-Turner and Trompenaars, 1993).

The orientation towards individualism in the culture of the UK and the USA is reflected in increasing fragmentation, litigiousness, short-termism, scientism, and standardisation. An individualistic orientation is linked to the concept of universalism where rules and theories are applied to all individuals. Individualism is oriented towards wealth creation, sequential time, consumption and entrepreneurial risk taking. Individualism has an analytical orientation towards economics, science, engineering and technology. US corporate culture is highly individualistic and scientific. The USA pioneered piecework, paying workers according to the number of pieces produced above a fixed minimum (Hampden-Turner and Trompenaars, 1993).

Americans prefer push strategies whereas the Japanese prefer pull strategies (Hampden-Turner and Trompenaars, 1993). A pull strategy is characterised by a spirit of partnership to achieve a future goal. A push strategy starts with the present and projects a schedule of sequential stages. Completion of each stage pushes the next stage into motion and if nothing goes wrong then the product emerges at the end on schedule. The entire schedule is in jeopardy with the lateness of any one stage, unless time for uncertainties is built into the contract, which is rare in cultures where time is money and competition is harsh. The push strategy is characteristic of the UK construction industry culture.

Societal cultural biases influence peoples' attitudes, exposure and responses to risks and accidents. Hampden-Turner and Trompenaars (1993) describe how the Japanese and American directors of the same company acted differently in response to an accident. The American director looked to pin the blame for the accident on an individual. The Japanese manager, valuing teamwork above retribution, accepted collective responsibility for accidents because everybody makes mistakes and identifying a guilty party might destroy team morale. It is more important to identify accident causes to learn to avoid similar problems in the future, than to apportion blame. This approach contrasts sharply with the blame approach prevalent in the UK construction industry.

Positive safety cultures are characterised by employees having a questioning attitude, a rigorous and prudent approach and positive communications (Meshkati, 1995). Communications in organisations are influenced by national culture as well as the organisation's specific culture. Hofstede (1994) and Meshkati (1995) consider that the degree to which employees will have a questioning attitude is greatly influenced by power distance, rule orientation and uncertainty avoidance of the social environment, and the openness of the organisation's culture. Changing from rule-based to knowledge-based functioning is influenced by the safety culture of the organisation and societal cultural orientation (Meshkati, 1995).

2.7.7   International studies of safety culture

International comparisons of organisational safety culture have tended to be limited to the nuclear industry (Fujita *et al.*, 1993; Meshkati, 1995; Rochlin and von Meier,

1994; Rosen, 1995). Rochlin and von Meier (1994), in a study of nuclear reactors in the USA and Europe, found that the unique historical, social, and cultural environment has both functional and operational consequences. A characteristic of collectively orientated cultures is employee involvement - e.g. suggestion boxes are well used in Japan and Scandinavian countries but not in the USA and Germany.   American culture views collectivist behaviour negatively, while Japanese culture views it positively (Fujita *et al.*, 1993). Characteristic orientations of US culture include linear engineering paradigms and a command and control style of regulation that is legalistic and adversarial.   Rosen (1995) reports significant national variations in the degree of reliance placed on staff to discharge their responsibilities.   Based on effective national training programmes, Switzerland, Sweden and Germany place a high degree of confidence on operating staff.   In the USA greater emphasis is placed on supervision and systems for oversight and review.

Rosen (1995), in a study of 20 nuclear plants, distinguishes three broad culturally different geographical areas; the Far East, the former Soviet - Union and Eastern Europe, and Western Europe and North America.   In the Far East politeness and respect towards superiors are characteristics of the work culture. Staff are expected to be supportive of their superiors and to work closely as a team. Most notably in Japan, decisions come about only after extensive discussion of alternatives and possible consequences.   With these culturally driven practices, management can easily emphasise each employee's responsibility.

Characteristics of safety culture in the countries of the former Soviet Union and Eastern Europe include authoritarian management styles and compartmentalisation of activities and responsibilities (Rosen, 1995). This results in a lack of knowledge and curiosity among power plant personnel, a lack of a questioning and a non-self-critical approach.   In the former Soviet Union, construction and production were emphasised over safety, resulting in designs lacking safety features commonly found in Western and Far Eastern plants. Operational safety practices were also weak at Soviet designed plants, while severe economic problems continue to limit improvements. The shortage of electricity in some countries results in considerable pressure to maximise production and continue operating plants with safety deficiencies.

Hofstede's dimensions of culture have been used in a cross-cultural study of safety culture in nuclear power plants owned by the same US corporation located in Mexico, Puerto Rico and the USA (Meshkati, 1999). The Mexican plant achieved the highest scores of cultural compatibility and safety performance, while the US achieved the lowest scores. It was found that even within a multinational corporation with strong efforts devoted to socialisation, national culture played a major role in influencing attitudes to work, risk, power and time (Meshkati, 1999). Open communications, procedure following and teamwork is influenced by power distance, collectivism and uncertainty avoidance. Zhang and Barrett (1997) use Hofstede's dimensions to examine the fit of quality management techniques in the Chinese construction industry and highlight the importance of societal culture.

At a tri-nation seminar on safety and health in the construction industry, cultural differences in approaches to accidents and safety management were highlighted. The American delegate considered that many accidents resulted from a 'get it done' attitude, by equipment malfunctions, and by momentary mental lapses among employees. Worker compensation costs were the driving force for safety and health in the US construction industry. The Canadian delegate considered the root cause of accidents to be the safety versus productivity conflict, while the key to success was training supervisors. The Mexican delegate considered that Mexican construction managers see safety and health as an integral part of the work process, but referred to the fact that it is difficult to separate workers from their community standards (Drysdale, 1995).

In a study of 30 construction sites in Botswana, Ngowi and Mothibi (1996) found that in spite of having the same level of basic training, employees from different cultural backgrounds viewed safety procedures differently. Site managers stated that safety gear provided to employees from impoverished backgrounds was often sold. Traditional construction techniques, such as using mud mixed by hand, proved to be obstacles to get workers to appreciate the need to wear gloves when working with concrete. Some local cultures were considered to be more emotional or more dominant, causing some difficulties for safety management (Ngowi and Mothibi, 1996).

Culture is increasingly becoming an issue of concern to construction academics. Langford (2000) highlights differences in expression, formal contract, individualism and perceptions of time as key societal cultural differences that impact upon construction project management. Phua (2002) explores the impact of collective bias in the Hong Kong construction industry, identifying positive cooperative operating within organisations, but non-cooperation with out-groups outside the organisation. Tijhuis (2002) highlights the impact of cultural issues when transferring Western construction technologies overseas. Decisions made during the procurement process significantly impact upon the success of construction projects. The failure to consider local cultural issues such as the lack of training opportunities, problematic infrastructure, local conflicts and attitudes to time can cause projects to fail. The procurement process needs to consider how to transform technologies to meet local conditions, expectations and needs.

2.8   COMPARING UK AND CARIBBEAN CONSTRUCTION INDUSTRY CULTURES

The UK construction industry produces many excellent buildings and structures, however, projects frequently overrun time and cost schedules, do not meet user requirements, and fail to last as long as they should (Ball, 1988). Industry characteristics include fierce competition, poor communications, little teamwork, contractual problems, litigation and poor safety management (Abeytunga, 1978; Latham, 1993). It has a macho culture where crisis management and conflict are commonplace and safety responsibilities are ignored (Latham, 1994; Sherrington, 1997). The public image of the industry is that it is a dirty, dangerous and hard, with long working hours and frequent discrimination against ethnic minorities and women (Strategic Forum for Construction, 2002). Productivity in the UK construction industry is 30% lower than in most other European countries (Atkin *et al.*, 1995; Financial Times, 1993).

The UK construction industry is dominated by 19th Century attitudes towards managing the labour force and pays scant regard to worker well-being. At site level UK construction workers experience poor health, safety and welfare provision as well as situational pressures such as bonus schemes, tight time schedules, poor designs and unsafe working conditions (Dawson *et al.*, 1985; Leather, 1987). They have to work long hours, rush work, and take short cuts and risks. They often mistrust managers, think they are forced to take risks to get the work done and to earn decent money. Workers tend to believe that management are more concerned with getting the work done than with managing safely. Risk taking, conflict, restricted participation and teamwork, and lack of competence are key negative features of the safety culture of the UK construction industry.

Standards of site safety often depend on the attitudes and experience of line managers. Managers often consider that the best way to motivate operatives is through traditional scientific management techniques (Honey, 1996). UK site managers commonly often consider that the primary cause of accidents is worker's stupidity, that accidents are part of life, and there is nothing you can do to stop them (Dawson, *et al.* 1985; Honey, 1996; Leather, 1987; Whittington, Livingston and Luca, 1992). Site managers use a form of sliding calculus in which they balance their assessment of the risks involved in a particular contravention against the benefits gained from it (Leather, 1987; Dawson *et al.*, 1988). They commonly ignore unsafe work practices particularly during the completion of urgent work. Managers in the British construction industry operate in an extremely competitive and pressured environment. Their reputations and futures depend on the successful completion of the project within the limits set by the contract.

> *Given these types of pressures and the enormous variety of tasks they are expected to undertake, it is not unexpected that considerations of health and safety are quite often of marginal significance, except when they directly affect, or are likely to affect, the progress of works.* Dawson, *et al.* (1988:110)

Many construction companies are family owned and characterised by tough managers who favour Fordist and Taylorist management techniques based on piece-rates and bonuses (Ball, 1988). In the construction industry, subcontractors, supervisors and workers often have to sort out how to execute the work. Skills, knowledge, and resources are often limited at this decision-making level, reducing the effectiveness of risk management. The one-off nature of most projects, wide use of subcontracting, and lack of worker loyalty, create a challenging environment in which to manage.

UK construction industry characteristics include endemic risk taking, cut-throat competition, price-based competitive tendering, low profit margins, penalty clauses and litigation (Latham, 1994). Complex contractual chains link many different firms, each with their own skills, methods of work and particular financial and time pressures. Fragmentation encourages each party to protect their own short-term interests. Price-based competitive tendering, programme pressures and use of penalty clauses set up pressures that conflict with good safety management (Leather, 1987).

Lowest-price competitive tendering and reactive quality control lead to what CIRIA (1991) refer to as the vicious circle of construction procurement (Figure 2.3). Contractors and designers are forced to under-price jobs to get work. Predicted costs are exceeded and cost-cutting exercises are put in place. To cope with the lack of finance, consultants and contractors reduce the quality of work and make claims. The client then institutes more reactive control measures. It is hard for a project to recover from cost-cutting exercises. Contractors hide delays caused by them in the hope that delays caused by the client or consultants, may be blamed to avoid liability. This vicious circle of construction procurement is the antithesis of the quality philosophy.

Fragmentation encourages mistrust and poor communication between designers, contractors and clients. Tight financial and time constraints restrict the efforts contractors will put into effective health and safety management. Those who design structures frequently fail to ensure that their designs can be constructed safely.

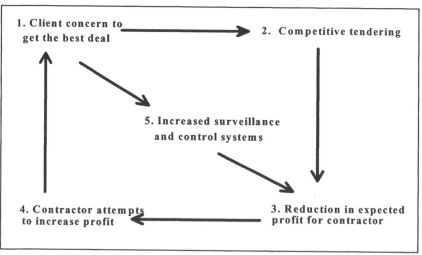

Figure 2.3   The Vicious Circle of Construction Procurement (CIRIA 1991)

Whittington, *et al.* (1992) identify the following fundamental problems of the industry:

- time and financial pressures
- poor design
- poor equipment
- inadequate safety skills and management training
- poor communications
- inadequate planning
- reliance on subcontractors

Latham (1993) likens the UK construction industry as a mighty machine that has grit in the engine instead of oil. Clients worry that contractors are cheating them and that they will not get the buildings they want. Professional consultants often fear that they will be held responsible by clients for any unseen extra expenditure on contracts. Contractors fear underpayment by clients, overcharging by specialist subcontractors, the rising cost of materials, and inflated demands from specialists. Latham (1993) describes a debilitating culture of conflict:

... The industry has deeply engraved adversarial attitudes. The culture of conflict seems to be embedded, and the tendency towards litigiousness is growing. ... disputes and conflicts have taken their toll on morale and team spirit. Defensive attitudes are common place. A conflict-ridden, adversarial project is unlikely to come out to cost and time, or to the quality required, and the client will be the loser in the end. (Latham, 1993:10)

Popular construction industry definitions reflect the uncertainty and adversity inherent in the UK construction industry. The following descriptions were contained in a poster on the wall in the office of a construction manager on a major civil engineering contract:

subcontractor:	a gambler who never gets to shuffle or deal or cut;
tender submission:	a poker game in which the losing hand wins;
tender sum:	a wild guess to 2 decimal points;
successful tenderer:	a contractor who is wondering what he left out;
architects estimate:	the cost of construction in heaven;
management contract:	the technique of losing your shirt under perfect control;
completion date:	the point at which liquidated damages begin;
liquidated damages:	the penalty for failing to achieve the impossible;
quantity surveyor:	people who go in after the war is lost and bayonet the wounded;
lawyers:	people who go in after the QS and strip the bodies.

The lack of both trust and money creates an adversarial culture and confrontational environment between clients, designers, contractors and workers

(Latham, 1994). The director of a national contracting firm summed up the culture of the UK construction industry as a 'macho culture that screws the subcontractor' and rewards 'crises management' (quoted in Latham, 1994). Health and safety matters are considered unimportant in relation to the fight for work completion to tight time schedules and financial survival. The construction industry in the UK is contractually one of the most complicated and conflict ridden in the world. Amec's Chairman and Chief Executive, Sir Alan Cockshaw, stated that:

> *I have always believed that this is the most adversarial country in the world for the way that contracts are organised. It has to change. The consequence is the clients don't get value. Co-operation within and without Britain is where a lot of the future lies.* (Cockshaw, 1995, *Contract Journal* 19 January,. p. 21)

2.9   THE SAFETY CULTURE OF THE UK CONSTRUCTION INDUSTRY

The characteristics of positive safety cultures align with many aspects of total quality management. At the organisational level, characteristics of positive safety cultures include belief in the prime importance of risk elimination and control, humanistic values, empowerment, participation, team working, monitoring and training. Many of the characteristics of the UK and Caribbean construction industries identified in this and other studies conflict with the essential characteristics of positive safety cultures. Key negative characteristics of the UK construction industry safety culture that emerge from this study are summarised in Table 2.2.

Research on safety management in the construction industry in Britain, includes work by Abeytunga (1978), Bomel (2001), Cooper *et al.,* (1991), Leather and Butler (1983), Dawson *et al.* (1988), Jannady (1996), Leather (1987, 1988), Whittington *et al.* (1992) and Wilson (1989). Construction managers often view safety as a cost that conflicts with production, take little direct interest in safety, rely on site supervisors to manage safety, and neglect safety when they feel strong programme and financial pressures (Leather, 1987). Time and cost pressures result in risks being ignored and the law being broken with the consent of supervisors and managers. Workers commonly view safety officers as largely powerless, perceiving safety policies to contain good words, but are ignored by those making management decisions (Jannady, 1996). Other key factors affecting safety in the construction industry are the frequent failure to provide effective control of site hazards, adequate safety training and to specify safety responsibilities.

Following Shillito's (1995) typology, the safety culture of the UK construction industry is a *rule book* orientated culture where safety management is poor and motivated more by the desire to comply with the legal rules and avoid prosecution, than to actively prevent loss. Workers are considered careless and are blamed and punished for mistakes and accidents. Training is scarce, scientific management styles prevail and many in the industry believe that safety conflicts with productivity and profitability.

Following Eckenfelder's (1997) typology, the safety culture of the UK construction industry is *immature*. At best, some of the larger contractors are at the *novice* stage of maturity, where efforts are made to integrate safety into business goals, safety awareness is growing, safety management is sometimes proactive, but empowerment and involvement of workers is limited. Many of the smaller contractors, which dominate the industry, are at *the darkness* or *ignorance stage* of safety culture maturity where safety is seen as a burden, safety management is purely reactive and safety performance measurement is absent.

Table 2.2   Negative Characteristics of the UK Construction Industry Safety Culture

INDUSTRY LEVEL	ORGANISATION LEVEL	SITE LEVEL
☠ frequent conflict and litigation	☠ domination by small companies with few resources	☠ job insecurity
☠ focus on cost		☠ dangers taken for granted
☠ complex regulatory system	☠ intense competition	☠ risks not controlled
☠ division between design and construction	☠ stress	☠ poor communications
	☠ production related bonus schemes and piece-rates	☠ blame culture
☠ mistrust and blame	☠ safety seen as a site matter	☠ tight deadlines, long hours, low pay, low productivity
☠ shorttermism	☠ limited safety management	
☠ absence of investment in the construction process	☠ safety thought to conflict with production	☠ rush to complete work
		☠ risk taking
☠ prevalence of subcontracting and risk avoidance	☠ safety training limited but improving	☠ short cuts common
	☠ regulations frequently ignored	☠ variable competence
☠ complex, fragmented relationships	☠ limited monitoring and reviewing performance	☠ lack of participative decision making
☠ lack of consideration of health and safety in design and planning	☠ poor organisational learning	

2.9.1   The Caribbean

The Caribbean construction industry is an important part of the region's economy, consisting of private and public sector clients, design, management and construction specialists. Each country has a dozen or so major construction and civil engineering contractors, which include branches of international contractors. UK contractors frequently operate within the Anglophone Caribbean region. New build work dominates the Caribbean construction industry, as the national development process continues.

In the early 1990s the Caribbean construction industry lacked formal safety management systems and effective health and safety regulation. However, relatively low-technology construction methods and societal values to life, time and work limited the degree of rushing and risk taking in the Caribbean (Peckitt, 2001). Bonus schemes are rarely employed and tight time schedules rarely set. The lack of formal safety management systems at organisational and national levels within the Caribbean safety culture is countered by positive factors manifest at the site level. The Caribbean construction industry safety culture was positively biased towards:

- good interpersonal communications
- not rushing work
- enjoying work and
- risk perception allied with locus of control

Managers and directors with experience of working both in Britain and the Caribbean considered that labourers in the Caribbean tended to be better workers, but that tradesmen were not as skilled as their UK counterparts. They stated that construction standards were better in Britain because the building process is more regulated and controlled. Emphasis was also placed on the lack of life pressures compared with more developed countries. When questioned regarding their thoughts on what were the key influences upon site safety culture in the Caribbean, most interviewees commented on the workers' aversion to risk. Common replies from project managers and construction company directors included:

- the guys don't want to die, they enjoy life and don't take risks
- man don't like pressure, and will walk off the job if he feels pressured
- the men are jack of all trades, they are strong, and they don't cut corners
- the extended family, social networks, and farming their own land takes the pressure off men to earn money all the time
- the guys don't rush or take risks, and there is no sense of machismo

Directors, managers and foremen stated that there were good relations on site between workmen and site managers. Caribbean managers often stressed the importance of open communications and the need to consult with workers. Site relations tended to be more paternalistic on Caribbean sites, with less emphasis on efficiency and speed (Peckitt, 2001). Caribbean foremen were described as experienced and dedicated men, who treated their workers with respect. Both foremen and managers across the Caribbean stated that you had to get the respect of the workers in order to get them to work hard and that workers do not like pressure and will walk off the job if shouted at or pressed to work faster or harder. A Guyanese site manager stated:

The guys are very innovative, they need minimum supervision and like learning new things. You have to know how to deal with people in order to apply pressure. You can't just be blunt and direct, you need to be tactful and get their respect.

A British project manager working in Trinidad stated:

Works look dangerous because there are very few physical precautions taken. The men seem more aware of what they are doing and the potential dangers. They don't expect any safety barriers or other equipment and therefore look after themselves. You cannot pressure the men to work faster.

A Portuguese site manager working in Guyana stated:

Workers here are very keen to learn new skills, but generally you don't get specialist craftsmen. Climate is important in dictating the pace of work. People don't like pressure and will walk off the job. African-Caribbean workers enjoy life and say that white men work to pay bills while black men work to enjoy life.

When questioned regarding the actions they would take to ensure that projects came in on time and on budget, Caribbean directors and project managers stated that they would increase the number of workers on site to speed up production, rather than putting more pressure on existing workers. Workers on job rates stated that they did not speed up work to earn more money, but worked longer hours instead. If they did not feel like working on any particular day, workers would often not bother to turn up to site for a week or two. Particularly in rural areas, they commonly work on a construction site for several weeks and then go missing for a week or two to enjoy the fruits of their labour, or work on their own houses or farmlands.

Compared with the UK there was less pressure on workers and a slower work pace in the Caribbean. Construction managers with experience of both cultures stated that works of similar sizes would take approximately twice as long in the Caribbean compared with Britain. Many factors contributed to extended project lengths in the Caribbean, including: materials shortages due to supply difficulties; inadequate infrastructure; decision changes by clients and architects; lack of finance; climatic influences; management practices; attitudes; and working methods.

Most Caribbean construction projects experience delays due to design changes and materials supply problems. In a study in Trinidad, clients were found to be responsible for 50% of contract delays, consultants for 20% and contractors for 30% (Lewis and Mugishagwe, 1993). Contractors were generally dissatisfied with consultants and 80% of contractors experienced contractual problems relating to poor briefs, drawings and specifications. However, in contrast with the UK the courts were rarely used to settle contractual disputes in the Caribbean in the early 1990s.

2.9.2   The safety culture of the Caribbean construction industry

Lewis and Mugishagwe (1993) highlight the low pay, inadequate design briefs, drawings and specifications, poor motivation, low productivity and skills shortages in the Caribbean construction industry.   Other factors that prevail include corruption, poor infrastructure, political influences on site planning and effects of 'boom and bust' cycles.   These practical challenges dictate a contingency style approach to project management and a relaxed attitude to time schedules.   There is little recourse to the legal system to enforce health and safety or contractual law. Caribbean construction contractors generally did not formulate or implement formal safety management systems and paid little regard to health and safety regulations (Peckitt, 2001).

As in the UK, poor levels of safety training, job insecurity, hazardous working conditions and long work hours characterize the Caribbean construction industry.   Subcontracting, poor quality management and devolving safety management to site level are common to both cultures and are characteristic features of much of the construction industry around the world.   At site level workers do not experience situational drivers for speed and risk taking, with the virtual absence of production-related bonus schemes and little time pressure (Table 2.3).   Workers have a strong sense of locus of control over safety, generally trusted management and looked out for each other.

Table 2.3   Negative Characteristics of the Caribbean Construction Industry Safety Culture

INDUSTRY LEVEL	ORGANISATIONAL LEVEL	SITE LEVEL
absence of investment in the construction process	domination by small companies with few resources	job insecurity
few specific regulations relating to construction	lack of formal safety management systems	long working hours
lack of inspector enforcement powers	few specialisms	multi-skilling
little enforcement action	little training	poor provision of PPE
limited resources	few resources devoted to safety	risks not controlled
limited use of technology	skills shortages	variable competence
unreliable infrastructure	safety seen as a site matter	little safety training
materials supply problems	legislation ignored (where present)	plant and equipment poorly maintained
lack of consideration of health and safety in design and planning		lack of participative decision making

The safety culture of the Caribbean construction industry displays features of the *behavioural* safety culture following Shillito's (1995) typology.   Workers tend to accept responsibility for their own safety and do not rush to achieve production targets.   However, at the organisational level safety is not actively managed.   In Eckenfelder's (1997) typology, the safety culture of the Caribbean construction industry is *immature* as safety management is reactive or absent, training is scarce, there are skills shortages and evidence of poor process management. A prevalent attitude in the industry is that safety expenditure reduces profitability. Little motivational influence is exerted by government through legislation and inspection activities.

The Caribbean construction industry faces less technological risk due to the comparatively limited use of chemicals, plant and machinery, the relatively low height of buildings and robust earthquake resistant designs, general absence of asbestos and easy to handle design of concrete blocks. Time is less of a pressure, production related bonus schemes are generally absent and work pace is relatively slow. The Caribbean construction industry is more labour intensive, with less use of mechanisation and less wastage due to economic pressures than in the UK. The culture of the Caribbean construction industry not as adversarial, formal, pressured and regulated than its UK counterpart.

People who had experience of both cultures highlighted the facts that the tropical climate dictated a slower work pace in the Caribbean, and that less pressure was applied by management to workers in the Caribbean compared with the UK. The absence of formal health and safety risk management systems was balanced to a degree by Caribbean workers' strong locus of control for their own safety and tendancy to be risk averse. Interviewees in the Caribbean stressed the importance to African-Caribbean workers of the values of freedom, love of life (*joie de vivre*) and social aspects of work when describing the important factors that impact upon construction site safety in the Caribbean. Those project managers and company directors who had experience of working both in the Caribbean and the UK frequently stated:

- everything takes twice as long in the Caribbean than it does in the UK
- there are few specialist trades on site therefore you don't get the same kind of competition on Caribbean sites
- site relations are generally less adversarial in the Caribbean than in the UK

African-Caribbean people dominate the manual trades in the region's construction industry. The degree of influence of Afrocentric values and their impact on modern Caribbean societies has provoked much debate because of the history of European hegemony. African-Caribbean culture emphasises collectivity, spirituality and *joie de vivre*. Time is not money and people are prepared to take their time to do things. Both Caribbean and European construction professionals working in the Caribbean stated: *people live to work in the Britain, while in the Caribbean they work to live.* Many Caribbean construction workers live in extended families, place great emphasis on human interaction, dance, music, fertility, and African inspired religious beliefs. Attitude scale items relating to interactions with fellow workers, communications on site and enjoyment of work were all scored highly by Caribbean workers (Peckitt, 2001).

There is a higher risk burden in the UK construction industry compared with its Caribbean counterpart.   There is also a greater degree of safety management by the larger contractors in the UK construction industry that reduces and controls many of the risks involved in the process. This positive aspect of the safety culture is promoted, at least in part, by the long-established system of standard setting, legislation and enforcement. However, the small contractors, who dominate the industry and who account for the majority of fatal accidents rarely have anything more than rudimentary safety management systems and a vague knowledge of health and safety law (Bomel, 2001).

2.10   SOCIETAL IMPACT ON SAFETY CULTURE

A major, although subtle, factor affecting the safety and performance of a technological system is the degree of compatibility between organisational culture and the national culture of the host country. (Meshkati, 1995:264)

The relative importance of the impact of the different levels of analysis of safety culture is important in understanding, predicting and controlling industrial risk (Reason, 1997). According to several studies, an operator's culturally-driven habit is a more potent predictor of behaviour than his/her intentions (Meshkati, 1995). However, current knowledge relating to the impact of societal factors on safety culture is limited due to a focus on organisational and individual factors. Reason (1997) identified individual, workplace, organisational and societal factors in the causation of major accidents. He speculated that of these different causative factors of accidents, organisational culture, processes and the workplace, are the key indicators of explanatory, predictive and remedial value. He ranked individual factors second to organisational factors, followed closely by regulation and society, to which he assigns little remedial value. Findings from this study of safety culture suggests that societal influences may be more important than envisaged by Reason.

Hofstede's dimensions of cultural orientation can be used to compare the biases inherent in UK and Caribbean cultures. Hofstede's sample contained an inadequate number of responses from Caribbean workers to be considered representative of the Caribbean as a whole. Hofstede's findings on UK culture are compared with the related findings reported by Peckitt (2001) relating to Caribbean orientations. The collective and time synchronous orientations of Caribbean societies, reflected in Afrocentric paradigms, contrast sharply with UK orientations towards individualism and perception of time as sequential.

How a culture conceives of time is fundamentally important to many attitudes and behaviours (Schein, 1985). The Afrocentric concept of time is based on natural events rather than imposed mathematical formulas; it is a phenomenal event. Time corresponds with the duration of an event, rather than the number of minutes. It does not matter how long it takes, the most important thing is participation in the event taking place (Alleyne, 1988). The concept of time being determined by events and participation in them has puzzled many outsiders who complain that Africans and African-Caribbeans do not keep time or just waste time sitting down doing nothing and seem lazy.

Life in the Caribbean in the 1990's was comparatively unhurried and relatively little importance was attached to strict timekeeping. In the construction industry this was reflected by strict time schedules not being assigned to many construction projects. Caribbean people generally do not experience anxiety or impatience when waiting, it is not important how long you spend at (say) the barbershop, what matters is the quality of social interaction. Communication, style and immediate gratification are important (Alleyne, 1988). Life is not all about work, and is not divorced from social activity.

On one site visited in Trinidad in 1992 an English company was involved in the construction of several large concrete silos by slip forming. The Trinidadian site manager expressed the view that British employees worked at a greater speed than Trinidadians, and as a result the accident rate of British workers was higher. He also stated that the stripping down of the flying form on completion of each silo would have taken up to three weeks if carried out by Trinidadians. The English firm carried out this operation within one week. However, during one of the dismantling phases the British foreman tried to release a jammed part of the structure by kicking it away. As he did this, the part gave way and as the crane took the load it jerked him off the platform that he was standing on. He was fortunate that one of the Trinidadians was quick enough to grab him and save him from falling off the silo.

Hofstede (1980) identified the UK as one of the most individualistic societies in his study. Emphasis on technology, material wealth and formal legal regulation characterise individualistic cultures. Individualistic cultures are biased towards risk taking and poor team working, both of which are negative aspects of the UK construction industry. Caribbean people recognise the increasing influence of the USA in the region, a growing trend towards individuality and a breakdown in the sense of community over the last 20 years. Alleyne (1988) describes a process as the confrontation of European individualism and traditional Jamaican collectivism. Those living in the urban Caribbean, in particular, feel the tremendous pull of modernisation, the pressure to get rich quick and consume, creating a cultural rootlessness (Alleyene, 1988). However, the collectivist bias is still strong.

Hofstede (1980) identified UK culture as orientated towards low power distance with emphasis placed on the legal system, democratic power and technology. The presence of factors associated with high power distance in the Caribbean highlight differences with the UK on this dimension. Differences in power distance between these cultures can at least in part account for differences in approaches to regulation and relationships within the construction process. Caribbean culture is less biased towards masculinity and low uncertainty avoidance than in the UK, which is reflected in the greater importance attached to communication, self-expression and team working in Caribbean society.

Table 2.4 summarises key influences upon the relative orientations of the two cultures identified in the literature and reported in Peckitt (2001). The factors in each category combine and interact in a variety of ways to form the cultural orientations of each region.

At the societal level in the Caribbean, the natural environment plays a significant role in shaping safety culture, influencing building designs, construction methods and working patterns. The relative lack of economic development impacts upon the infrastructure, legislature, methods of work and use of technology. The orientation towards collectivism, enjoyment of life, spirituality and synchronous time influence project planning, management styles and working patterns. Competency, competition, societal values and the socio-technical work environment influence risk taking. Afrocentric attitudes to work, life and time together with the tropical climate do not promote rushing (Peckitt, 2001).

In Caribbean culture time is considered synchronous, the past connects with the present to influence the future, activity is unhurried, people take their time, and limited importance is attached to strict punctuality. Life is about enjoyment and people work to live. The collective spirit is demonstrated in the Rastafari concept of *I and I is one*. Caribbean cultures place high value on expressiveness through conversation, art, music and dance. Nature is respected and considered an essential aspect of existence rather than the European view that it is something to defy, control and exploit (Ani, 1994; Asante, 1990). Clean, sharp lines and divisions characteristic of European culture dissolve into the Afrocentric understanding of the oneness of man and nature (Asante, 1992).

Table 2.4   Key influences upon Societal Cultural Orientation in the UK and the Caribbean

	UK	Caribbean
Situation	temperate climate	tropical climate
	industrialised	agricultural, industrialising
	technology dominates life	nature dominates life
	former colonial power	former colonies
Cognition	universalistic, segmented	particularistic, holistic
	domin nature	live with nature
	scientific knowledge	intuitive knowledge
	inner-directed	outer-directed
	time sequential	time synchronous
	low uncertainty avoidance	medium uncertainty avoidance
	material wealth	*joie de vivre*
	bulldog spirit	Anansi smartness
Behaviour	individualistic, hierarchist	egalitarian, fatalistic expressive
	competitive, assertive	medium / high power distance
	low power distance	work to live
	live to work	slow pace of life
	fast pace of life	

Compared with the Caribbean, the UK construction industry is more positive in relation to regulation, resources, training and formal safety management.   Negative influences within the UK construction industry safety culture include adversity, complex subcontracting relationships, risk taking and rushing. The low uncertainty avoidance orientation of the UK is associated with individualism, risk taking and blame (Hofstede, 1980). British concepts of time and risk ensure that construction clients and contractors frequently focus too much on costs and progress at the expense of quality and safety, resulting in frequent accidents, disputes, litigation, poor quality and rushed work.

Positive attitudes towards risk taking, materialism, individualism, hierarchy and technology are central tenets of the UK industrial culture. The dominant rational positivistic paradigm has created a culture of technocracy and severe managerial and communications problems (Smith, 1998). These problems are not conducive to development of positive risk management culture, which requires humanistic management styles, open communications and commitment to loss control.   The tendency for individualistic cultures to blame individuals for accidents has its roots in the societal orientation towards individuality, which is a feature of modernity (Rosen, 1995).

2.11   CONCLUSION

The safety culture paradigm provides a global characteristic through which deep-rooted preconditions of accidents can be identified (Pidgeon, 1991). Despite its lack of precisely quantifiable parameters, the safety culture paradigm directs consideration of factors that have been overlooked by previous safety paradigms. Societal cultural biases can have a significant impact upon safety culture. For example, despite the common occurrence of unprotected drops in the Caribbean, mediating cultural factors reduce the potential for falls, corresponding with Hinze's (1996) *Distractions Theory* and Leather's (1987) PAS *Model*. The relatively slow pace of work, worker agility, and sense of locus of control over their own safety are cultural factors which help explain why falling accidents appear to be less of a problem compared with the UK (Peckitt, 2001).

There are common features of the construction industry that impact negatively on safety culture.   The fragmentation of the industry and nature of procurement promotes distrust between organisations.   Job insecurity, long working hours, lack of training and poor health and safety risk management appear endemic in the construction industry.   However, there are other factors that influence the culture of the industry.

Societal orientations to power relationships, time, human relations, materialism and risk taking emerge as important, but subtle factors that impact on safety culture. The British individualistic and hierarchical bias promotes a blame centred, legalistic bias towards health and safety regulation and management. The perception of time as sequential and as a cost combined with strong individualism, ensures construction clients and contractors focus on time and cost, resulting in frequent disputes and litigation. The bias towards low uncertainty avoidance limits the detail of safety management systems and procedures in both cultures.

To improve the health and safety risk management culture of the British construction industry there is a need for a shift away from the cultural antecedents of poor performance. Risk taking as a cultural value needs to be tempered by the risk management philosophy that promotes partnering, team working and humanistic management styles. Time needs to be seen as an ally as well as a cost and scarce resource. This requires a cultural shift towards collective, synthesising, particularistic and feminine values.

Change initiatives in the UK construction industry prompted by the reports of Latham (1994) and Egan (1998) are resulting in positive action to alter procurement methods to increase trust and team working, to improve competence and the respect for workers. However, there is still much room for improvement in the procurement process to remove adversity and increase efficiency and quality (Jagger *et al.*, 2002). Relational contracting, based on win-win scenarios and cooperative working, could well offer the new procurement paradigm that will propel the cultural shift needed in the industry (Rahman, *et al.*, 2002). True partnerships will allow the building of trust and understanding that can remove the adversity, distrust and shorttermism currently hampering the industry.

2.12   REFERENCES

Abeytunga, P.K., 1978, The role of the first line supervisor in construction safety: The potential for training, University of Aston in Birmingham.

Adams, J., 1995, *Risk* (London: UCL Press).

Alleyne, M., 1988, *The Roots of Jamaican Culture* (London: Pluto Press).

Andriessen, J.H., 1978, Safe behaviour and safety motivation. *Journal of Occupational Accidents* (Amsterdam: Elsevier Scientific Publishing Co.), pp. 363-376.

Ani, M., 1994, *Yurugu: An African-centred Critique of European Cultural thought and Behaviour* (New Jersey: African World Press Inc.).

Asante, M.K., 1990, *Kemet, Afrocentricity and Knowledge* (New Jersey: African World Press Inc).

Asante, M.K., 1992, *Afrocentricity* (New Jersey: African World Press).

Ball, M., 1988, *Rebuilding Construction: Economic Change in the British Construction Industry* (London: Routledge).

Back, W. and Woolfson, C., 1999, Safety culture - A concept too many? *Health and Safety Practitioner*, January, pp. 14-16.

Bird, F.E. and Loftus, R.G., 1976, *Loss Control Management, Loganville*, (Georgia: Institute Press).

Bomel Ltd, 2001, *Improving Health and Safety in Construction, Phase 1: Data Collection, Review and Structuring*, Contract Research Report 386/2001 (Sudbury: HSE Books).

Bond, M.H., and Hofstede, G., 1989, The cash value of Confucian values. *Human Systems Management,* 8, pp. 195-199.

Building, 2001, *If You Think It Is Bad Here....,* Anon, 9 February.

CIRIA, 1991, *Roles Responsibilities and Risks in Management Contracting,* Special Publication 81 (London: CIRIA).

Cockshaw, A., 1995, Amec chairman quoted in *Contract Journal* 19, January.

Cooper, M.D., 1998, *Improving Safety Culture: A Practical Guide* (J. Wiley and Sons).

Cooper, M.D., Phillips, R.A. Robertson, I.T. and Duff, A.R., 1991, Improving safety on construction sites by the utilization of psychologically based techniques: Alternative approaches to the measurement of safety behaviour. Paper given at the *5th European Congress of the Psychology of Work and Organization,* Rouen, France.

Coote, J.A. and Lee, T.R., 1993, *Employee Perceptions of Safety at Sellafield: Initial Results of the Safety Survey* carried out in 1991/92 BNFL.

Cotgrove, S., 1982, *Catastrophe or Cornucopia: The Environment Politics and the Future,* Chichester, Wiley.

Cox, S. and Flin, R., 1998, Safety culture: Philosopher's stone or man of straw? *Work and Stress,* 12. Safety culture - special issue, pp. 187-201.

Dawson, S., Clinton, A., Bamford, M. and Willman, P., 1985, Safety in construction: self-regulation, industrial structure and workforce involvement. *Journal of General Management,* 10, 4, pp. 21-38.

Dawson, S., Willman, P., Bamford, M. and Clinton, A., 1988, *Safety at Work: The Limits of Self-Regulation* (Cambridge University Press).

Dester, W.S. and Blockley, D.I., 1995, Safety, behaviour and culture in construction. *Engineering, Construction and Architectural Management,* 1, pp. 19-26.

Douglas, M., 1992, *Risk and Blame: Essays in Cultural Theory* (London: Routledge).

Douglas, M. and Wildavsky, A., 1982, *Risk and Culture,* University of California.

Drysdale, D.T., 1995, A summary of the trinational seminar on safety and health for the construction industry. *American Industrial Hygiene Association Journal,* 56, pp. 929-934.

Eckenfelder, D.J., 1997, It's the culture stupid. *Occupational Hazards.* June 1997. pp. 41-44.

Egan, J., 1998, *Rethinking Construction* (HMSO).

Erickson, J.A., 1997, The relationship between corporate safety culture and performance, *Professional Safety.,* ASSE, May 1997, pp. 29-33.

Financial Times, 1993, *Building and Development Economics in the EC,* Report.

Fujita, Y., Toquam, J.L., Wheeler, B., Tani, M. and Mouri, T., 1993, Ebunka: Do cultural differences matter. In *Proceedings of IEEE Nuclear Safety Conference,* 92CH3233-4, pp. 188-194.

Glendon, A.I., 2000, Safety culture. In W. Karwoski (ed.) *International Encyclopedia of Ergonomics and Human Factors* (London: Taylor & Francis), pp. 1337-1340.

Glendon, A.I., and Stanton, N.A., 2000, Perspectives on safety culture, *Safety Science*, 34, **1-3**, pp. 193-213.

Glendon, A.I. and McKenna, E.F., 1995, *Human Safety and Risk Management*, (London: Chapman Hall).

Goszczynska, M., Tyska, T. and Slovic, P., 1991, Risk perception: a comparative study with 3 other countries. *Journal of Behavioural Decision Making*, 4, pp. 79-193.

Guest, D.E., Peccei, R. and Thomas, R., 1994, Safety culture and safety performance: British Rail in the aftermath of Clapham Junction disaster. Paper presented at the *Bolton Business School Conference on Changing Perceptions of Risk*, Bolton, February 23-March 1.

Guldenmund, F.W., 2000, The nature of safety culture: a review of theory and research. *Safety Science*, 34, pp. 215-257.

Hampden-Turner, C. and Trompenaars, F., 1993, *The Seven Cultures of Capitalism* (New York: Doubleday).

Hale, A.R., 2000, Culture's confusions (editorial), *Safety Science*, 34, pp. 1-14.

Hall, S., 1997, Cultural representations and signifying practices. In *Representation Introduction to Book 2 Cultural Studies MA Course* (London: Sage/Open University).

Haralambos, M. and Holborn, M., 1991, *Sociology Themes and Perspectives*, 3rd ed. (London: Collins Educational).

Health and Safety Commission, 1993, *ACSNI Human Factors Study Group Third Report, Organising for Safety* (London: HMSO).

Health and Safety Executive, 1989, *Human Factors in Industrial Safety*, Health and Safety Series booklet HS(G), 48 (HMSO).

Health and Safety Executive, 1994, *Fatal Injuries in the Construction Industry 1986/87 to 1991/2*, Statistical Services (stavdg\94\15\2) Bootle.

Hinze, J.W., 1996, The distractions theory of accident causation. In *CIB W99, Implementation of Safety and Health on Construction Sites*, edited by L.M. Alvez Dias, and R.J. Coble, Lisbon, Balkema, Rotterdam, pp. 375-373.

Hofstede, G., 1980, *Cultures' Consequences*, International Differences in Work-related Values, 5, Cross-cultural Research and Methodology Series, (Sage Publications).

Hofstede, G., 1994, *Culture and Organisations: Software of the Mind* (London: Harper Collins Publishers).

Hofstede, G. and Bond, M.H., 1988, The Confucius connection: from cultural roots to economic growth. *Organizational Dynamics*, 16, pp. 4-21.

Honey, R., 1996, Motivation is of the key to site safety. *Construction News*, 29 August.

House, R., Javidan, M., Hanges, P. and Dorfman, P., 2002, Understanding cultures and implicit leadership theories across the globe: an introduction to project GLOBE. *Journal of World Business,* 37, pp. 3-11.

International Atomic Energy Authority, 1998, *Developing Safety Culture in Nuclear Activities Practical Suggestions to Assist Progress,* SRS, 11, IAEA Vienna.

Jannady, T.M.O., 1996, Factors affecting the safety of the construction industry. *Building Research and Information,* 24, **2**, pp. 108-112.

Jagger, D.M., Ross, A., Love P.E.D. and Smith, J., 2002, Towards Achieving More Effective Procurement through Information. In M. Lewis (ed.), *Procurement Systems and Technology Transfer CIB W92,* University of the West Indies, Trinidad & Tobago, pp. 179-193.

Langford, D., 2000, *The Influence of Culture on Internationalisation,* Keynote address, Meeting of CIB TG29, Botswana.

Latham, M., 1993, Trust and Money, *Interim Report of the Joint Government/ Industry Review of Procurement and Contractual Arrangements in the United Kingdom Construction Industry* (London: HMSO).

Latham, M., 1994, *Constructing the Team* (London: HMSO).

Leather, P.J. and Butler, R., 1983, Attitudes to safety among construction workers a pilot survey, BRE Report, Dept. of Behaviour in Organisations, University of Lancaster.

Leather, P.J., 1987, Safety and accidents in the construction industry: a work design perspective, *Work and Stress,* 1, **2**, pp. 167-174.

Leather, P.J., 1988, Attitudes towards safety performance on construction work: an investigation of public and private sector differences. *Work and Stress,* 2, **2**, pp. 155-167.

Levine, J.B., Lee, J.D., Ryman, D.H. and Rahe, R.H., 1976, Attitudes and accidents aboard an aircraft carrier, Report No. 75-68. *Aviation and Space Environment Medicine,* 47, **1**, pp. 82-85.

Lewis, T.M. and Mugishagwe, D.D., 1993, Operational problems facing a sample of construction firms in Trinidad and Tobago (unpublished Paper), Civil Engineering Dept, St Augustine Campus, University of the West Indies, Trinidad & Tobago.

Lockhart, R.W. and Smith, K.U., 1986, Theory and practice of involving workers in management programs, *First International Congress on Industrial Engineering and Management,* Paris.

MacKenzie, J., Gibb, A.G.F. and Bouchlaghem, N.M., 1999, Communication of health and safety in the design phase, In *Proceedings of the Second International Conference of CIB Working Commission W99, Implementation of Safety and Health on Construction Sites,* edited by A. Singh, J. Hinze, and R.J. Coble., Honolulu, Hawaii, 24-27 March, pp. 419-426.

Meshkati, N., 1995, The culture context of safety culture: a conceptual model and experimental study. In *Safety Culture in Nuclear Installations Conference*

Proceedings of the International Topical Meeting on Safety Culture in Nuclear Installations, 24-28 April, Vienna, Austria, pp. 261-270.

Meshkati, N., 1999, The cultural context of nuclear safety culture: A conceptual model and field study. In *Nuclear Safety: A Human Factors Perspective* edited by J. Misumi, B. Wilpert and R. Miller, pp. 5-24 (London: Taylor and Francis Ltd).

Ng, S.H., 1980, *The Social Psychology of Power*, European monographs in social psychology 21 series editor- Henri Tajfel (London: Academic Press).

Ngowi, A.B. and Mothibi, J., 1996, Culture and safety at work site - A case study at Botswana. In *CIB W99, Implementation of Safety and Health on Construction Sites*, edited by L.M. Alvez Dias and R.J. Coble., Lisbon, Balkema, Rotterdam, pp. 417-429.

Ordraguez, Y.A., Jaselskis, J.E. and Russell, J.S., 1996, Relationship between project performance and accidents. In *CIB W99, Implementation of Safety and Health on Construction Sites* edited by L.M. Alvez Dias and R.J. Coble, Lisbon, Balkema, Rotterdam, pp. 251-257.

Peckitt, S.J., 2001, Construction industry safety culture: A comparative study of Britain and the Caribbean, Aston University.

Peckitt, S.J., Glendon A.I. and Booth R.T., 2002, A comparative study of safety culture in the construction industry of Britain and the Caribbean. In *Proceeding of the International Symposium of the Working Commission CIB W92, Procurement Systems and Technology Transfer*, edited by M. Lewis, The Engineering Institute, University of West Indies, Trinidad & Tobago, pp. 195-220.

Pedersen, A., 1995, Safety culture in nuclear installations. In *Safety Culture in Nuclear Installations Conference Proceedings of the International Topical Meeting on Safety Culture in Nuclear Installations*, 24-28 April, Vienna Austria, pp. 271-288.

Phua T.K., 2002, Effects of cultural differences on project participants' behaviour in construction projects. In *Proceeding of the International Symposium of the Working Commission CIB W92, The Case of Hong Kong in Procurement Systems and Technology Transfer*, edited by M. Lewis, The Engineering Institute, University of West Indies, Trinidad & Tobago, pp. 497-511.

Pidgeon, N.F., 1991, Safety culture and risk management in organisations. *Journal of Cross Cultural Psychology*, 22, **1**, pp. 129-140.

Pidgeon, N.F., 1998, Safety culture: Key theoretical issues. *Work and Stress*, 12, **3**, pp. 202-216.

Pidgeon, N.F. and O'Leary, M., 1994, Organisational safety culture: implications for aviation practice. In *Aviation Psychology in Practice*, edited by N. Johnston, N. McDonald, and R. Fuller, pp. 21-43.

Rahman, M., Kumaraswamy, M. Rowlinson, S. and Palaneeswaran, E., 2002. In *Procurement Systems and Technology Transfer, CIBW92, Transformed Culture and Enhanced Procurement through Relational Contracting and*

Enlightened Selection edited by M. Lewis, University of the West Indies, Trinidad & Tobago, pp. 383-401.

Reason, J., 1995, A systems approach to organisational error. *Ergonomics*, 38, **8**, pp. 1708-1721.

Reason, J.T., 1997, *Managing the Risks of Organisational Accidents* (Aldershot: Ashgate).

Reason, J.T., 1998, Achieving a safe culture: theory and practice. *Work and Stress*, 12, pp. 293–306.

Roberts, K.H., 1990, Some characteristics of one type of HRO. *Organisation Science*, 1, 160 pp.

Roberts, K.H., 1992, Structuring to facilitate mitigating decisions in reliability enhancing organizations. In *Top Management and Effective Leadership in High Technology Firms*, edited by L. Gomez-Melia and M.W. Lawless. JAI Press. Greenwich, CT.

Roberts, K.H. and Rousseau, D.M., 1989, Research in nearly error free, high reliability organisations: having the bubble. *IEEE Transactions*, 36, pp. 132-139.

Rochlin, G.I., 1993, Defining high reliability organisations. In *New Challenges to Understanding Organisations*, edited by K.H. Roberts (New York: Macmillan).

Rochlin, G.I., 1999, The social construction of safety. In *Nuclear Safety: A Human Factors Perspective*, edited by J. Misumi, B. Wilpert and R. Miller (London: Taylor and Francis Ltd), pp. 5-24.

Rochlin, G.I. and von Meier, A., 1994, Nuclear power operations: A cross-cultural perspective. *Annual Review. Energy Environment*, 19, pp. 153-87.

Rosen, M., 1995, Safety culture: An international perspective. In *Proceedings of the 5th International Nuclear Safety Forum*, Tokyo, 14 March, International Atomic Energy Commission.

Rousseau, D.M., 1990, Assessing organizational culture: The case for multiple methods. In *Organisational Climate and Culture*, edited by B. Schneider (San Francisco: Josey-Bass).

Sayers, D., 1994, On the road to a new safety culture. *Occupational Health and Safety*, Nov/Dec, 10, **6**, pp. 36-38, Canada.

Schein, E.H., 1985, *Organisational Culture and Leadership* (San Francisco: Jossey-Bass).

Schein, E.H., 1990, Organizational culture, *American Psychologist*, 45, **2**, pp. 109-119.

Schein, E.H., 1997, *Three Cultures of Management: The Key to Organisational Learning in the 21st Century Internet document*
http://learning.mit.edu/res/wp/10011.html

Sherrington, M., 1997, Don't write CDM off. *Construction News*, 7 February, 24pp.

Shillito, D.E., 1995, Grand unification theory or should safety, health, environment and quality be managed together or separately? In *Process Safety and*

Environmental Protection Transactions of the Institution of Chemical Engineers, part 3, 73, **B3**, pp. 194-202, August.

Smallwood, J.J., 1996, The influence of management on occupational health and safety. In *CIB W99, Implementation of Safety and Health on Construction Sites*, edited by L.M. Alvez Dias and R.J. Coble, Lisbon, Balkema, Rotterdam, pp. 215-226.

Smith, D., 1998, Paranoia is a healthy state of mind: Management process in systems failure. *The Safety and Health Practitioner,* February, pp. 28-30.

Strategic Forum for Construction, 2002, *Accelerating Change, A Report by the Strategic Forum for Construction chaired by Sir John Egan*, Rethinking Construction, CIC, London.

Suraji, A., Duff, R., and Peckitt, S.J., 2001, Development of a causal model of construction accident causation. *Journal of Construction Engineering and Management*, ASCE, Jul/Aug 2001, 127, **4** , pp. 337-344.

Tam, C.M. and Chan, A.P.C., 1999, Nourishing safety culture in the construction industry of Hong Kong. In *Proceedings the Second International Conference of CIB Working Commission*, W99, *Implementation of Safety and Health on Construction Sites*, edited by in A. Singh, J. Hinze and R.J. Coble, Honolulu, Hawaii, 24-27 March, pp. 117-122.

Thompson, N., Ellis, R. and Wildavsky, A., 1990, *A Cultural Theory* (San Francisco: Westview Press Boulder).

Tijhuis W., 2002, Improving technology-transformation in the construction industry by understanding culture's influences. In *Proceeding of the International Symposium of the Working Commission CIB W92, Procurement Systems and Technology Transfer*, edited by M. Lewis, The Engineering Institute, University of West Indies, Trinidad & Tobago, pp. 513-530.

Triandis, H.C., 1989, The self and social behaviour in differing cultural contexts. *Psychological Review,* 96, **3**, pp. 506-520.

Waring, A.E., 1992, Organisational culture, management and safety, *Paper presented at the British Academy of Management 6th Annual Conference*, 14-16 September, Bradford University.

Waring, A.E. and Glendon, A.I., 1998, *Managing Risk: Critical Issues for Survival and Success into the 21st Century* (London: ITBP).

Whittington, C., Livingston, A. and Lucas, B.A., 1992, *Research into Management, Organisational and Human Factors in the Construction Industry*, HSE Research Report No 45/1992.

Williams, A., Dobson, P., and Walters, M., 1989, *changing culture*, Institute of Personnel Management, Britain.

Wilson, H.A., 1989, Organizational behaviour and safety management in the construction industry. *Construction Management and Economics*, 7, pp. 303-319.

Zhang, Z. and Barrett, P., 1997, *Culture and Chinese Construction Quality*, Research Centre for the Built and Human Environment, University of Salford, http://www.scpm.salford.ac.uk/buhu/bizfruit/1997/papers/zhang.htm.

CHAPTER 3

Construction Site Safety in China

S.X. Zeng, C.M. Tam and Z.M. Deng

INTRODUCTION

Construction is one of the most hazardous industries due to its unique nature. Measured by international standards, construction site safety records in China are poor. This chapter aims to examine the status of safety management in the Chinese construction industry, explore the risk-prone activities on construction sites, and identify factors affecting construction site safety. The findings reveal that the behaviours of contractors in terms of safety management are of grave concern, including the lack of provision of personal protective equipment, regular safety meetings, and safety training. The main factors affecting safety performance include 'poor safety awareness of top management', 'lack of training', 'poor safety awareness of project managers', 'reluctance to input resources to safety' and 'reckless operations'. The study also suggests that the government should play a more critical role in stricter legal enforcement of safety legislation and organising safety training programmes.

All over the world, construction is one of the most hazardous industries due to its unique nature (Jannadi and Bu-Khamsin, 2002). Construction safety is always a grave concern of both practitioners and researchers. A number of causes influencing safety performance in the construction industry have been identified including workers' attitudes (Hinze, 1981); construction company size, safety policy, project coordination, and economic pressure (Hinze and Raboud, 1988); management training (Gun, 1993; Jaselskis and Suazo, 1994); and safety culture (Glendon and Stanton, 2000; Tam and Fung, 1998; Tam, Fung and Chan, 2001). Measures taken to prevent occupational injuries and improve safety performance have been extensively explored (Laufer and Ledbetter, 1986; Harper and Koehn, 1998). Some of these studies (Mattila and Hyodynmaa, 1988; Fellner and Sulzer-Azaroff, 1984; Laitinen and Ruohomaki, 1996) reveal that when goals are posted and feedback is given, safety performance is significantly better than when no feedback is given. Hakkinen (1995) advocated a training programme called 'one hour for safety management' for top management. The application of the programme was successful in drawing management's attention to safety issues. Research indicates that 83% of projects have achieved the zero accident goals after applying the 'Zero Accident Program' (Center to Protect Workers' Rights, 1993; Hinze and Wilson, 2000).

As regards construction safety in China, the record is poor in terms of international standards. In 1999, 923 site accidents above Grade IV [1] were recorded in countryside construction, in which 1,097 construction workers lost their lives (China Statistical Yearbook of Construction, 2001). This chapter describes the findings from a structured questionnaire survey and interviews on the subject of safety management in the Chinese construction industry with the following objectives:

- to examine the status of safety management in the industry
- to explore the risk-prone activities on construction sites
- to identify the factors affecting construction site safety and
- to propose suggestions for improving safety performance

3.1   BACKGROUND OF CONSTRUCTION SAFETY IN CHINA

3.1.1   Role of government in construction safety

The Ministry of Construction takes overall responsibility in overseeing the construction industry in China. It takes the leading role in implementing the new strategies and policies, including preparing development programmes, regulating construction markets and construction institutions, and monitoring construction safety. The role of the central Ministry is mirrored by the provincial construction departments and those of the independent municipalities. They are charged with the responsibility of regulating construction safety (see Figure 3.1).

Empowered by the relevant legislations on construction safety, such as 'Construction Law', 'Inspection Standards for Construction Safety' and 'Inspection Standards for Labour Protection in Construction Enterprises', the Ministry of Construction annually hires about 50 safety auditors conducting nationwide safety inspections. The scope of the audits includes the safety management system of construction firms, labour protection measures, safety pitfalls on construction sites and so on.

3.1.2   Safety management system of construction firms

Protection of labour from occupational diseases and accidents in the construction industry of China is defined by law; for example, for construction sites having 50 employees or more, main contractors have to nominate a full-time safety inspector; for sites with an area exceeding 10,000m^2 there must be 2-3 safety inspectors;

---

[1]   The Regulation on Procedures for Reporting and Investigation into Serious Accidents in Construction was issued by the Ministry of Construction of China in 1989. According to the regulation, accidents on construction sites are classified into four grades as listed in Table 3.5.

wherever the site exceeds 50,000m^2, the main contractor has to establish a safety management team.

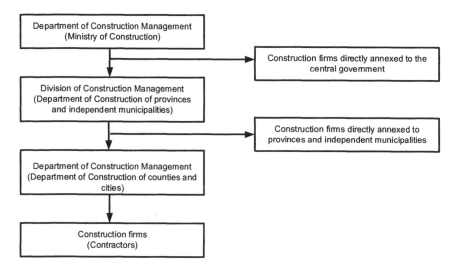

Figure 3.1   Governance hierarchy for construction safety management

3.1.3   Construction project supervisors' inspection of safety matters

Since 1989 China has begun adopting the 'Construction Supervision Scheme'. One of the project supervising engineer's responsibilities is to monitor construction safety. Since the issuance of the 'Regulation on Construction Project Supervision' in 1996, the construction supervision scheme has been extensively practised in China. According to the system, the role of the supervisors is to enhance construction supervision by introducing checks and controls at various construction stages on behalf of the clients. Under the 32nd clause of the current Construction Law, issued in 1997, supervisors' duties are to ensure construction works in compliance with the construction regulations, to supervise execution of the work, to monitor construction safety, to prepare supervision plans and to notify the government in case of any violation of the relevant statutory legislations.

In spite of the well-defined roles of the above parties in construction safety, the safety performance on construction sites is still disappointingly poor in China. Occupational accidents have not been effectively prevented.

3.2   RESEARCH METHODOLOGY

In order to explore the status of construction safety management in China, a survey of 200 large and medium-sized construction firms listed in the Dictionary of

Quality System Certificated Enterprises was conducted. Structured questionnaire surveys and interviews were used to collect the necessary information and data. Questionnaires were sent to safety representatives of the construction firms listed in the Dictionary of ISO 9000 Certified Enterprises. The areas of investigation of the questionnaires are summarised as follows:

- safety management system
- safety behaviours and safety measures
- impacts of site accidents on companies
- factors affecting safety management and
- government support

Sixty completed questionnaires have been received with a response rate of 30%. In addition, interviews were conducted with government officials of the construction departments in charge of construction safety, focusing on safety policies and procedures issued by the government, safety standards, and factors affecting safety on construction sites.

In the survey, all the 60 construction firms were ISO 9000 certified. They fall into two categories of ownership: 52 state-owned (87%) and 8 share-holding enterprises (13%). Out of all the responding firms, 52 firms (87%) employ over 1,001 people, 4 firms (6.5%) between 501 and 1,000, and 4 firms (6.5%) below 500. In China, all large construction firms were state-owned establishments under the traditional planned economy system. Since the adoption of the reform and opening policies in 1978, the traditional planned economy system has been gradually replaced by the market economy in China. A great number of peasants were liberated from the traditional cultivation and farming works and organised themselves into rural-village-enterprises and the rural construction teams (RCT). This is closely associated with rapid economic expansion, which has resulted in high volumes of construction activities and renders China the largest construction market in the world. As at 1999, the proportion of RCT has reached 51% of all construction firms, and that of state-owned enterprises (SOE) at 10%, urban collective-owned (UCO) at 26%, and others (including share-holding, foreign-funded enterprises) at 13% (see Figure 3.2), which represents a great change in the form of ownership of construction enterprises in China.

Previous research has revealed that there is a relationship between sizes of firms and accident rates (Hinze and Raboud, 1988). A study by McVittie, Banikin and Brocklebanh (1997) indicated that accident rates decrease as the sizes of firms increase. The underlying factors include the degree of planning and organization at large firms versus that of small firms, the presence of in-house health and safety expertise or resources, the degree of unionisation, access to and use of external support services relating to health and safety, levels of government inspection and the effects of economies of scale. Comparing SOEs with RCTs in China, the average numbers of employees are 735 and 150 respectively (China Statistical Yearbook of Construction, 2001). However the fatality rate for the former is three times as that of the latter (China Statistical Yearbook of Construction , 2001). This trend contradicts to the findings of McVittie, Banikin and Brocklebanh (1997), representing on exceptional difference in construction safety in China compared to the west.

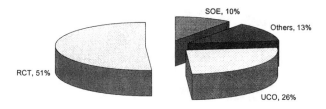

Figure 3.2   Categories of construction firms of various ownerships in 1999

3.3   ANALYSIS AND RESULTS

3.3.1   Safety manual and procedures

The purpose of the safety manual is to communicate a firm's safety policy, identify the safety factors, define responsibility and control the safety management system. As the backbone of the management system, the manual defines the safety procedures and instructions and identifies the specific requirements. The respondents were asked whether they have safety manuals and safety procedures. Sixty-two percent of the respondents claimed that they did not have documented safety manuals. With respect to safety procedures, all the respondents opined that they had documented procedures for safety management on construction sites, which formed part of the procedural process control (Clause 4.9) of ISO 9000:1994.   However, the majority of the respondents (92%) claimed that not every worker knew the procedure.

3.3.2   Provision of personal protective equipment

The status of the provision of personal protective equipment (PPE) for workers is illustrated in Figure 3.3.

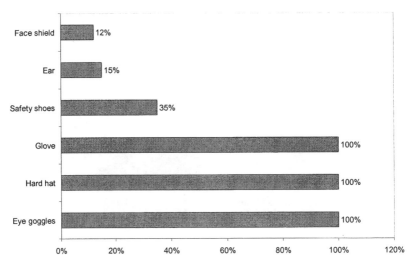

Figure 3.3   Personal protective equipment provided by contractors

Figure 3.3 shows that the most common PPE provided are gloves, hard hats and eye goggles. However, many workers opined that hard hats are not convenient for their operations. Face shields, ear plugs and safety shoes are not common with only 12%, 15% and 35% of contractors claiming the provide these.

3.3.3   Safety meetings and training

Regular safety meetings are necessary for communicating safety information to all parties. 36% of the respondents claimed that they had regular safety meetings, and the others indicated that safety issues were discussed and presented at other meetings, such as construction planning meetings. However 87% argued that top management seldom attended the safety meetings.

The respondents were asked whether they had provided safety training for their first-line workers. Twenty-four percent of the respondents claimed the provision of systematic training; 65% offered occasional training; and the others (11%) rarely provided any training. In the construction industry, construction workers have high mobility and they switch from one company to another frequently. The transient nature of the construction workforce makes it difficult to train workers. The lack of effective labour training is a major concern in safety management.

3.3.4   Impacts of site accidents

The questionnaire has highlighted the impacts of site accidents to construction firms including 'increase in cost', 'interrupting construction schedule', 'impairing reputation of firms', 'imposing psychological burden on workers', and others. Figure 3.4 shows the results.

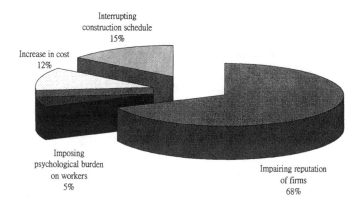

Figure 3.4   Impacts of site accidents

Figure 3.4 shows that 68% of the respondents considered 'impairing reputation of firms' as the most serious impact of site accidents. 'Interrupting construction schedule', 'increase in cost' and 'imposing psychological burden on workers' were indicated by 15%, 12% and 5% of the respondents respectively.

3.3.5   Risk-prone operations on sites

Construction sites exhibit unique hazardous characteristics; for example, workers are crowded together on sites, operating at height and outdoors, and the use of heavy machine and equipment is common. The questionnaire explored the nature of risk-prone operations on construction sites. The respondents were asked to choose the most hazardous site operations (see Figure 3.5).

Figure 3.5 indicates that 'falling from height' is considered most risky (92%) (Larsson and Field, 2002). The other operations in descending order of risk-proneness are 'hit by falling materials', 'collapse of earthwork, 'use of heavy machinery', and 'electrocution'. The results are comparable with the safety statistical data of the construction industry in China (as listed in Table 3.1).

Table 3.1 shows that 'falling from height' scores high both in accident and fatality records. This type of accident reaches 50% of the total accidents, with respect to fatalities, 524 construction workers (48%) lost their life due to falling from height in 1999 (see Table 3.2).

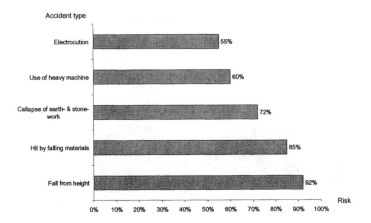

Figure 3.5   Risk-prone accidents on construction sites

Table 3.1   Safety accidents in the construction industry (in 1999)

Accident category	Cases of accidents	Fatality	Severe injury
Falling from height	466 (50)	524 (48)	133 (44)
Electrocution	120 (13)	124 (11)	4 (1)
Hit by falling materials	115 (12)	116 (11)	45 (15)
Collapse of earthwork	87 (9)	148 (13)	36 (12)
Use of heavy machine	63 (7)	71 (6)	38 (13)
Lifting of weights	32 (3)	45 (4)	18 (6)
Toxic and suffocation	16 (2)	29 (3)	2 (1)
Use of motor	8 (1)	8 (1)	3 (1)
Fire and explosions	5 (1)	20 (2)	3 (1)
Others	11 (2)	12 (1)	17 (6)
Total	923 (100)	1097 (100)	299 (100)

The figure in parentheses indicates the percentage of the total.
Source: China Statistical Yearbook of Construction (2001), pp. 105.

Table 3.2   Fatal accidents due to fall from height (in 1999)

Type	Fatality	Severe injury
Hole and edge	182 (35)	24 (18)
Scaffolding	133 (25)	51 (39)
Crane	78 (16)	13 (10)
Tower crane	35 (7)	9 (7)
Formwork	34 (6)	9 (7)
Construction machine	10 (2)	10 (7)
Earthmoving	7 (1)	3 (2)
Building demolition	7 (1)	–
Others	38 (7)	14 (10)
Total	524 (100)	133 (100)

The figure in parentheses indicates the percentage of the total.
Source: China Statistical Yearbook of Construction (20001), pp. 105.

3.4 ROOT CAUSES AFFECTING SAFETY PERFORMANCE

There are various factors influencing safety management in the construction industry. These factors can be grouped into people's role, organization, management, technology, industrial relationship and so on. Due to the difference in culture, management and the market structure, these factors have diverse influence on construction safety. The related literature to date on safety management is tabulated in Table 3.3.

Table 3.3 Previous researches on safety management

Areas	Items	Relative researches
People's role	Role of leaders	Hakkinen (1995); Koehn, Kothari and Pan (1995); Levitt and Parker (1976); Tam and Fung (1998); Wentz (1998)
Organisation and management	Worker's behaviour	Hinze (1981); Yu (1990)
	Training	Gun (1993); Hakkinen (1995); Hale (1984); Krause (1993); Tam and Fung (1998)
	Safety systems	Hale, *et al.* (1997); Hale and Hovden (1998); Hinze (1981); Jaselskis, Anderson and Russell (1996); Tam, Fung and Chan (2001)
Apparatus and equipment	Equipment	Jaselskis and Suazo (1994); Krause (1993); Larsson and Field (2002)
Technology	Technology control	Blank, Laflamme and Anderson (1997); Lingard and Holmes (2001); Jannadi and Assaf (1998)
Industrial relationship	Market	Hinze and Raboud (1988); Kartam, Flood and Koushki(2000)
	Safety regulations	Gun (1993); Seppala (1995)

Based on the above, the questionnaire has incorporated 25 factors affecting construction safety. The respondents were asked to provide their opinions on the importance of the factors affecting construction site safety by scores from 1 to 5, where '1' represents the least important and '5' the most important. To determine the relative ranking of the factors, the scores were then transformed to importance indices based on the following formula (Tam *et al.*, 2000).

$$\text{Relative importance index} = \frac{\sum w}{A \times N} \qquad (1)$$

where w is the weighting given to each factor by the respondents, ranging from 1 to 5, A is the highest weight (i.e. 5 in the study) and N is the total number of samples.

Based on equation (1), the relative importance index (RII) can be normalised to fall within 0 to 1. Table 3.4 shows the relative importance index of each factor affecting construction site safety.

Table 3.4   Relative importance index of factors affecting construction site safety

Ranking	Factors affecting site safety	Relative importance index
1	Poor safety awareness of firm's top leaders	0.93
2	Lack of training	0.90
3	Poor safety awareness of project managers	0.89
4	Reluctant safety	0.86
5	Reckless operation	0.86
6	Lack of certified skilled labour	0.84
7	Poor equipment	0.82
8	Lack of first aid measures	0.81
9	Lack of rigorous enforcement of safety regulations	0.74
10	Lack of organisational commitment	0.71
11	Low education level of workers	0.68
12	Poor safety conscientiousness of workers	0.65
13	Lack of personal protective equipment	0.62
14	Ineffective operation of safety regulation	0.59
15	Lack of technical guidance	0.55
16	Lack of strict operational procedures	0.55
17	Lack of experienced project managers	0.54
18	Shortfall of safety regulations	0.53
19	Lack of protection in material transportation	0.53
20	Lack of protection in material storage	0.51
21	Lack of teamwork spirits	0.50
22	Excessive overtime work for labour	0.49
23	Shortage of safety management manual	0.48
24	Lack of innovation technology	0.43
25	Poor information flow	0.40

The respondents ranked 'poor safety awareness of firm's top leaders' and 'poor safety awareness of project managers' the first and the third, with a relative importance index of 0.93 and 0.89. This indicates that leaders play a very important role in construction safety management. The top management sets up appropriate environments for safety by defining the safety policy and allocating resources. The attitude of the top leaders plays an important role in cultivating a good safety culture (Seppala, 1995). However, in practice, not all business leaders pay great attention to safety management because other business objectives such as profitability, schedule and quality are always competing for their time and resources (Hakkinen, 1995). This can be reflected from the questionnaire response that only a small proportion of top management attended safety meetings. As contractors have to finish the work within a specified period of time, at an agreed price and at a certain standard of workmanship, most people focus on the immediate problems and view their top priorities as meeting the production schedule, quota and cost targets, and quality requirements. Only after achieving these objectives will they give consideration to safety (Tam, Fung and Chan, 2001).

'Lack of training' is ranked the second with a relative importance index of 0.90. Training programmes help personnel carry out various activities effectively, establish a positive safety attitude, and integrate safety with construction and quality goals. In fact, the percentage of construction workers being trained is very low in China. Statistics reveal that only 3% of workers have been trained and certified; 7% trained under short-term programmes; and 90% without any training at all (Zhang, 2001). In addition, one of the characteristics of the Chinese construction industry is the existence of a large number of peasant-workers, who receive little education and are unskilled, untrained, and inexperienced (education level of these workers, which is ranked eleventh, is shown in Figure 3.6).

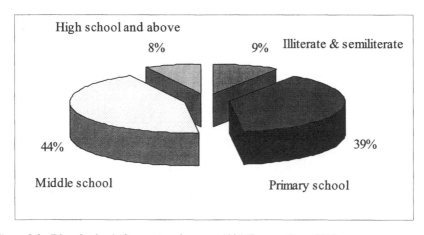

Figure 3.6   Education level of peasant-workers as at 1996 (Source: Zhang, 2001)

They come from poor provinces and are ready to take up any job to earn a reasonable living for their families. Due to the relatively low requirements for skills in construction, the industry has been overwhelmed with peasant-workers.

'Reluctance to input of resources to safety' is graded fourth, with a relative importance index of 0.86. This is closely associated with the operational nature of construction firms in China. Almost all Chinese construction firms of different sizes compete for the same jobs in the construction market and manage similar projects that results in 'excessive competition' and thin profit margins, sometimes even losses due to unhealthy levels of competition. This vicious cycle hinders technology and productivity improvement, in which the state-owned construction firms are the direct victims (see Figure 3.7).

Meanwhile government policy has burdened SOEs as they need to shoulder extra welfare costs and retirement pensions of workers, including social and welfare amenities.

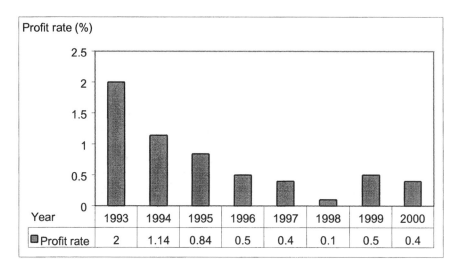

Figure 3.7   Profit rate of construction state-owned firms (1993-2000)
Source: China Statistical Yearbook of Construction (2001)

Additionally, delay in payment is a common problem in China. In 1999, the value of arrears in payment reached 222,140 million RMB (Yuan) (27,090 million US$), which was 19.1% of total construction output. This phenomenon is attributed to two main reasons. First, all construction firms can and are willing to tender for all kinds of jobs, which results in acute competition in the construction market. Under this circumstance, construction firms, which are desperate for jobs, would use their own funds or even borrow money to start the project from the banking system on behalf of the clients. Second, the operational environment for the Chinese state-owned construction firms is far from perfect. Although the National Construction Law was issued in November 1997, the existing legislative and regulatory frameworks are incomplete and, often, high-level government officials have the final say on any decisions. As a result, the clients and contractors are on an equal footing. The legitimate interests of construction firms cannot be protected when they come into conflict with other, higher-level benefits. As a result, it is not easy for construction firms to spare extra resources for safety management.

The respondents graded 'reckless operation' fifth, with a relative importance index of 0.86. The reckless operations largely occur during building demolition. According to the China Statistical Yearbook of Construction, the fatality rate was 46 (4%) resulting from reckless operations in 1999.

'Non-certified skilled labour' is ranked sixth, with a relative importance index of 0.84. In construction, some activities demand a high level of skill, such as tower crane and gantry operations, and framework and scaffold erection etc. In 1999, the number of fatalities was 102 (9%) and 46 (4%) resulting from gantry and framework erection respectively.

The respondents graded 'poor equipment' seventh, with a relative importance index of 0.82. In 1999, 95 fatalities (8%) resulted from the problems of construction equipment. Construction equipment is considered to be one of the weakest links in the Chinese construction industry. As there are no plant-hiring services offered in China, the construction firms have to own their construction equipment. Most of the equipment that is not fully utilised places a heavy burden on firms. Although around 30% of construction equipment is old and obsolete, it is still being used because most state-owned firms lack money to replace them. As a result, site operations are still rather primitive due to the shortage of practical hand tools. The abundant supply of cheap labour further augments the situation rendering the construction industry lagging behind in technology.

The respondents graded 'lack of first aid measures' eighth, with a relative importance index of 0.81. In general, construction firms have no first aid measures for emergency situations in China. This is related to the factors on 'lack of attention from leaders' and 'poor safety conscientiousness of managers'.

The respondents ranked 'non-rigorous enforcement of safety regulations' ninth, with a relative importance index of 0.74. While legislation on construction safety has been issued, the state administration still retains considerable powers, and government officials can influence legal enforcement in local areas. Violation of construction safety laws often goes unpunished. For example, the qualifications and certification of safety inspectors are stipulated by construction regulations; however, it is not uncommon for contractors to appoint people without the stipulated competence for safety management. Meanwhile, many clients do not consider past safety records of contractors in the tender prequalification process of project procurement.

The respondents graded 'lack of organisational commitment' tenth, with a relative importance index of 0.71. Although all respondents claimed to have organisational commitment, however, as indicated by an interviewee, an official of the Ministry of Construction, many contractors just put their commitments on paper but actually behave differently. For a construction project, the project manager's safety responsibility should at least cover the following (MacCollum, 1995):

1. Preparation of a project safety plan.
2. Review of plans and specifications to identify the location and the nature of potential hazards.
3. Review of specifications to identify appropriate safety standards and special safety conditions.
4. Requiring construction superintendents to prepare a written safety plan for each major phase of work, including hazard analysis for high-risk activities.
5. Insisting upon immediate reporting of all injuries, deaths, and property damage as a result of accidents.
6. Employment of qualified and certified safety inspectors and personnel.

Sometimes, the safety management system requires site supervision staff to carry out a lot of documentation and paperwork which may create frustrations leading to completion of the inspection forms without making any inspections at all.

In addition, the current labour union does not play an active role in defending the rights of labour in the construction industry of China when compared with those in industrialized countries, which are powerful and can insist contractors provide safe working conditions and safety equipment to protect their workforces' rights and health (Kartam, Flood and Koushki, 2000).

3.5   DEMANDS FOR GOVERNMENTAL SUPPORT

The government should play an important role in safety management in the construction industry (Kartam, Flood and Koushki, 2000). In this survey, respondents were asked to provide their opinions on the ways that the government can support safety management. Figure 3.8 shows the results.

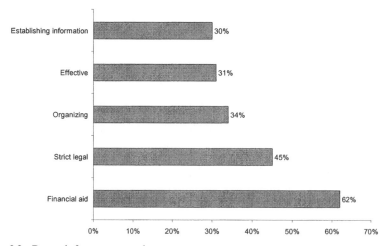

Figure 3.8   Demands for governmental support

Figure 3.8 shows that 62% of the respondents indicated 'financial aid' as the major tool. According to an interviewee, an official in the Ministry of Construction in charge of construction safety, the provision of financial aid is not feasible under the current government policy. Thirty-one percent of the respondents felt that government should adopt a more effective inspection regime. Currently the inspections are not conducted effectively as it is impossible to have sufficient auditors to visit all workplaces at all times and be there when new hazards arise (MacCollum, 1995). Under current circumstances, the most feasible way that the government can manage is the provision of support in safety training and legal enforcement.

3.6   CONCLUSIONS

Construction is one of the most hazardous industries due to the nature of the industry. By international standards, the construction site safety record of China is poor. Every year about 1,000 workers lose their lives on construction sites and there are numerous occupational accidents. The safety management systems adopted by most firms are of grave concern:

- most contractors do not have any documented safety manuals
- only a small percentage of contractors provides necessary PPE for their workers
- top management seldom attends the safety meetings and
- only a small number of contractors offer systematic safety training

Sixty-eight percent of the respondents considered 'impairing reputation of firms' as the most serious impact of site accidents. Although there are many factors affecting contractors' safety performance, the main factors affecting safety performance include:

- poor safety awareness of firm's top management
- lack of training
- poor safety awareness of project managers
- reluctance to assign resources to safety and
- reckless operations

Based on the analysis, it can be concluded that the government should play a more active role in stricter legal enforcement and organising training programmes for improving safety performance.

Table 3.5   Classification of Serious Accidents in Construction in China

Accident grade	Demarcation
I	Over 30 fatalities; or Over 3.0 million RMB Yuan in direct economic losses
II	10-29 fatalities; or 1.0-3.0 million RMB Yuan in direct economic losses
III	3-9 fatalities; or over 20 severe injuries; or 0.3-1.0 million RMB Yuan in direct economic losses
IV	2 fatalities; or 3-19 severe injuries; or 0.1-0.3 million RMB Yuan in direct economic losses

Source: Lei and Qian (1999)

3.7   REFERENCES

Blank, V.L.G., Laflamme, L. and Andersson, R., 1997, The impact of advances in production technology on industrial injuries: a review of the literature. *Safety Science*, 26, **3**, pp. 219-234.

Center to Protect Workers' Rights., 1993, An agenda for change. *Report of the National Conference on Ergonomics, Safety, and Health in Construction*, July 18-22, Washington, D.C.

China Statistical Yearbook of Construction, 2001 (Beijing: China Statistical Press).

Fellner, D.J. and Sulzer-Azaroff, B., 1984, Increasing industrial safety practices and conditions through posted feedback. *Journal of Safety Research,* 15, pp. 7-21.

Glendon, A.I. and Stanton, N.A., 2000, Perspectives on safety culture. *Safety Science* 34, pp. 193-214.

Gun, R.T., 1993, The role of regulations in the prevention of occupation injury. *Safety Science*, 16, pp. 47-66.

Hakkinen, K., 1995, A learning-by-doing strategy to improve top management involvement in safety. *Safety Science*, 20, pp. 299-304.

Hale, A.R., 1984, Is safety training worthwhile? *Journal of Occupational Accidents*, 6, pp. 17-33.

Hale, A.R., Heming, B.H.J., Carthey, J. and Kirwan, B., 1997, Modeling of safety management. *Safety Science*, 26 **(1/2)**, pp. 121-140.

Hale, A.R. and Hovden, J., 1998, Managing and culture: the third age of safety. A review of approaches to organizational aspects of safety health and environment. In *Occupational Injury: Risk, Prevention and Injury*, edited by A. Williamson, and A.M. Feyer, (Taylor & Francis).

Harper, R.S. and Koehn, E., 1998, Managing industrial construction safety in Southeast Texas. *Journal of Construction Engineering and Management*, 124, **6**, pp. 452-457.

Hinze, J., 1981, Human aspects of construction safety. *Journal of the Construction Division*, ASCE, 107, pp. 61-72.

Hinze, J., Raboud, P., 1988, Safety on large building construction projects. *Journal of Construction Engineering and Management*, ASCE, 114, pp. 286-293.

Hinze, J., Wilson, G., 2000, Moving toward a zero injury objective. *Journal of Construction Engineering and Management*, ASCE, 126, **5**, pp. 399-403.

Jannadi, M.O., Assaf, S., 1998, Safety assessment in the built environment of Saudi Arabia. *Safety Science*, 29, **1**, pp. 15-24.

Jannadi, O.A. and Bu-Khamsin, M.S., 2002, Safety factors considered by industrial contractors in Saudi Arabia. *Building and Environment*, 37, **5**,pp. 539-547.

Jaselskis, E.J., Anderson, S.D. and Russell, J.S., 1996, Strategies for achieving excellence in construction safety performance. *Journal of Construction Engineering and Management*, 122, **1**, pp. 61-70.

Jaselskis, E.J. and Suazo, G.A.R., 1994, A survey of construction site safety in Honduras. *Construction Management and Economics*, 12, pp. 245-255.

Kartam, N.A., Flood, I. and Koushki, P., 2000, Construction safety in Kuwait: Issues, procedures, problems, and recommendations. *Safety Science*, 36, pp. 163-184.

Koehn, E., Kothari, R. and Pan, C.S., 1995, Safety in developing countries: professional and bureaucratic problems. *Journal of Construction Engineering and Management*, ASCE, 121, 3, pp. 261-265.

Krause, T.R., 1993, Safety and quality: two sides of the same coin. *Occupational Hazards*, April, pp. 47-50.

Laitinen, H. and Ruohomaki, I., 1996, The effects of feedback and goal setting on safety performance at two construction sites. *Safety Science*, 24, 1, pp. 61-73.

Larsson, T.J. and Field, B., 2002, The distribution of occupational injury risk in the Victorian construction industry. *Safety Science*, 40, pp. 439-456.

Laufer, A. and Ledbetter, W.B., 1986, Assessment of safety performance measures at construction sites. *Journal of Construction Engineering and Management*, 112, 4, pp. 530-542.

Lei, Y.J. and Qian, K.R., 1999, *Manual of Supervision on Construction* (Beijing: China Architecture & Building Press).

Levitt, R.E. and Parker, H.W., 1976, Reducing construction accidents - top management's role. *Journal of the Construction Division*, 102.

Lingard, H. and Holmes, N., 2001, Understanding of occupational health and safety risk control in small business construction firms: barriers to implementing technological controls. *Construction Management and Economics*, 19, pp. 217-226.

MacCollum, D.V., 1995, *Construction Safety Planning* (Van Nostrand Reinhold).

Mattila, M. and Hyodynmaa, M., 1988, Promoting job safety in building: an experiment on the behavior analysis approach. *Journal of Occupational Accidents*, 9, pp. 255-267.

McVittie, D., Banikin, H. and Brocklebanh, W., 1997, The effects of firm size on injury frequency in construction. *Safety Science*, 27, 1, pp. 19-23.

Seppala, A., 1995, Promoting safety by training supervisors and safety representatives for daily safety work. *Safety Science*, 20, pp. 317-322.

Tam, C.M., Deng, Z.M., Zeng, S.X. and Ho, C.S., 2000, Quest for continuous quality improvement for public housing construction in Hong Kong. *Construction Management and Economics*, 18, 4, pp. 437-446.

Tam, C.M. and Fung, I.W.H., 1998, Effectiveness of safety management strategies on safety performance in Hong Kong. *Construction Management and Economics*, 16, 1, pp. 49-55.

Tam, C.M., Fung, I.W.H. and Chan, A.P.C., 2001, Study of attitude changes in people after the implementation of a new safety management system: The supervision plan. *Construction Management and Economics*, 19, 4, pp. 393-403.

Wentz, C.A., 1998, *Safety, Health and Environmental Protection* (New York: McGraw-Hill).

Yu, Y.F., 1990, The role of workers' behavior and accomplishment in the prevention of accidents and injuries. *Safety Science*, 12.

Zhang, X., 2001, Analysis of professional quality in the construction industry, *Construction Economics*. 2, pp. 16-19.

CHAPTER 4

A Study of the Present Situation and Reform of Construction Safety Supervision in Shenzhen

X.W. Lu and D.X. Ji

INTRODUCTION

Ever since the establishment of the Shenzhen Special Economic Zone (SSEZ) in March 1979, the local construction industry has made significant progress and the scale of construction has been greatly expanded. As a result of such developments, employment opportunities in the construction industry flourished. Most of the construction workers were poorly educated and had little awareness of the safety hazards on site. Consequently, during the construction boom period, a high accident ratio resulted. During the period 1982 to 1986, statistics indicated that 102 were killed, 213 seriously injured and 509 slightly wounded in Shenzhen construction sites. In 1985, 41 persons were killed. During the following years, about 15 to 20 persons lost their lives working on construction sites. While the accident statistics in the industry have improved over the years, construction site safety continues to be a major issue on the industry agenda.

4.1   CURRENT SITUATION OF CONSTRUCTION SAFETY SUPERVISION IN SHENZHEN

In the early period of SSEZ establishment, supervision of construction safety was inadequately managed. From March 1979 to December 1983, there was only one part-time technical officer in charge of safety for the government. From January 1984 to March 1988, there was one full-time technical officer appointed to be in charge of construction quality and safety. Their duty was to manage the documentation and statistics, as well as to provide an annual report on the construction industry's safety developments. From 1988, a temporary Construction Safety Supervision Group was established to enforce the safety legislation on construction site safety. In 1989, *Standard of construction safety inspection* was issued by the Construction Department of the Central Government and in the following year, a campaign was conducted to promote safety awareness in the industry. From February 1991 to October 1994, a series of documents and technical standards were issued by the municipal government. Together, this formed the foundation of construction safety supervision in Shenzhen.

The Shenzhen Construction Safety Supervision Office was established on 28th, November 1994, and began formal operations on 1st, January 1995. Since then, district offices have been established, as well as some town offices. There are more than 150 officers employed in these offices forming the network of construction safety supervision. The work of the Office has made significant improvements to the industry's site safety record and has continually raised the awareness of construction workers on site safety issues. In recent years, with the scale of construction investment expanding, the construction safety record has improved year by year. From 1994 to 1997, the ratio of implementing safety standards increased by 10 to 15% annually, and the death ratio decreased gradually to the lowest point in 1997. In 1995, a countrywide report revealed that Shenzhen was ranked second, and in 1997, the leader among a total of 44 cities.

From 1995, SSEZ has followed the principle 'to tackle a problem not only on the surface, but also at the root', and made effort to look for innovative ways for '3 stage supervision', 'Special supervision for dangerous operations' and so on. The campaigns have been carried out as follows:

Firstly, '3 stage supervision'. This includes the following stages. Before on-site construction starts, the site safety plan must be examined and approved; during construction, site safety standards must be implemented and; after the construction is completed, an overall evaluation of the project site safety performance is made, which influences the contractors' capacity to work in future projects. According to the principle 'to aim at safety and put prevention first', the key point of our work is precaution. We stress on the examination of prerequisites and implementation of safety standards, trying to eliminate the hidden dangers and hazards. At the same time, we advise the contractors to eliminate these hidden dangers by on-site spot checks, and trying to help the companies to become more self-regulating.

Second, on training and education in construction site safety, in 1996 the *Regulation of training and education about construction safety* was enacted to improve the workers' performance, providing for the aim, scale and requirement of training and education, and enforcing rules about certificate of 'safety operation', examination of education, and keeping record of training. From 1997, a series of textbooks for training have been compiled, among which are textbooks about worker training, electrical safety, piling machines, vertical transport equipment, and machine disassembling and assembling. A pamphlet for training of construction safety supervision was also provided for the entire Guangdong province, authorised by the construction committee of Guangdong. With these textbooks, the training and education work has a plan, basis, examination and effect, making way for safety professionalism in the industry. At the end of 2000,

116 safety related training courses have been offered, and more than 6000 technical officers trained and certified.

Third, on construction safety campaigns. From 1995, we published free periodic newsletters on construction safety mainly to provide information, knowledge and reports about construction safety. It is made available to all organisations in the industry. In June 1998, a video CD-ROM on construction safety education was prepared and delivered to construction sites. In addition, competitions were organised and posters promoting a safety culture were offered to the industry.

Fourth, on promoting the standardisation and legitimisation of construction safety supervision. From 1995, a series of district standards and laws have been laid down, including *Standard of construction safety inspection for special building machines in SSEZ, Standard of construction safety inspection for SSEZ, Instruction standard of construction safety inspection for SSEZ and Supervision and safety code for climbing frames in temporary support in SSEZ* etc. At the same time, *Bylaw of construction safety supervision in SSEZ* was issued on 13th February 1998, standardising the supervision of construction site safety and providing legal safeguards.

Fifth, on special treatment to accident prevention. Protection of the edge, fall accidents, and especially dangerous operations (including assembling and disassembling of vertical transport equipment, piling machines, climbing frames in temporary support, etc.) have been dealt with, and the ratio of serious accidents has been reduced.

Sixth, we conduct testing of PPE, regulate the use of safety protection articles on site. Every year, we provide assistance to the Construction Department, the Industrial and Commercial Bureau and the Technical Supervision Bureau to inspect construction safety protection articles and facilities, preventing unauthorised personnel from entering construction sites, therefore hidden dangers are eliminated. This is regarded as the daily responsibilities of safety officers, which has had a strong influence on improving site safety conditions.

4.2   CONSTRUCTION SAFETY SUPERVISION REFORMS IN SHENZHEN

General reform principles are: to limit the government, to entrust the authority to basic units, to supervise what should be done, to regulate the supervision, to generally optimise technical standards and regulations, to advise construction sites to achieve 'civilised sites', to impel workers to join the insurance plan, to advance the training plan which aims at the person is the core, to accelerate safety culture achievement.

Measures are listed as follows:

Firstly, based on 'the person is the core', the workers are encouraged to improve safety performance. They are an integral part of the construction process and they need to recognise their importance by practising good on-site safety practices to ensure good safety performance. Statistics suggest that 85% of accidents are caused by human beings. *Domino*'s theory confirms that humans play a leading role, and dangerous human activity is the critical factor in the chain which causes accidents, and if this chain doesn't exist, no accidents will take place. Chairman *Jiang Zemin* points out that the economy must be developed, based on the progress of science and technology and the improvement of the worker's quality. With the advancements made to the economic system and the establishment of a market-based economy, the government entrusts the companies with part of its authority, which requires the companies to have more self-restraint to improve workers' performance.

In order to accelerate the campaign based on 'the person is the core', the first thing is to standardise and regulate the training and education, then the problem can be sorted out, not only on the surface but also at the root. Today, some companies pay little attention to the training and education of the workers. It seems that these companies have got more economic benefit, but there are more hidden dangers in these companies, and the result is unimaginable.

Thus the general principle of construction safety supervision should be followed, with an emphasis on company responsibility, guiding, supervising and checking. The standard textbooks should be used with the same requirement and the same standard. Rules should be enforced and records kept. The training and education plan should be carried out in the following procedure: firstly, to train the piling workers, electricians, workers whose job is to assemble and disassemble the vertical transport equipment, both part-time and full-time employers in charge of safety; then, to train three kinds of workers; finally, to enforce the regulations. Workers who do not pass the required safety examinations are not certified work on the construction site.

Second, technical standards and regulations about construction safety should be drawn up. The supervision of construction safety started quite late in China, thus the construction of technical standards and regulations need to catch up with the development of the construction industry. At present there are only seven standards, which require no construction safety supervision at all and, furthermore, three of these standards were worked out in the 'Great leap' period, which need serious amendment in line with the development of today's science and technology. The relevant standards and regulations should be consummated within 2 to 3 years.

Third, construction sites will be advised to be 'civilised sites' in order to advance the level of supervision. In 1996, the Central Construction Department called on us to take part in this campaign. In September 1998, 'civilized sites' were established, which stipulated one construction site being a 'civilized site'. This was a prerequisite of its safety. The measure has changed the dirty, disturbed and bad appearance of construction sites, and taken on a new look. In the future, this campaign will be promoted with the best and the related supervised investment in order to make more construction sites 'civilized sites'.

Fourth, a kind of injury insurance, caused by accidents, will be launched. *The construction law* states that all companies in the industry must pay the insurance for injury caused by accidents, which is compulsory. Shenzhen must learn from advanced countries in this aspect and make this work conform with the international standards.

Fifth, the research and application of new technology will be promoted. 'Science and technology is the most important productive force'. The companies should be encouraged and helped to engage in the research and further application of new technology. Each year, guidelines will be worked out for research, trying to address relevant problems, such as the safety devices on scaffolding and hoists, and safety devices on the tower crane and hand-operated construction equipment.

Sixth, campaigns to promote a safety culture within the general working population should be encouraged accelerating the development of a safer work environment. It is necessary to form a kind of atmosphere where everybody pays attention to safety and practises good safety management in the construction industry. In the future, besides publications on safety, every means to educate the workers, such as posters, pamphlets, short and simple artistic creations, exhibitions, videos, competitions and speeches through which the workers can easily learn safety knowledge and technology, should be pursued.

The construction safety situation in Shenzhen will strive to become better and better in the future!

4.3   REFERENCES

The Construction Department of P.R. of China, 1989, *Standard of Construction Safety Inspection.*
Shenzhen Construction Safety Supervision Center, 1997, *Standard of Construction Safety Inspection for SSEZ.*
SSEZ Construction Bureau, 1998, *Collection of Law and Regulations about Construction in SSEZ.*

4.4   APPENDIX

Table 4.1   Law and regulations

Year	Law and regulations
1989	Standard of construction safety inspection
1991	Contemporary regulations of construction safety punishment
1994	Qualification supervision of construction enterprise in SSEZ
1994	Contemporary regulations of civilisation and environment in construction sites in SSEZ
1996	Standard of construction safety inspection for special building machine in SSEZ
1996	Contemporary regulations of construction safety training and education in SSEZ
1996	Standard of construction safety inspection for SSEZ
1996	Instruction standard of construction safety inspection for SSEZ
1996	Supervision and safety code for climbing frames in temporary support in SSEZ
1996	Supervision and safety code for piles in SSEZ
1998	Byelaw of construction safety supervision In SSEZ
1998	Regulations of construction sites civilisation in SSEZ
1998	Standard of construction sites civilisation in SSEZ

Table 4.2   The present situation in SSEZ

No.	Description
1	3 stage supervision
2	Training and education
3	Propaganda about construction safety
4	Promote the standardisation and legitimisation of construction safety supervision
5	Special treatment of serious accidents
6	Promote the test of safety protection staff and facilities

Table 4.3   Reform of construction safety supervision in SSEZ

No.	Context
1	The person is the core
2	Draw up the technical standards and regulations
3	Civilised construction sites
4	Injury insurance
5	Research and application of new technology
6	Safety culture

CHAPTER 5

Analysis of Construction Site Injuries in Palestine

Adnan Enshassi and Peter E. Mayer

INTRODUCTION

Construction sites are considered dangerous places. A large number of people die on them every year. Many site injuries result from people falling from structures like roofs and scaffolds, or being hit by falling objects. Many others are caused by the misuse of mechanical plant and site transport, including hoists (Frayer, 1995). There is nearly always keen competition for new contracts and site personnel are often under pressure to work to tight time and cost constraints. It is hardly surprising that safety is often neglected.

In spite of the low attention often given to construction sites injuries in many countries, the statistics continue to be alarming. For instance, fatal accidental injury rates in the United Kingdom and Japan are reported to be four times higher in the construction industry, when compared to the manufacturing industry (Bentil, 1990). Construction is often classified as a high-risk industry because it has historically been plagued with much higher and unacceptable injury rates when compared to other industries. In the United States, the incidence rate of accidents in the construction industry is reported to be twice that of the industrial average. According to the National Safety Council, there are an estimated 2,200 deaths and 220,000 disabling injuries each year (National Safety Council, 1987).

Over the years, construction has been a major contributor to the national economy. Presently, however, its share of the GNP is about 5-7 percent. In the 1970s this percentage was roughly 7-10 percent (Koehn and Pan, 1996). In Palestine, construction share of the GDP in 1994 is about 16.8, in 1992 the percentage was about 13.0 (Table 5.1). Therefore, it is crucial to take safety factor into consideration in the construction industry in Palestine in order to minimise the number of injuries on construction fields. This paper presents an analysis of statistical data on construction site injury between 1999 to 2000 published by the Ministry of Labour.

5.1   SAFETY MANAGEMENT

The attitudes and behaviour of construction workers and managers towards safety
is undoubtedly a major factor in the poor accident record of the industry in
Palestine. Many workers and managers see construction as a rough job for tough,
self-reliant people. It is widely believed that building to tight deadlines at low cost
is incompatible with high safety standards (Abd. Ghani, 1996).

Construction workers accept that their work is demanding and risky,
although they usually underestimate the risk. Group norms may cause individuals
to ignore safety measures for fear of appearing cowardly to their workmates
(Laney, 1982). Many managers and workers resent outside pressures on them to
comply with safety regulations and sometimes make collusive arrangements to
avoid them. Unfortunately, in Palestine safety regulations did not exist till now
(Enshassi, 1998).

If the construction site is to become safer, the major task is to change
people's attitudes. This may mean changing contractors' and clients' attitudes too.
They may have to accept that there is a safety premium to pay on the cost of
construction; that if getting a rock-bottom price means that people will be killed or
seriously injured, then the price is too low.

The management of labour safety and health at a construction site is the
responsibility of the employer. The implementation of a specific safety plan is the
responsibility of various sections and departments of a company. Depending upon
the scope and nature of the organisation, the employer should direct the labour
safety and health management committee to perform the following duties (Labour
Safety and Health Act, 1991):

- Prepare a thorough occupational accident prevention plan and direct
  implementation on sites.
- Plan and monitor labour safety and health management in all divisions.
- Plan and check regularly the labour safety and health installations.
- Direct and supervise safety personnel in conduction inspection tours,
  periodic inspections, and priority inspections of the work environment.
- Plan and conduct labour safety and health education and training.
- Respond to and supervise occupational accident investigations and
  compile occupational accident statistics.
- Provide the employer with data and suggestions related to labor safety
  and health management.

The construction industry should be interested in safety regulations and
implementation for humanitarian and economic reasons. In humanitarian terms, if
accidents occur, a construction firm and/or its employees may suffer pain and/or
loss of life due to an accident. In addition, its safety record will be lower and there
may be a loss of the firm's reputation. In economic terms, an effective safety

management programme may be profitable for a construction firm. This is because accidents have high direct (medical, worker's compensation) and indirect (reduction of productivity, job schedule delays, rework) costs.

Table 5.1   Economic Structure (Percentage of GDP)

Industry category	Year 1992	Year 1993	Year 1994
Construction	13.0	16.4	16.8
Agriculture	35.5	29.0	33.3
Industry	7.4	8.4	8.2
Government	10.5	11.8	11.2
Other GDP	33.6	34.4	30.5
Employment in West Bank and Gaza Strip	204.0	232.0	249.0
Employment in Israel	116.0	83.0	68.0

Source: IMP and PECDAR

Table 5.2   Construction Injuries as compared to Total Industrial Injuries

Year	Total industrial injuries (A)	Construction injuries (B)	(B/A) x 100% (%)
1994	22	1	4.5
1995	32	1	3.1
1996	23	1	4.3
1997	48	6	12.5
1998	35	9	25.7
1999	75	31	41.3
2000	247	89	36
Average percentage construction injuries to total industrial I injuries			18.2%

Source: Ministry of Labour, PNA, Annual Reports

5.2   ANALYSIS OF THE CONSTRUCTION SITE INJURIES

The analysis was based on the statistical data published by the Ministry of Labour in Palestine between 1994 and 2000. Table 5.2 shows that from 1994 to 2000 the total industrial injuries increased from 22 cases to 247 cases. In addition, the number of reported construction site injuries increased from 1 case to 89 cases. The site injuries account for an average of (18.2%) of the annual total industrial injuries. However, it has been found that the majority of contractors in Palestine did not report site injuries to the Ministry of Labour (Enshassi, 1997). Two main reasons were found for this. Firstly, the employees have no insurance, and, secondly, the employers are concerned at the effect on their reputation if they report the site injuries. Therefore, it is uncertain whether the number of injuries published by the Ministry of Labour is correct.

It can be seen from Table 5.3 that construction site injuries (28.6%) are the highest among other major industrial injuries. It can be observed from Tables 5.2 and Table 5.3 that construction site injuries have dramatically increased in 1999 (41.3%) and in 2000 (36%) compared to 1994 (4.5%). This can be traced to the increase of building and construction activities aimed at the reconstruction of newly developing Palestine after the peace accord. The International and Arab communities have contributed to the Palestinian National Authority in 1994 US$511,738m, in 1995 US$429,995m and in 1996 US$516,276m in order to reconstruct Palestine (PECDAR, 1997).

Due to the urgent need for improving the infrastructure in Palestine, an emergency programme was established to reconstruct and build as much as possible in a short duration. This situation puts a lot of pressure on clients and contractors as well, in order to complete a number of projects (e.g. housing, schools, roads, water and sewage) in a short time. There is still an absence of safety codes and regulations in Palestine. This has led to an increase in construction site injuries.

According to a survey conducted by the author (Enshassi, 1997), the number of construction site injuries is far more than the figure published by the Ministry of Labour. It has been observed that when an accident happens on a construction site, the employer does not report it to the concerned department. He tries to solve the problem internally according to the culture and attitudes of the people involved. This unhealthy system has contributed negatively to the safety requirements on the site. In addition, nobody can learn what the cause of the accident is and how it can be avoided in the future. Moreover, no reliable statistical data is available regarding the real number and nature of injuries on construction projects.

Most reported construction site injuries in Palestine occurred to the production-related workers and equipment operators. The type of accidents mainly are the result of construction workers, being struck by falling objects and stepping on, striking against or being struck by objects. The statistics show that the reasons for accidents were falling (18%), equipment and plant (7%), material dropped from a height (9%), misuse of manual equipment (3%), steel works (6%), nails (49%) and other miscellaneous (8%) causes. The nature of the injuries mainly consists of fractures, wounds, cuts, sprains and pains, confusions, crushings and superficial injuries. Most frequent site injuries occur as a result of poor usage of safety belts, safety helmets and boots, and over exertion of construction workers on site.

Table 5.3   Total Major Industrial Injury according to Type and Industry

Year	All industries	Manufacturing & processing	Wood workshops	Tiles & blocks factories	services	construction	Plastic factories	Electricity	Sanitary services	Others
1994	22	7	2	3	1	1	1	--	--	7
1995	32	11	6	2	1	1	1	--	--	10
1996	23	10	1	1	1	1	--	--	--	9
1997	48	6	8	5	7	6	1	3	8	4
1998	35	6	5	6	--	9	--	2	2	5
1999	75	8	6	12	1	31	2	5	4	6
2000	247	32	15	19	12	89	15	16	15	34
Total	482	80	43	48	23	138	20	26	29	75
%	100	16.6	8.9	10	4.8	28.6	4.1	5.4	6	15.6

Source: Ministry of Labour, PNA, Annual Reports

5.3   DEGREE OF INJURY

Construction site injuries consist of minor injury, disablement and fatality. Table 5.4 shows the degree of injuries for all industries. In 1994, it was found that 7 cases were minor injuries, 13 cases were disablement and 2 cases were fatalities. In 2000, it can be seen that 42 cases were minor injuries, 198 cases were disablement and 7 cases were fatalities. This indicates that safety regulations and measures were not effectively implemented by employees.

Table 5.5 shows the percentage of construction injuries over the total industrial injuries. In 1994, one minor injury case was found, and no single disablement or fatality was found. In 2000, 51 minor injury cases were found, 34 disablement cases were observed and 4 fatal cases were found. This indicates that the rate of injury has increased for all degree of injuries. It indicates that safety procedures were not a priority for most employees in Palestine.

By comparing the degree of construction injuries to the total industrial injuries (see Tables 5.4 and 5.5) the percentage of minor injury, disablement and fatality of construction over all industries gradually increased from 1994 to 2000. The percentage of minor injuries increased from 4.5% to 20.7% and disablement rose from 0% to 13.7%, and similarly fatalities rose from 0% to 1.6% of the total industrial fatalities. Unfortunately, the Ministry of Labour has no data regarding the financial implication of construction injuries. However, insurance claims, delay in construction works and compensation paid to the affected workers or families have financial implications as a result of construction site injuries.

Adnan Enshassi and Peter E. Mayer

Table 5.4   Degree of Industries for All Industries

Year	Total industrial injury	Minor injury		Disablement		Fatality	
		No.	%	No.	%	No.	%
1994	22	7	32	13	59	2	9
1995	32	15	46	16	50	1	4
1996	23	13	57	10	43	-	-
1997	48	23	48	23	48	1	4
1998	35	20	57.2	14	40	1	2.8
1999	75	37	49	35	47	3	4
2000	247	42	17.2	198	80	7	2.8

Source: Ministry of Labour, PNA, Annual Reports

Table 5.5   Percentage of Construction Injuries over the Total Industrial Injuries
(According to the Degree of Injury)

Year	Total Industrial Injury	Construction Injury	Minor Injury		Disablement		Fatality	
			No.	%	No.	%	No.	%
1994	22	1	1	4.5	-	-	-	-
1995	32	1	1	3.1	-	-	-	-
1996	23	1	-	-	1	4.3	-	-
1997	48	6	3	6.25	3	6.25	-	-
1998	35	9	3	8.6	5	14.3	1	2.8
1999	75	31	16	21	13	17.3	2	3
2000	247	89	51	20.7	34	13.7	4	1.6

Source: Ministry of Labour, PNA, Annual Reports

5.4   CONCLUSIONS

The analysis of the data presented in this paper brings the following findings:

- Construction site injuries account for an average of (18.2%) of the annual total industrial injuries, this is relatively high.
- The number of construction site injuries (28.6%) is the highest among other major industries. This is quite alarming.
- The majority of contractors in Palestine did not report construction site injuries to the Ministry of Labour. Contractors try to solve the workers injury problems internally according to the local culture.
- Safety codes and regulations still do not exist in Palestine. Therefore, there is an urgent need to establish and implement safety codes.
- Most frequent construction site injuries occur as a result of poor usage of safety belts, safety helmets and boots and over exertion of construction workers on site.

- The common types of injury were fractures, wounds and cuts and sprains and strains.

The findings are quite alarming to those concerned with site safety in Palestine, even more so if we consider the unreported injuries. It is important to eliminate the cause of accidents in order to minimise the number of injuries. Construction site safety precautions should be promoted in all construction sites.

5.5   THE WAY FORWARD

The job safety plan must be based on an analysis of project risks and a preventive programme specifically modelled to combat the particular and peculiar hazards of that project. In addition, construction workers often tend to be unmindful of safety regulations. It is recommended that construction workers should be constantly reminded of safety problems through their supervisors. Clear instruction is required concerning how specific tasks should be carried out so as to minimise the chances of accidents and injuries. The success of a safety programme is contingent upon the advance recognition of the hazards present and personal safety instruction to the tradesmen before hazardous work is initiated.

In Palestine, it is the responsibility of the contractor to provide a safe and healthy working environment. This must be accomplished in conformity with established standards that are designed to prevent the risk of injury caused by tools, machinery, equipment, etc. It is recommended that safety codes and regulations should be created and updated frequently. The purpose of these codes is to prevent injuries, fatalities, and structural failures. They also indirectly reduce the expenses and assist in maintaining the reputation of construction firms.

The Ministry of Labour should put more emphasis on safety requirements and implementation. The Ministry must set up a safety committee, the main functions of this committee should be:

- To monitor working arrangements on site with regard to health and safety.
- To develop site safety rules, safe systems of working and guidelines for hazardous operations.
- To study accident trends and safety reports.
- To investigate the causes of serious accidents.

Effective implementation programmes should focus on both the physical and the behavioural sides of safety and health. An adequate safety programme can assist in minimising construction site injuries which in turn can reduce financial loss, increase productivity, enhance profit and, above all, save the lives of work teams and so to help the industry and nation as a whole.

There is a pressing need to make students of civil engineering and, architecture in Palestine more aware of the importance of health and safety. This can be accomplished by incorporating courses on construction safety in the core curricula of such programmes. In addition, there is a need to conduct training programmes on a regular basis for construction engineers and workers about safety rules and regulations. Construction site safety should be seriously considered as one of the keys to a successful construction industry.

5.6   ACKNOWLEDGEMENT

The authors would like to express their appreciation to George Forster Fellowship Programme of Alexander von Humboldt Foundation in Germany for funding this research.

5.7   REFERENCES

Abd. Ghani, K., 1996, Construction site injuries: the case of Malaysia. In *Proceedings of the First International Conference of CIB Working Commission W99*, Lisbon, Portugal.

Bentil, K., 1990, The impact of construction related accidents on the cost and productivity of building construction projects. In *Proceedings of the CIB 90*, 6, Management of the Building Firm, Sydney.

Enshassi, A., 1998, Construction safety: issues in Gaza Strip. *International Journal of Building Research and Information*, 25, **6**.

Enshassi, A., 1997, Construction site injuries in Palestine: causes and effects, unpublished paper.

Frayer, B., 1995, *The Practice of Construction Management* (London: Collins).

Koehn, E. and Pan, C.S., 1996, Comparison of construction safety and risk between USA and Taiwan. In *Proceedings of CIB W65*, The Organization and Management of Construction, UK.

Laney, J.C., 1982, *Site Safety* (London and New York: Construction Press).

Labour Safety and Health Act, 1991, *Council of Labor Affairs*, Executive Yuan, Taiwan, Republic of China.

National Safety Council, 1987, OSHA investigates Connecticut Apartment project collapse. *OSHA Up to Date*, XVI, **5**, Chicago, Illinois.

PECDAR, 1997, *Monthly Information Bulletin*, Jerusalem, 1, 7.

Section 2

Safety Management Issues

CHAPTER 6

An Overview of Safety Management Systems

Steve Rowlinson

INTRODUCTION

Safety management is the key to ensuring that safety measures are implemented on construction sites. With the implementation of performance based safety legislation in many countries the onus has moved away from enforcement of prescriptive regulations to setting up of an infrastructure within the company to manage safety issues in a spirit of self regulation. The following chapters address a number of key issues relating to safety management and the examples used are drawn from four continents. Three chapters deal with the related issues of safety culture, relational contracting and subcontractor management as means to effective safety management. The first of these identifies the goals of a safety management system in terms of safety culture and performance measurement and the following chapters discuss organisational mechanisms for achieving these goals. The next chapter moves down to the individual level and deals specifically with the issue of the performance based approach to safety management and its impact on construction managers. Three subsequent chapters address issues of safety management through the medium of project management and project planning, emphasising the need for communication. The final paper deals with the practical implementation of the 5S system of safety management.

6.1   SAFETY CULTURE AND CLIMATE

Mohamed discusses safety culture, safety climate and safety performance measures. He starts from the premise that there is a need for information to stimulate appropriate action and cites the view that organisational learning emphasises the need for performance measurement (Brignall and Ballantine, 1996). He states that performance measurement is the process of quantifying past action (Neelly, 1998). To facilitate such a process, performance measurement systems are developed as a means of monitoring and maintaining organisational control, i.e. ensuring that an organisation pursues strategies that lead to the achievement of goals and objectives (Nanni, Dison and Vollmann, 1990). He points out that although accident statistics are widely used throughout the construction industry, Laitinen, Marjamaki and Paivarinta (1999) state that it is almost impossible to use accidents as a safety indicator for a single building construction site. This is because of random variation, where many sites will have no accidents, and it is not possible to determine whether these sites with zero accidents are safer than sites with, for example, four or five accidents. Glendon and McKenna (1995) identify a number of reasons why accident data, or similar outcome data, are poor measures of safety performance. In discussing how safety can be managed he states that safety culture concerns the determinants of the ability to manage safety (top-down organisational attribute approach), whereas safety climate concerns workers' perception of the role safety plays in the workplace (bottom-up perceptual approach). The top-down approach should include observable measures such as management commitment, participation and accountability, procedures and policies, communication, etc. On the other hand, the bottom-up approach should include a different set of observable measures such as workers' constructive involvement, proactive reporting, individual attitude, group behaviour, working relationships with supervisor and co-workers, etc. He goes on to argue that this dual perspective of safety should be adequate to comprehensively assess and benchmark safety performance. This chapter gives a company wide perspective on how safety can be managed and how a culture and a climate can be nurtured. The following chapter takes this analysis down to the level of the individual manager and how the managers' attitudes can affect performance.

Haupt discusses the performance approach to construction worker safety and health which involves identifying and satisfying user requirements that must result from applying a safety standard, regulation or rule without setting out the specific technical requirements or methods for doing so (Haupt, 2001). This proactive approach describes what has to be achieved to comply with the regulations leaving the means and methods of doing so up to the contractor. It demands the involvement in the safety and health effort of all participants in the construction process from owners to construction workers. In terms of this performance approach contractors are not solely and exclusively responsible for construction worker safety. Consequently, the performance approach to construction worker safety and health demands a paradigm shift from the traditionally prescriptive approach (Haupt and Coble, 2000).

In the United States the prescriptive approach prevails. It describes means, as opposed to ends, and is primarily concerned with the type and quality of materials, method of construction, and workmanship. This approach attempts to standardise the work process using prescriptive rules and procedures usually backed by the monitoring of compliance and by sanctions for non-compliance (Reason, 1998). On the other hand, the performance approach requires a culture change that relies on a continuous and long-term commitment to understanding, evaluating and improving construction activities and processes (Coble and Haupt, 1999; 2000). This chapter discusses the results of an attitudinal survey of construction managers in the United States regarding a performance approach to construction safety.

The next chapter deals with mechanisms for fostering a safety climate and culture throughout construction project teams, within the typical temporary multi-organisations of a typical construction project. Heath begins his treatise as follows 'To build is to be robbed': Johnson's edict still carries much weight in the modern age. The Latham Report (1994) paints a picture of distrust and conflict, not just between client and contractor but between the design and construction team and within the construction team itself.

The UK Government has attempted since the Latham Report was published in 1994 to change the ethos of the construction industry and make it less conflict orientated. Legislation followed on from Latham in the form of the Housing Grants, Construction and Regeneration Act 1996, and further initiatives in the form of The Egan Report and most recently the Movement for Innovation (M4i). The focus for these initiatives was the performance of the construction industry, principally its inability to satisfy customer expectations, the apparent lack of teamwork between the various contracting parties and the lack of post contract liability. In short, the emphasis was on value for money.

At the same time, initiated by European social legislation, the Construction (Design Management) (CDM) Regulations has provided a focus on health and safety in the design, construction and maintenance phases of construction projects allocating specific responsibilities to those named as function holders under the CDM Regulations.

The issue for consideration within this chapter is whether both sets of initiatives can change the culture and focus of the construction industry and act in concert to promote health and safety on construction sites and improve the performance and image of the industry in this respect.

The next chapter takes these arguments a step further and looks at a crucial element of the supply chain, subcontracting practices. Henry Tang (CIRC, 2001) in his review of the construction industry in Hong Kong stated: 'To improve construction quality, non-value adding multi-layered subcontracting must be eradicated.' and 'To improve safety performance, we need to foster a safety culture within the industry. We advocate the wide adoption of a preventive approach.' Wong et al look at this issue from the dual viewpoint of subcontracting and safety improvement. They report that the safety record of the construction industry was poor and much worse than other industries in Hong Kong. The reasons of the poor safety record may correlate with many factors such as complexity of the work or system, risk nature of works, management style, safety knowledge and commitment, and personal behaviour (Stranks 1994). The multi-layered subcontracting practice is a unique determinant of safety performance in the Hong Kong construction industry. When a principal contractor secures a project from a developer, usually it would break down the project activity into different trades and then sublet each category to individual subcontractors with the lowest bid (Lee, 1991). These subcontractors would normally further subcontract their work without the consent of their principal contractor to several smaller firms in order to minimise their overheads.

The problems of multi-layered subcontracting practice have long been an issue and a controversial subject in the industry. Recently researchers (Linehan 2000 and Tong, 2000) have suggested having legislation for restricting the subcontracting practice in the construction industry for better safety performance. Although such restrictions may help to improve the safety performance, it would be extremely difficult to implement in the workplace due to the traditional work practice. On the other hand, some people including Lee (1999) and So (2000) disagreed with the idea of restricting the multi-layered subcontracting practice. They considered that the subcontracting practice has great market values for the industry, which ought not be interfered by restriction in the form of legislation. This chapter reviews these arguments and puts them into a context of international best practice.

6.2   SAFETY MANAGEMENT AND PLANNING

The next three chapters deal with the impact of various aspects of planning on the safety management process. Cameron and Duff report that the performance standards traditionally associated with the delivery of construction projects are time, cost and quality. However, in recent think-tanks such as The Construction Task Force (www.construction.detr.gov.uk/cis/rethink/index.htm), The Construction Best Practice Programme (www.cbpp.org.uk), The Key Performance Indicators Zone (www.kpizone.com), and The Movement for Innovation (www.m4i.org.uk) have set additional targets, challenging construction to plan more thoroughly and improve performance in many areas, including a greatly increased attention to health and safety.

Today's thinking seriously challenges the old model of time/cost/quality trade-off, which suggested that an improvement in one must lead to deterioration in at least one of the others. It now extends the total quality management philosophy that 'quality is free' and embraces the premise that delivery in one area, safety, can actually lead to benefits in other areas, time and cost. The importance of effective construction planning and control in the communication and avoidance of health and safety risks cannot be overstated but the fundamental premise which underlies this chapter, is that this need not and should not be a separate exercise aimed solely at health and safety. This chapter argues that effective management should embrace all production objectives, as an integrated, holistic process, and deliver construction, which satisfies all these objectives, not one at the expense of the others.

Murray, using the South African construction industry as his example, argues that governments in developed countries attempt to address the poor safety record of the construction industry by implementing legislation which punishes infracting companies or persons, by the application of fines (New Civil Engineer International 2000) and even prison sentences (Engineering News Record 2001a).

A complementary approach is to exhort the construction industry to improve its safety record (Engineering News Record, 2001b).

In developing countries, appropriate legislation quite often exists but it is often not applied in the case of infraction of laws, probably because construction workers are often migrant, low-paid, are not afforded union-protection and are held in low esteem by the society in which they live. Another factor which influences attitudes to safety and health is cost: a worker from a developing country who earns US$100 an hour and who dies in an accident at age 25 would be assessed for future loss of future earnings of US$80,000, a US worker at least 20 times that amount.

At the same time, in the corporate world, in Murray's experience, issues of safety and health are rarely really seriously addressed by owners, engineers and contractors during the construction planning and execution processes. The general approach seems to be top-down where head-office inspectors visit and monitor sites or provide guidelines. Two examples of this are the approaches taken by AMEC (AMEC, 2000) and Skanska (Särkilahti, 2001). A more suitable approach would be the bottom-up one where construction site managers automatically implement safety and health planning in their day-to-day activities. For this to happen, these areas would probably need to receive additional exposure in undergraduate construction management courses.

In this chapter the construction planning processes is examined and the application of the Project Management Body of Knowledge (PMBOK) processes and areas of knowledge, suitably adapted to include issues of safety, health and environment and social issues, will be described. The application of PMBOK processes includes applying risk management techniques to the analysis of safety and health and to do this it would be necessary to develop registers of indicators which would alert management to the need for risk analysis. A project Plan of Action or Business Plan including the above processes is shown to assist Safety and Health officials and others in evaluating a contractor's approach to safety and health matters.

Saurin *et al.* discuss the phenomenon that in spite of the high costs of construction related accidents (Hinze, 1991), many construction companies in Brazil adopt as their only safety management strategy the minimum requirements set out by the governing safety regulations. In Brazil, the main regulation related to construction industry is the NR-18 standard (Work Conditions and Environment in the Construction Industry). However, compliance with these standards might not be sufficient to guarantee acceptable safety performance, since they cover only minimal acceptable preventative measures. This statement is also applicable to regulations on safety and health management systems, such as OHSAS 18001 (Occupational Health and Safety Management Systems). Like quality assurance standards, OHSAS 18001 also does not establish performance targets, but is concerned with the compliance of safety management procedures.

Safety planning appears as a core requirement of OHSAS 18001 as well as of NR-18. On the one hand, OHSAS 18001 safety planning requirements consist of a risk management cycle, involving the continuous risk identification, evaluation and control. On the other hand, NR-18 (like similar regulations in many other countries) requires the elaboration of a safety and health plan, named PCMAT (Plan of Conditions and Work Conditions in the Construction Industry), which has a minimal mandatory scope. Since NR-18 was established in 1995, most companies produce the PCMAT only to avoid fines from government inspectors and do not effectively use it as a mechanism for managing site safety. Its main shortcomings are highlighted in the chapter as follows:

a. PCMAT implementation is usually regarded as an extra activity to managers, since it is not integrated to routine production management activities. NR-18 does not require its integration to other plans, except for site layout planning.
b. Safety experts (usually external to the company) produce it and normally there is no involvement of production managers, representatives of subcontractors or workers.
c. PCMAT does not usually take into account the uncertainty of construction projects. A fairly detailed plan is produced at the beginning of the construction stage and it is not usually updated during the production stage. The lack of update is caused by another common problem: the absence of formal control.
d. PCMAT emphasises physical protections, normally neglecting the necessary managerial actions (for instance, implementing proactive performance indicators) for achieving a safe work environment; and
e. PCMAT does not induce risk elimination through preventive measures at the design stage.

6.3   DAY-TO-DAY SAFETY MANAGEMENT

The final chapter deals with the implementation of a safety management system through a structured approach to day to day safety management, borrowing heavily from a concept developed in Japan. Yang, Zhang and Zhan note that one particularly worrying observation is that almost identical situations continue to cause death and injury in the construction industry. The industry seems unable to learn from its past mistakes. There is a tendency to blame external factors for the poor safety record. Factors such as:

•   The transient nature of the industry
•   The complete disregard for safety of many of its employees
•   The need to use a partially completed permanent structure, or a regularly changing temporary platform, to access works at a higher level
•   The constantly changing hazards as the project is constructed

- In general every corporation has its own safety policy. A corporate safety policy will include
- Policy statement: states what the organisation will do
- Operation of the policy: explains how the organisation will ensure that the policy is adhered to
- Organization: states who is responsible for safety at different levels of the organisation; explains how the policy affects departments, sections or projects within the organisation
- Communication: explains how the policy will be communicated throughout the organisation, and how senior management will be briefed on its implementation.

There is a danger that, once compiled, the safety policy is filed out of sight. The safety policy must be a working document, used regularly to monitor site practice. If practice begins to device from the policy, the major contractor must act immediately. This chapter describes the system in place in China which attempts to address these fatal flaws and assesses its effectiveness.

6.4   REFERENCES

AMEC, 2000, *Safety, Health and Environment*, www.amec.com.

Brignall, A. and Ballantine, B., 1996, Performance measurement in service business revisited. *International Journal of Service Industry Management*, 7, **1**, pp. 6-31.

CIRC (Construction ndustry Review committee), 2001. *Construct for Excellence*, The Government of Hong Kong Special Administrative Region.

Coble, R. and Haupt, T.C., 1999, Safety and health legislation in Europe and United States: A comparison. In *Proceedings of International Conference of CIB Working Commission W99 and Task Group, Safety Coordination and Quality in* Construction, edited by A. Gottfried, L. Trani and L.A. Dias, 36, Milan, Italy, 22-23 June, pp. 159-164.

Coble, R.J. and Haupt, T.C., 2000, Performance vs. prescription based safety and health legislation - A comparison. *The American Professional Constructor*, 24, **2**, pp. 22-25.

Engineering News Record, 2001a, Contractors face criminal charges in workers' deaths. *Engineering News Record*, USA, August 27, 2001. p. 7.

Engineering News Record, 2001b, Summit tackles work zone safety. *Engineering News Record*, USA, July 23, p. 10.

Glendon, A.I. and McKenna, E.F., 1995, *Human Safety and Risk Management*. (London: Chapman & Hall).

Haupt, T.C., 2001, *The Performance Approach to Construction Worker Safety and Health*, University of Florida, Unpublished Ph.D. dissertation.

Haupt, T.C. and Coble, R.J., 2000, International safety and health standards in construction. *The American Professional Constructor,* 24, **1**, pp. 31-36.

Hinze, J., 1991, *Indirect Costs of Construction Accidents* (Austin: The Construction Industry Institute).

Laitinen, H., Marjamaki, M. and Paivarinta, K., 1999, The validity of the TR safety observation method on building construction. *Accident Analysis and Prevention,* 31, **5**, pp. 463-472.

Latham, M., 1994, *Constructing the Team, Final Report on Joint Review of Procurement and Contractual Agreements in the UK Construction Industry,* (London: HMSO).

Lee, H.K., 1991, *Safety Management - Hong Kong Experience,* Lorraine Lo Concept Design, Hong Kong.

Lee, S.S., 1999, An article of interview by Labour Department. *A Hong Kong Journal of Construction Safety Newsletter,* Labour Department of HKSAR, 4, p. 4.

Linehan, A.J, 2000, Subcontracting in the construction industry. *A speech presented in Safety and Health Conference,* Hong Kong, 22-23 March.

Nanni, A.J., Dixon, J.R. and Vollmann, T.E., 1990, Strategic control and performance measurement. *Journal of Cost Management,* Summer Edition, pp. 33-42.

Neelly, A., 1998, *Measuring Business Performance* (London: Economist Books).

New Civil Engineer International, 2000, Heathrow Express report slates poor risk management. *New Civil Engineer International,* London, August, p. 5.

Reason, J., 1998, Organizational controls and safety: the varieties of rule-related behavior. *Journal of Occupational and Organizational Psychology,* 71, **4**, pp. 289-301.

Särkilahti, S., 2001, Skanska OY's Integrated Management System. In *Proceedings of the International Conference on Costs and Benefits Related to Quality and Safety and Health in Construction,* Barcelona, Spain.

So, K.C., 2000, The Multi-layers subcontracting system. An article of interview from the Hong Kong newspaper of *Hong Kong Economic Post,* 12 May.

Stranks, J., 1994, *Health & Safety in Practice Management System for Safety* (U.K. Bell & Bain Ltd).

Tong, Y.C., 2000, Scrapping of rotten subcontracting system urged. An article of interview by *South China Morning Post.*

CHAPTER 7

Safety Culture, Climate and Performance Measurement

Sherif Mohamed

INTRODUCTION

The need for information to stimulate appropriate action and organisational learning emphasises the need for performance measurement (Brignall and Ballantine, 1996). Literally, performance measurement is the process of quantifying past action (Neelly, 1998). To facilitate such a process, performance measurement systems were, and continue to be, developed as a means of monitoring and maintaining organisational control, i.e. ensuring that an organisation pursues strategies that leads to the achievement of goals and objectives (Nanni, Dixon and Vollmann, 1990).

Although accident statistics are widely used throughout the construction industry, Laitinen, Marjamaki and Paivarinta (1999) state that it is almost impossible to use accidents as a safety indicator for a single building construction site. This is because of random variation, where many sites will have no accidents, and it is not possible to determine whether these sites with zero accidents are safer than sites with, for example, four or five accidents. Glendon and McKenna (1995) identify a number of reasons why accident data, or similar outcome data, are poor measures of safety performance. The main problems are that such data are insufficiently sensitive, of dubious accuracy, retrospective, and ignore risk exposure. The use of workers, compensation statistics such as EMR (Experience Modification Rate) was also criticised for being sensitive to factors unrelated to safety. Hinze, Bren and Piepho (1995) outline scenarios where by changing some variables that are unrelated to safety performance, such as labour cost or company size, the EMR value was drastically altered, regardless of lack of change in actual accidents.

To overcome the disadvantages of adopting reactive measures, it has been suggested to use behavioural observation measures (Peterson, 1998; Laitinen, Marjamaki and Paivarina, 1999). These measures are based on random samples of workers behaviour which is then evaluated to be safe or unsafe (Tarrants, 1980). The advantage of using a behavioural observation method in measuring safety performance is that it does not just focus on non-compliant behavior, but also acknowledges safe behaviour. However, this method is not without its drawbacks. One of its major disadvantages is that no allowance is made for the severity of the safety breach. This could mean that a site could still end up with a high safety score but could have had a number of serious safety breaches at the same time. If using this to compare with other sites, a site with a few serious safety breaches could score higher than a site with more safety breaches but of a less serious nature.

In addition to the above, current performance measures offer too little in reflecting management commitment and corporate culture within which safety is supposed to permeate all levels of the organisation. Moreover, the traditional assumption that safety is the sole responsibility of the contractor (Hinze and Wiegand, 1992) is no longer valid, especially after the introduction of the Construction, Design and Management (CDM) regulations in many developing countries. The fundamental principle on which these regulations are based is that all project participants (client, architects, designers, subcontractors, etc.) who contribute to safety on a project are to be included in considering safety issues systematically, stage by stage, from the outset of the project (Baxendale and Jones, 2000). For example, once a decision is taken to commission a project, the client, together with the designer, must apply the CDM regulations by appointing a planning supervisor that has to ensure that all design work has been considered from a safety perspective. The planning supervisor is also responsible for assessing the safety competence of principal contractors and for observing their performance. As can be seen, these regulations bring safety, on an obligatory basis, into the planning and design of construction work (Baxendale and Jones 2000). Accordingly, project participants are drawn into the sphere of responsibility for safety plan implementation, thus changing the safety norms (Langford, Rowlinson and Sawacha, 2000). As such, performance measures need to reflect the safety aspects of these work phases too.

7.1   SAFETY CULTURE AND CLIMATE

The term 'safety culture' first made its appearance in the 1987 OECD Nuclear Agency report (INSAG, 1988) on the 1986 Chernobyl disaster. Gaining international popularity over the last decade, this term is loosely used to describe the corporate atmosphere or culture in which safety is understood to be, and is accepted as, the number one priority (Cullen 1990). Numerous definitions of safety culture abound in the academic safety literature, with all of them identifying it as being fundamental to an organisation's ability to manage safety-related aspects of its operations (Glendon and Stanton 2000).

Cooper (2000) argues that defining the *product* of safety culture is very important to clarify what a safety culture should look like in an organisation. He adds that this also could help to determine the functional strategies required to developing this *product*, and it could provide an outcome measure to assess the degree to which organisations might or might not possess a 'good' safety culture. This outcome has been severely lacking in construction, hitherto.

Although Blockley (1995) advocates that the construction industry would be better characterised as one with a poor safety culture and that attempts to improve the safety record will not be fully effective until the safety culture is improved, progress over the last decade on defining and measuring the safety culture concept in construction appears to have been somewhat slow. Confusion between the 'culture' and 'climate' terms might have contributed to such a slow progress. While these two terms have been used interchangeably due to the relationship and some overlap between them, climate refers only to the people's perception of the value of safety in the work environment. According to Cooper and Philips (1994), safety climate is concerned with the shared perceptions and beliefs that managers and workers hold regarding safety in the workplace (i.e. safety climate is, to some degree, dependent on the prevalent safety culture). It can be, therefore, argued that safety climate is largely a product of safety culture, and the two terms should not be viewed as alternatives.

Safety culture concerns with the determinants of the ability to manage safety (top-down organisational attribute approach), whereas safety climate concerns with the workers' perception of the role safety plays in the workplace (bottom-up perceptual approach). The top-down approach should include observable measures such as management commitment, participation and accountability, procedures and policies, communication, etc. On the other hand, the bottom-up approach should include a different set of observable measures such as workers' constructive involvement, proactive reporting, individual attitude, group behaviour, working relationships with supervisor and co-workers, etc. This dual perspective of safety should be adequate to comprehensively assess and benchmark safety performance. A number of previous studies in construction have, either directly or indirectly, addressed some elements of the safety culture and climate concepts. However, these studies are relatively few compared to the many that have focussed upon safety performance records, type and rate of accidents, and associated cost and lost time.

In recent years, there has been a movement away from safety measures purely based on retrospective data or 'lagging indicators' such as accident rates, towards so called 'leading indicators', such as measurements of safety climate (Flin et al. 2000). The shift of focus has been driven by the awareness that organisational, managerial and human factors rather than purely technical failures are prime causes of accidents (Weick, Sutcliffe and Obstfeld, 1999; Langford, Rowlinson and Sawacha, 2000).

Zohar (1980) uses the term 'Safety Climate' to describe a construct that captures employees' perceptions of the role safety plays within the organisation. It is regarded as a descriptive measure reflecting the workforce's perception of, and attitudes towards, safety within organisational atmosphere at a given point in time (Gonzalez-Roma et al., 1999). Budworth (1997) refers to measuring safety climate as taking the 'safety temperature' of an organisation. Flin et al. (2000) reviewed existing safety climate measures in an attempt to establish a common set of organisational, managerial and human factors that are being regularly included in measures of safety climate. Based on this particular review, the most frequently measured factors are related to management, risk, and safety arrangements (systems). Work pressure and competence are the other two emerging, although less frequently used factors in the literature.

As mentioned earlier, the construction industry unfortunately continues to rely heavily on traditional measures such as accident, and workers' compensation statistics. This implies that the issue of measuring safety culture and climate in construction is in its infancy and needs to be addressed. As Grubb and Swanson (1999) note there is virtually no research examining work organisations factors such as safety climate in construction. Attention must also be given to the managerial, organisational and human factors. However, construction organisations lack the insight for the development of effective performance measures and metrics needed to achieve a comprehensive safety management system.

7.2   Safety culture determinants

As mentioned above, safety culture relates to the determinants of the ability to manage safety, which, in turn, reflect the effectiveness and efficiency of the safety managerial and operational procedures. To be able to measure how good a safety organisational culture is, one would need to develop a set of performance measures and metrics in the context of:

1. Assessing safety performance objectively as well as subjectively, where a mix of quantitative and qualitative performance measures is adopted. This is to enable organisations to regularly evaluate safety performance and identify areas of potential improvement.
2. Distinguishing between metrics at the managerial (strategic) and operational levels. Using a classification based on these two levels, each metric can be assigned to a level where it would be most appropriate.
3. Using leading performance measures such as measurements of safety climate to capture the perception, inter alia, of both management and workers.

As with any benchmarking exercise, care must be taken to concentrate on meaningful measures that are:

1. understandable (can be expressed in clear terms to avoid misinterpretation or vagueness);
2. attainable (can be met with reasonable effort);
3. valid (can capture and reflect the main features of the process/aspect to be measured); and most importantly
4. client-focused.

The measures should also be incorporated in a performance measurement framework that provides more than a group of isolated and eventually conflicting measures and strategies. This could be achieved by utilising the strategic management tool, known as the Balanced Scorecard (BSC).

The BSC was first introduced by Kaplan and Norton (1992) to allow managers to look at their business performance from four important perspectives (*financial, customer, internal business and innovation and learning*) shown in Figure 7.1. The BSC attempts to integrate all the interests of key stakeholders; i.e. owners, customers, employees, etc., on a scorecard. The term 'balanced' in the name reflects the balance provided between short- and long-term objectives, between lagging and leading indicators, and between external and internal performance perspectives.

To effectively develop a safety balanced scorecard that can meet its potential in measuring safety culture, the four traditional perspectives, defined by Kaplan and Norton (1992), were put in a different light as shown in Figure 7.2. The figure shows, still four perspectives, but with slightly different names and content to meet the nature of the task.   A brief description of each one of the four perspectives is given below.

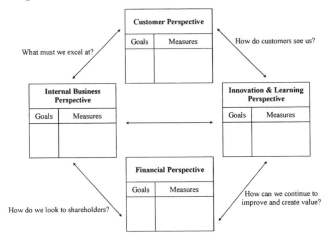

Figure 7.1 The Balanced Scorecard Framework (Kaplan and Norton, 1992)

7.2.1   Management perspective

Despite notions that culture cannot easily be created or engineered (Schein, 1990), in practice, the creation or enhancement of a safety culture is dependent upon the deliberate manipulation of various organisational management characteristics and activities thought to impact upon safety management practices. Management's commitment and involvement in safety is the factor of most importance for a satisfactory safety level (Jaselskis, 1996). Therefore, this strategic perspective reflects the following: What must Management excel at to achieve a zero-accident culture?

A focus on this perspective should lead to measures that would likely relate to such elements as management safety policy, commitment, accountability, and leadership. Using these elements in the BSC results in a number of criteria reflecting management control activities such as directing, leading, planning and co-coordinating. The process of deciding which measures of these criteria to adopt is a valuable one because it forces management to be very explicit about their safety-related management control activities and the relationship between them. Structuring this perspective according to a number of management control activities would also provide a focus on the goals of different activities necessary to accomplish the overall objective (i.e. achieving a zero-accident culture).

To reflect this perspective and to avoid developing an incoherent measurement system, it is crucial to incorporate measures that emanate from the organisation's safety policy. As Keegan, Eiler and Jones (1989) argue that the process of deciding what to measure must start with looking to the organisation's business strategy, defining the objectives, and then determining how it could be translated into divisional goals and individual management activities.

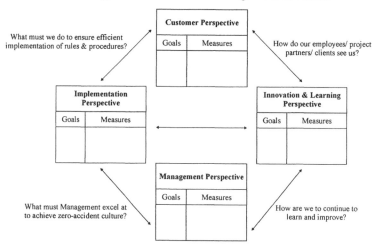

Figure 7.2   The Proposed Safety Balanced Scorecard Framework

7.2.2   Operational perspective

Although the BSC is a strategic rather than a diagnostic information system (Simons, 1995), Kaplan and Norton (1996) take the view that it is primarily a mechanism for strategy implementation, not for strategy formulation. To account for the latter viewpoint, this perspective poses the question: *What must we do to ensure efficient implementation of safety rules and procedures?* The main objective in this operational perspective is to enhance the integrity of the safety management system through addressing operational activities such as having and maintaining safer workplaces, improving working relationships, being proactive in reporting and detecting hazards, etc. Despite the fact that some of these activities may not always be measurable or conducive to quantification, they affect the achievement of the goals stipulated in the management perspective similar to how internal business processes determine the success of a business. As can be seen, this perspective is concerned with implementing action plans (means to the ends stipulated in safety rules and procedures).

7.2.3   Customer perspective

The shift in philosophy that takes place when safety management's viewpoint is embedded within the BSC framework is that the customer perspective is expanded to include the employees, project partners as well as clients. Stated another way, the customer perspective advocated herein should incorporate measures to capture how internal as well as external customers perceive the endeavours to achieve a zero-accident culture, as being promoted by the organization. Ideally, these measures should show all those involved how the safety culture is performing and foster incentives to work together (employees, project partners and clients). As a safety management's customer is viewed in three dimensions, it is important to focus on each. Employees represent the first dimension, as they are the source of achieving safety goals. This dimension reflects the following: *How do our employees perceive the role safety plays on site and how do they view our efforts?* The second dimension deals with project partners where measures should reflect: *How do our project partners see us dealing with safety in addressing specific project objectives?* Finally, the third dimension concerns clients where it would be important for the organisation to reflect: *compared to competing organisations, how do our clients see us in the context of safety?*

7.2.4   Learning perspective

If the measuring process remains static, then its potential to affect a positive outcome for the organisation is limited. Market, technology, project, client and other factors will often lead to changes in the type of information that needs to be collected to evaluate safety performance. Therefore, this strategic perspective reflects the following: *How are we to continue to learn and improve?* The focus herein is on the future as opposed to current safety performance levels. This perspective adds a dynamic element to the measurement framework. It recognises that organisations must continually learn and improve to achieve better safety performance levels. It is in this perspective, that organisations should incorporate human resource management measures, thereby recognising that people are true drivers of learning and improvement. This provides the rationale for investments in developing individuals' skills and capabilities, information systems, and enhanced organisational procedures (motivation, empowerment, etc.).

7.3   PERFORMANCE MEASURES

Although the BSC is undoubtedly valuable, its adoption is often constrained by the fact that it is simply a framework. It suggests some areas in which measures of performance might be useful, but provides little guidance on how the appropriate measures can be identified, introduced and ultimately used to manage performance. Therefore, performance measures that can relate to a perspective must be linked to the goals of this particular perspective. Developing such measures and linking them to goals is challenging but essential to ensure the effectiveness of the BSC approach. Another important facet of the BSC framework is that the measures represent a chain of cause and effect. Therefore, the proposed BSC should be viewed as a set of hypotheses about cause and effect relationships that affect safety performance. For example, the decision to commit extra resources (Management perspective) for training employees to identify site hazards (Learning perspective) should lead to minimised number or severity of hazard-related incidents (Operational perspective), which should reduce the number of near-misses, leading to improved employee and client perceptions (Customer perspective). It is worth pointing out that other cause-and-effect relationships, in the BSC, do not have to follow the order of this particular causality chain.

Given the overall framework of BSC application to measure safety culture, performance measures can be developed to fit into each perspective. In practice, an organisation should concentrate on a shortlist of measures that best capture and communicate the goals of each perspective. Also, organisations need to select only measures that are more suitable to the nature of their business. To measure training, for example, an organisation may decide to keep a record of the number of hours spent by individuals on training (hours/employee), whereas another organisation might be more interested in the level of competency related to the training. Once the measures are established, necessary information is then obtained, and the key linkages among the measures are identified. Proposed goals and measures for each perspective are briefly described below.

7.3.1   Management perspective

As mentioned earlier, this perspective concerns with the overall strategic objective of achieving a zero-accident culture and should represent a top-down driven strategy on safety, as part of an organisation's overall strategy for business. As such, goals in this perspective should reflect business, procurement, human resource management, and finance strategies. This multitude of strategies, in combination or alone, influences safety performance levels, be it voluntarily or involuntarily. It is essential therefore that there is coherence between these strategies, so that employees receive a consistent message on safety as a strategic issue. Typical goals would include accident elimination, reduction of number of incidents, improved productivity (as a result of less accident-related disruption to work), enhanced business image, accident-related cost reduction (compensation, insurance claims, etc.), highly competent workers (via a diligent recruitment policy and effective training programmes), and more safety-aware subcontractors (via a rigorous evaluation and selection process). Having quite diverse goals is necessary to provide valid benchmark standards through the assessment of a wide range of management issues across a number of processes (Fuller, 1997). Once these goals are identified and agreed upon, a set of measures should be developed and put in place to give insight into current practice and progress rate towards meeting performance targets.

Traditional lagging performance measures (e.g. accident and incident statistics) still have an important role to play in this perspective, however, these measures are merely one part of a whole. An additional set of proactive measures is much needed to reflect the effectiveness of management activities (commitment, proactiveness, leadership, etc.) as related to the goals identified above. In the first instance, such proactive measures may seem unfamiliar to the construction industry due to their focus on behaviours, attitudes and situations. Although understandable, there are a variety of well-tested instruments (quantitative and qualitative data collection tools) that can be used to measure the psychological, behavioural and situational aspects of safety culture. These are briefly presented below.

To measure management commitment, for example, a series of questions that measure people's commitment along various dimensions of safety have been successfully used to survey individuals within organisations (Guldenmund, 1998). There are also alternative measures for capturing the psychological aspects, which include group interviews and discussion groups (Johnson, 1992). The behavioural aspects of safety culture can be examined via peer observations, self-report measures and/or outcome measures (Cooper et al., 1994). Other behavioral measures that encompass leadership behaviours are also well documented (Komaki, 1998). The situational aspects of safety culture tend to be reflected in organisation's policies, operating procedures, management systems, control systems, communication flows, and workflow systems (Thompson and Luthans, 1990). As such, this wide range of cultural influences should be measured via audits of safety management systems (Glendon and McKenna, 1995). Details of all these instruments are beyond the scope of this chapter, but the premise is that these developed instruments, in the safety and organisational culture literature, would serve as thoughtful background for the application of the BSC to measure organisational safety in construction.

7.3.2   Operational perspective

Rules and procedures are the core component of safety management systems. A successful safety management system programme is based upon the premise that safety is both a management responsibility and a line function. While top management helps formulate the safety policy, its actual success depends upon the ability of site management and supervisory personnel to ensure that rules and policies are adhered to during daily operations (Agrilla, 1999). Consequently, this perspective concerns with the efficient implementation of safety rules and procedures on site. It also encompasses the ability to address specific project objectives in relation to safety, appraisal of physical work environment and workers' constructive involvement. One of the greatest strengths of the BSC is the salience that it gives operators, as they become more aware of the linkage between policy, procedures and performance targets. This, in turn, would help them identify opportunities for meaningful safety performance improvements outside the realm of compliance.

Goals in this perspective include higher degree of compliance, higher level of workforce proactiveness, more efficient site layout planning, efficient communication/feedback systems, safer workplaces, and better workers/

supervisors relationships. Considering these goals, measures would likely relate to elements such as process improvement, frequency of suggestions to improve safety, safety meetings, plan reviews, extent of accident analysis tasks, and ratio of recommended/completed remedial actions, degree of employee empowerment and constructive involvement.

7.3.3   Customer perspective

Safety culture within the organisation must be a philosophy, not just a set of guiding rules and procedures. Ideally, evaluating safety culture should be a two-way process, in which, feedback from employees and clients are obtained on both the goals and measures being used to measure its attainment. This comprehensive role can be clearly seen in Figure 7.3, which demonstrates the role of the BSC as the 'King Pin' of the safety management system.   The customer perspective represents the product of the safety culture. It can be used to assess how employees and external parties perceive safety on construction sites as a product of the prevailing organisational safety culture, thus setting an important indicator of the extent to which individuals are actually implementing the safety management system. This, in turn, would indicate whether additional opportunities are present for improving safety performance and enhancing safety culture.

Goals in this perspective should mainly focus on safety climate on sites. Descriptive measures would be chosen to capture client as well as employees' opinions reflecting their perception of, and attitudes towards, safety within organisational atmosphere at a given point in time. These will include several standard measures such as employee attitude and response to management, and customer satisfaction. Ojanen, Seppala and Aaltonen (1988) argue that the only way to measure safety climate is by surveys. The reader is referred to Flin *et al.*'s (2000) study where contemporary safety climate surveys and measures were critically reviewed, in an attempt to establish a common set of organisational, managerial and human factors that are being regularly included in measures of safety climate. The last section of this chapter empirically examines the relationship between safety climate determined by the factors identified by Flin *et al.* (2000), and safe work behaviour in construction site environments.

7.3.4   Learning perspective

The BSC will often identify gaps between the targets and existing performance. Using it to identify strategic initiatives and related measures, these can be then addressed and closed by initiatives such as managers and workers training and development. This perspective addresses the increasingly important issue of learning and improvement. Strategy, goals and measures should not be set in stone; the process of strategy development and performance improvement should be an evolving one, as shown in Figure 7.3. Measures should focus on such issues as encouraging bottom-up information flow and feedback, enhancing skills through education and training, improving supervisor/workers relationships, aligning incentive and reward schemes that are related to superior safe behaviour, empowering workers, etc. Such measures should facilitate periodic review of performance and progress made in meeting strategic objectives. Based on this review, programmes should be designed to target identified problems. For example, research has shown that a safety behaviour modification programme can be used successfully in giving feedback to employees about their performance, thus increasing safe behaviour (Cooper *et al.*, 1994). Causal relationships between measures should also be validated at defined intervals. The outcome of the review may necessitate the modification of action plans and revision of the scorecard.

7.4   SAFETY CLIMATE DETERMINANTS

In this section, the research model shown in Figure 7.4 was developed and empirically tested using a survey, which contained multiple measurement items relating to each of the constructs in the research model. The broad model's hypothesis is that safe work behaviours are consequences of the existing safety climate which, in turn, is determined by a number of independent constructs, i.e. management commitment, communication, safety rules and procedures, supportive and supervisory environments, workers' involvement, personal risk appreciation, work hazards appraisal, work pressure and competence. Whenever possible, appropriate scales that had demonstrated good psychometric properties in previous studies were employed. A brief description of each construct is given below.

7.4.1   Commitment

The role Management plays in promoting safety cannot be overemphasised. Management's commitment is a central element of the safety climate (Zohar, 1980). Management's role has to go beyond organizing and providing safety policy and working instructions. Several studies show that the management's commitment and involvement in safety is the factor of most importance for a satisfactory safety level (Jaselskis, 1996). Langford, Rowlinson and Sawacha (2000) found that where employees believe that the management cares about their personal safety, they are more willing to co-operate to improve safety performance.

7.4.2   Communication

Management is expected to use a variety of formal and informal means of communication to promote and communicate its commitment to safety (Baxendale and Jones, 2000). Simon (1991) suggests that both management communication and employees' feedback are critical for suggesting safety improvements and reporting near misses as well as unsafe conditions and practices.

7.4.3   Safety rules and procedures

Rules and procedures are the core component of safety management systems. A major factor influencing safety level is the extent to which workers perceive safety rules and procedures as promoted and implemented by the organisation (Cox and

Cheyne, 2000). Hood (1994) states that problems related to safety can frequently be traced to inconsistently applied or non-existent operating procedures.

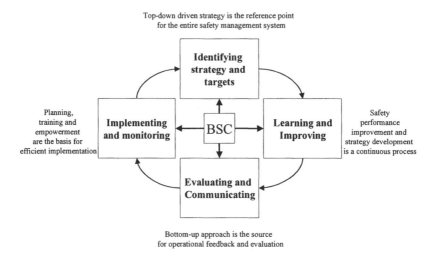

Figure 7.3   A Safety Management System for Strategic and Operational Implementation

7.4.4   Supportive environment

Supportive environment refers to the degree of trust and support within a group of workers, confidence that people have in working relationships with co-workers, and general morale. Having a supportive work environment demonstrates workers' concern for safety and fosters closer ties between them. Co-workers' attitude towards safety has been widely included in safety climate studies (Goldberg, Dar-El and Rubin, 1991).

7.4.5   Supervisory environment

A successful safety management system programme is based upon the premise that safety is both a management responsibility and a line function. While managers help develop and implement the programme, its actual success depends upon the ability of supervisory personnel to insure that the programme is carried out during daily operations (Agrilla, 1999). Langford, Rowlinson and Sawacha (2000) indicate that the more relationship-oriented supervisors are, the more likely it is that operatives will perform safely.

7.4.6   Workers' involvement

Anecdotal evidence suggests that it is not just management participation and involvement in safety activities which is important, but the extent to which management encourages the involvement of the workforce (Niskanen, 1994). Moreover, management must be willing to devolve some decision-making power to the workforce by allowing them to become actively involved in developing safety interventions and safety policy, rather than simply playing the more passive role of recipient (Williamson, *et al.*, 1997). Worker's involvement includes such issues as procedures for reporting injuries and potentially hazardous situations.

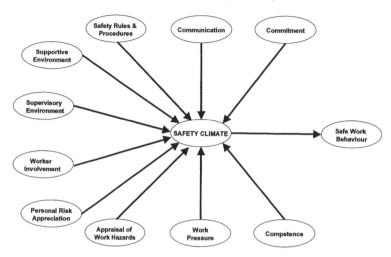

Figure 7.4   Research Model Linking Safety Climate Determinants to Safe Work Behaviour

7.4.7   Personal appreciation of risk

Cox and Cox (1991) argue that employees' attitudes towards safety are one of the most important indices of safety climate. Attitudes towards safety have been found to be associated with personal risk perception (Rundmo, 1997).   Individuals, however, differ in their perception of risk and willingness to take risks as demonstrated by March and Shapira (1992).

7.4.8   Appraisal of physical work environment and work hazards

The construction site is one of the primary resources available to the contractor. The aim in site layout planning and facilities is to produce a working environment that will maximise efficiency and minimise risks (Gibb and Knobbs, 1995). Aspects of site layout planning that need to be addressed include; access and traffic routes; material and storage handling; site offices and amenities; construction plant; fabrication workshops; services and facilities; and the site enclosure (Anumba and Bishop, 1997). Previous research shows that tidy and well planned (layout) sites are more likely to provide a high level of safety performance (Sawacha, Naoum and Fong, 1999).   For the purpose of this study, workplace hazards were defined as tangible factors that may pose risks for possible injuries or ailments. Within this definition, hazards do not always result in accidents, but they lurk in work environments, waiting for the right combination of circumstances to come together.

7.4.9   Work pressure

This construct deals with the degree to which employees feel under pressure to complete work, and the amount of time to plan and carry out work (Glendon, Stanton and Harrison, 1994). Ahmed et al. (1999) identify the tight construction schedule as the most serious factor that adversely affects the implementation of construction site safety in Hong Kong. This is supported by another study (Sawacha, Naoum and Fong, 1999), which found that productivity bonus pay could lead workers to achieve higher production through performing unsafely. Langford, Rowlinson and Sawacha (2000) state that supervisors are likely to turn a blind eye to unsafe practices on a site due to the pressure to achieve targets set by agreed programmes. They also argue that such ingrained practice of the industry (i.e. valuing expediency over safety) have to be overcome in order for safety management to be effective.

7.4.10   Competence

The essence of this construct is the workforce's perception of the general level of workers' qualifications, knowledge and skills, with associated aspects related to selection and training. Many researchers agree that training, especially in hazard-detection is a major factor influencing safety levels (Simon, 1991; Jaselskis, 1996). In summary, it is the workers' confidence that they have the skill to perform a particular job safely.

7.4.11   Safe work behaviour

As discussed earlier, traditional measures of safety performance are poor safety indicators due to their reliance on some form of accident or injury data. In view of this shortcoming, the research model adopts safe work behaviours (observable actions) as the safety indicator. This is based on the assumption that unsafe behaviour is intrinsically linked to workplace accidents (Thompson, Hilton and Witt, 1998). It is also supported by findings from studies and models developed based on the unsafe behaviour concept (Krause, 1997).

7.5   RESEARCH MODEL FINDINGS

A questionnaire survey was used in order to facilitate collection of information from construction sites. The questionnaire contained a total of 82 statements about safety issues at organisational, group and individual levels. Construction workers (contractors as well subcontractors' workforce) were the survey sample which included carpenters, steel fixers, equipment operators and electricians representing ten different organisations. To avoid the problem of bias, it was decided to administer the survey with no less than four employees working for the same organisation. A total of 68 questionnaires representing ten organisations, on six different construction sites in the south of State of Queensland – Australia, were completed.   Details of the questionnaire and statistical analysis are published elsewhere (Mohamed, 2002).

The research model was tested using the total sample (combining all responses solicited from all sites) in an attempt to seek correlation between safety climate and its determinants, and work behaviour. Generally speaking, support was found for the influence of management, safety and risk systems on safety climate. The empirical results indicate a significant relationship (positive association) between safety climate and safe work behaviour.

The path from the commitment construct to safety climate was found to be significant verifying previous research and further emphasising the importance of managers being committed and personally involved in safety activities to emphasise safety issues within the organisation. Also, the path from the communication construct to safety climate is positive, supporting the finding by Edmondson (1996) who found that handling errors in a negative way engendered a negative climate that in turn influenced the willingness of both management and employees to communicate freely and discuss mistakes and problems. Therefore, one can conclude that both commitment and communication are prerequisites to creating and sustaining a positive safety climate in construction site environments.

The results also suggest that safety rules and procedures, supportive and supervisory environments have influence on the safety climate. Supportive environment, however, seems to be perceived as of relatively more importance than the supervisory environment. This is not very surprising as a construction worker that continually interacts with co-workers also relies on them to a greater extent to provide a safer work environment. This finding suggests that workers in more positive safety climates are more likely to have above-average working relationships with managers, supervisors and co-workers.

The ability to identify hazards proved to have a strong positive relationship with safety climate. This finding was expected and corresponded to the findings of Mohamed (2000) who found that the ability to detect potential hazards has an important role in affecting the overall safety performance for an organisation and is dependent upon the intensity of management activities toward safety. Similarly, Salminen (1995) strongly emphasised the role of detecting hazards and suggested increased training in detecting these hazards.

The research model also predicted that the relationship between the competence and safety climate constructs is a significant one. It is, therefore, likely that a high level of competence will create a positive safety climate.   Laukkanen (1999) reports that skilled and experienced construction workers have fewer stress symptoms and are less prone to hazards than the inexperienced ones.

The expected influence of work pressure on safety climate was not supported, as the work pressure construct was not significantly related to safety climate. The absence of a significant direct relationship may be partially explained by the way in which work pressure was defined (i.e. a condition created by imbalance between high demand and low control at the work environment) in this study. Indicators used to measure this construct dealt solely with the psychological aspects of working under pressure (perceiving the conflicting safety and production requirements). However, 'working under pressure is the *norm* in the construction industry', this is a concern that was frequently expressed by informants during the course of data collection. As a result, and in view of the statistical inter-relation found between the work pressure and personal appreciation of risk constructs, an attempt was made to investigate whether an indirect relationship between work pressure and safety climate exists. The direct link between these two constructs was dropped from the model and a new link between work pressure and personal appreciation of risk was introduced. In doing so, the new link implies that the higher the perception of pressure to value expediency over safety, the higher the level of worker's willingness to take risk and consequently, the less positive safety climate becomes. The refined model was then analysed using the same data. Not surprisingly, support was found for the indirect influence work pressure has on safety climate. The amount of variance explained in the safety climate was slightly reduced. However, all paths remained largely unchanged, suggesting that the newly introduced link does not significantly affect the model.

7.6   CONCLUDING REMARKS

This chapter presented an overview of the concepts of safety culture and safety climate as applied to construction, with special emphasis given to developing and adopting proactive safety performance measures. It was advocated that the well-known performance measurement framework - the balanced scorecard (BSC) with slight modifications - could offer a tangible focal point for the complex process of getting an entire construction organisation to proactively manage safety. This framework would also offer the advantage of providing a mix of objective and subjective safety performance measures, and adding value by making both relevant and balanced information available to all stakeholders. Organisations need though to select only a handful of measures that reflect their policy and line of business, without leading to 'information overload'.

It is also important to note that the fundamental premise of the proposed performance measurement approach is that current results in all perspectives need to be displayed and reviewed, on a regular basis, in order to obtain an appropriate overview of an organisation's safety performance. Management should, therefore, set targets and assign responsibility for particular goals. Periodic reviews then will hold individuals accountable and ensure action. Because the BSC presents measures that management and employees can influence directly by their actions, this approach to performance measurement is expected to encourage behavioural changes aimed at achieving zero-accident culture. The proposed BSC is expected to create an environment which is conducive to learning organisations by testing, and providing feedback on, hypotheses regarding cause-and-effect relationships. All together should lead to measuring and benchmarking organisational safety culture.

Finally, the relationship between determinants of safety climate and safety climate, and that between safety climate and safe work behaviour on construction sites was examined. A research model was developed and tested using a survey, which contained multiple measurement items relating to each of the constructs in the model. Support was found for the influence of management, safety and risk systems on safety climate. The empirical results indicate a significant relationship (positive association) between safety climate and safe work behaviour. Positive safety climates seem to result from management's showing a committed and non-punitive approach to safety, and promoting a more open, free-flowing exchange about safety-related issues. Contrary to our expectations, this study indicated that work pressure has no significant direct relationship with safety climate. Instead, it has an indirect negative effect on safety climate through its impact on workers' willingness, under pressure, to take timesaving short cuts.

7.7   REFERENCES

Agrilla, J.A., 1999, Construction safety management formula for success.   In *Proceedings of 2nd International Conference of CIB Working Commission W99*, Hawaii, pp. 33-36.

Ahmed, S.M., Tang, S.L. and Poon, T.K., 1999, Problems of implementing safety programmes on construction sites and some possible solutions.   In *Proceedings of 2nd International Conference of CIB Working Commission W99*, Hawaii, pp. 525-529.

Anumba, C.J. and Bishop, G., 1997, Safety-integrated site layout and organization, *Proceedings of Annual Conference of the Canadian Society for Civil Engineers*, III, pp. 147-156.

Baxendale, T. and Jones, O., 2000, Construction design and construction management safety regulations in practice - progress and implementation. *International Journal of Project Management*, 18, **1**, pp. 33-40.

Blockley, D., 1995, Process re-engineering for safety.   In *Proceedings of risk engineering and management in civil, mechanical and structural engineering*, Institution of Civil Engineers, London, pp. 51-66.

Brignall, A. and Ballantine, B., 1996, Performance measurement in service business revisited. *International Journal of Service Industry Management*, 7, **1**, pp. 6-31.

Budworth, N., 1997, The development and evaluation of a safety climate measure as a diagnostic tool in safety management. *Journal of the Institution of Occupational Safety and Health*, 1, pp. 19-29.

Cooper, M.D., 2000, Towards a model of safety culture. *Safety Science*, 36, pp. 111-136.

Cooper, M.D. and Philips, R.A., 1994, Validation of a safety climate measure.   In *Proceedings of the Annual Occupational Psychology Conference*, British Psychological Society, Birmingham, UK, 3-5 January.

Cooper, M.D., Phillips, R.A., Sutherland, V.J. and Makin, P.J., 1994, Reducing accidents using goal setting and feedback: A field study. *Journal of Occupational and Organisational Psychology*, 67, pp. 219-240.

Cox, S.J. and Cheyne, A.J.T., 2000, Assessing safety culture in offshore environments. *Safety Science,* 34, pp. 111-129.

Cox, S.J. and Cox, T.R., 1991, The structure of employee attitude to safety: A European example. *Work and Stress*, 5, pp. 93-106.

Cullen, W.D., 1990, *The Public Inquiry into the Piper Alpha Disaster* (London: HMSO).

Edmondson, A.C., 1996, Learning from mistakes is easier said than done: Group and organizational influences on the detection and correction of human error. *Journal of Applied Behavioural Science*, 32, pp. 5-28.

Flin, R., Mearns, K., O'Connor, P. and Bryden, R., 2000, Measuring safety climate: identifying the common features. *Safety Science*, 34, pp. 177-192.

Fuller, C.W., 1997, Key performance indicators for benchmarking health and safety management in intra- and inter-company comparisons. *Journal of Benchmarking for Quality Management and Technology*, 4, **3**, pp. 165-174.

Gibb, A.G.F. and Knobbs, T., 1995, Computer-aided site layout and facilities. In *Proceedings of 11th Annual Conf. of ARCOM*, U.K., pp. 541-550.

Glendon, A.I. and McKenna, E.F., 1995, *Human Safety and Risk Management* (London: Chapman & Hall).

Glendon, A.I. and Stanton, N.A., 2000, Perspectives on safety culture. *Safety Science*, 34, pp. 193-214.

Glendon, A.I., Stanton, N.A., and Harrison, D., 1994, Factor analysing a performance shaping concept questionnaire. In *Contemporary Ergonomics*, edited by S.A. Robertson (London: Taylor & Francis), pp. 340-345.

Goldberg, A.I., Dar-El, E.M. and Rubin, A.E., 1991, Threat perception and the readiness to participate in safety programs. *Journal of Organisational Behaviour*, 12, pp. 109-122.

Gonzalez-Roma, V., Peiro, J., Lloret, S. and Zornoza, A., 1999, The validity of collective climates. *Journal of Occupational and Organisational Psychology*, 72, pp. 25-40.

Gru bb, P.L. and Swanson, N.G., 1999, Identification of work organization risk factors in construction. In *Proceedings of 2nd International Conference of CIB Working Commission W99,* Hawaii, pp. 793-797.

Guldenmund, F.W., 1998, The nature of safety culture: A review of theory and research. In *Proceedings of the 24th International Congress of Applied Psychology, Safety Culture Symposium*, August, San Francisco, CA.

Hinze, J. and Wiegand, F., 1992, Role of designers in construction worker safety. *Journal of Construction Engineering and Management,* ASCE, 118, **4**, pp. 677-684.

Hinze, J., Bren, D.C. and Piepho, N., 1995, Experience modification rating as measure of safety performance. *Journal of Construction Engineering and Management,* ASCE, 121, **4**, pp. 455-458.

Hood, S., 1994, Developing operating procedures: 9 steps to success. *Accident Prevention*, May-June.

INSAG, 1988, *Basic Safety Principles for Nuclear Power Plants* (Safety Series No. 75-INSAG-3), International Nuclear Safety Advisory Group, International Atomic Energy Agency, Vienna.

Jaselskis, E.J., 1996, Strategies for achieving excellence: contractor safety performance. *Journal of Construction Engineering and Management*, ASCE, 122, **1**, pp. 61-70.

Johnson, G., 1992, Managing strategic change: strategy, culture and action. *Long Range Planning*, 23, pp. 9-19.

Kaplan, R.S. and Norton, D.P., 1992, The balanced scorecard-measures that drive performance. *Harvard Business Review*, 70, **1**, pp. 71-79.

Kaplan, R.S. and Norton, D.P., 1996, *The balanced scorecard - translating strategy into action*. Harvard Business School Press, Boston, MA.

Keegan, D.P., Eiler, R.G. and Jones, C.R., 1989, Are your performance measures obsolete? *Management Accounting*, June, pp. 45-50.

Komaki, J.L., 1998, *Leadership from an Operant Perspective* (London: Routledge).

Krause, T.R., 1997, *The Behaviour-Based Safety Process Managing Involvement for an Injury-free Culture*, 2nd edition, Van Nostrand Reinhold, NY.

Laitinen, H., Marjamaki, M. and Paivarinta, K., 1999, The validity of the TR safety observation method on building construction. *Accident Analysis and Prevention*, 31, **5**, pp. 463-472.

Langford, D., Rowlinson, S. and Sawacha, E., 2000, Safety behaviour and safety management: its influence on the attitudes of workers in the UK construction industry. *Engineering Construction and Architectural Management*, 7, **2**, pp. 133-140.

Laukkanen, T., 1999, Construction work and education: occupational health and safety reviewed. *Construction Management and Economics*, 17, **1**, pp. 53-62.

March, J. and Shapira, Z., 1992, Variable risk preferences and the forces of attention. *Psychological Review*, 99, **1**, pp. 172-183.

Mohamed, S., 2000, Empirical investigation of construction safety management activities and performance in Australia. *Safety Science*, 33, **3**, pp. 129-142.

Mohamed, S., 2002, Safety climate in construction site environments. *Journal of Construction Engineering and Management*, American Society of Civil Engineers, 128, **5**, pp. 375-384.

Mohamed, S., 2003, Scorecard approach to benchmarking organizational safety culture in construction. *Journal of Construction Engineering and Mmanagement*, American Society of Civil Engineers, 129, **1**, pp. 80-88.

Nanni, A.J., Dixon, J.R. and Vollmann, T.E., 1990, Strategic control and performance measurement. *Journal of Cost Management*, Summer Edition, pp. 33-42.

Neelly, A., 1998, *Measuring Business Performance* (London: Economist Books).

Niskanen, T., 1994, Safety climate in the road administration. *Safety Science, 7*, pp. 237-255.

Ojanen, K., Seppala, A. and Aaltonen, M., 1988, Measurement methodology for the effects of accident prevention programs. *Scandinavian Journal of Work, Environment and Health*, 14, pp. 95-96.

Peterson, D., 1998, What measures should we use and why? *Professional Safety*, 42, **10**, pp. 37-40.

Rundmo, T., 1997, Associations between risk perception and safety. *Safety Science*, 24, pp. 197-209.

Salminen, S., 1995, Serious occupational accidents in the construction industry. *Construction Management and Economics*, 13, **4**, pp. 299-306.

Sawacha, E., Naoum, S. and Fong, D., 1999, Factors affecting safety performance on construction sites. *International Journal of Project Management*, 17, **5**, 309-315.

Schein, E.H., 1990, Organisational culture. *American Psychologist*, 45, pp. 109-119.

Simon, J.M., 1991, Construction safety performance significantly improves. *Proceedings of 1st International Conference of Health, Safety and Environment*, Netherlands, pp. 465-472.

Simons, R., 1995, Control in an age of empowerment. *Harvard Business Rev.*, 73, **2**, pp. 80-88.

Smith, D., Hunt, G. and Green C., 1998, *Managing safety the BS8800 way*, British Standards Institution, London.

Tarrants, W., 1980, *The Measurement of Safety Performance*, Garland STPM, New York.

Thompson, K.R. and Luthans, F., 1990, Organisational culture: A behavioural perspective. In *Organisational Culture and Climate*, edited by B. Schneider, San Francisco, CA., pp. 319-344.

Thompson, R.C., Hilton, T.F. and Witt, L.A., 1998, Where the safety rubber meets the shop floor: a confirmatory model of management influence on workplace safety. *Journal of Safety Research*, 29, **1**, pp. 15-24.

Trethewy, R.W., Cross, J. and Marosszeky, M., 1999, Safety measurement, a 'positive' approach towards best practice. In *Proceedings of the 2nd International Conference on Construction Process Re-engineering*, Sydney.

Weick, K., Sutcliffe, K. and Obstfeld, D., 1999, Organizing for reliability: processes of collective mindfulness. *Research in Organizational Behavior*, 21, pp. 81-123.

Williamson, A.M., Feyer, A., Cairns, D. and Biancotti, D., 1997, The development of a measure of safety climate: the role of safety perceptions and attitudes. *Safety Science, 25*, **1-3**, pp. 15-27.

Zohar, D., 1980, Safety climate in industrial organisations: Theoretical and applied implications. *Journal of Applied Psychology*, 65, **1**, pp. 96-101.

CHAPTER 8

Attitudes of Construction Managers to the Performance Approach to Construction Worker Safety

Theo C. Haupt

INTRODUCTION

The construction industry has a dismal track record with respect to the safety and health of its workers. Each year hundreds of accidents on construction sites either maim or kill many workers. Uncontrolled exposure to occupational health hazards curtails the working life of construction workers by several years as a result of occupational diseases. Workers in many trades die 8 to 12 years earlier, on average, than do many white-collar workers (Center to Protect Workers' Rights, 1993). Construction workers are two to three times more likely to die on the job than workers in other industries while the risk of serious injury is almost three times higher (Site Safe, 2000). The industry has not been unresponsive to this state of affairs, recognising that consistent excellence in worker safety requires a concerted effort on the part of everyone engaged in construction related activities. However, very few companies achieve consistently high levels of compliance with safety procedures.

For a long time legislators held that if they standardised the construction process and specified in exacting terms how contractors had to address any given conditions, they would be able to reduce the number of accidents on construction sites. Safety legislation described means of compliance and was concerned with the type and quality of materials used, the methods of construction employed, and the way construction activities were carried out. They concentrated on unsafe conditions in regulations while largely ignoring unsafe worker behaviour. The focus of implementation and enforcement of these regulations was on compliance rather than on proactive preventive measures. Noncompliance was usually punished in the form of fines. Unfortunately, simply providing and enforcing prescriptive rules and procedures is not enough to foster safe behaviour in the workplace (Reason, 1998).

Theo C. Haupt

For more than 50 years this prescriptive approach was instrumental in bringing down the number of accidents on construction sites. However, it became problematic to refine it to keep pace with innovation, better construction techniques and new materials and the process of making amendments was often extremely tedious and time-consuming.

More recently, legislators recognised these shortcomings of the prescriptive approach and opted in favour of a proactive approach that rather describes what has to be achieved to comply with safety regulations while leaving the means and methods of doing so up to the contractor. This performance approach demands the involvement in the safety and health effort of all participants in the construction process from owners to construction workers. In terms of this performance approach where it has been introduced, contractors are no longer solely and exclusively responsible for the safety and health of their workers. The approach involves the identification and satisfaction of worker safety and health requirements that must result from applying safety standards, regulations or rules without setting out the specific technical requirements or methods for doing so (Haupt, 2001). Consequently, this emerging approach to construction worker safety and health demands a complete paradigm shift from the traditionally prescriptive approach (Haupt and Coble, 2000). It requires a culture change that relies on a continuous and long-term commitment to understanding, evaluating and improving construction activities and processes (Coble and Haupt, 1999; 2000). This chapter discusses the results of an attitudinal survey of construction managers in the United States regarding the performance approach to construction safety.

8.1   CHANGE NATURE AND NEED

Nobody or organization is exempt from the effects of change. In particular, organisations have to cope with the effects of the globalisation of the economy, new market opportunities, technological advancements, emergence of new management approaches and paradigms, and appropriate responses to the needs of workers.

Bennis (1993:19) suggests that,

if change has now become a permanent and accelerating factor in American life [and elsewhere], then adaptability to change becomes increasingly the most important single determinant of survival. The profit, the saving, the efficiency, and the morale of the moment become secondary to keeping the door open for rapid readjustment to changing conditions.

Weatherall (1995) claims that continuing change will be the constant in this present century. It has been described as being *pervasive, important and most frustratingly, elusive* (Weston, 1998, p. 78). Change is a process of transition and transformation of people and systems that may be either temporary or permanent.

Humans have an almost natural tendency to resist change (Marshall, 1994), especially transformational change (Almaraz, 1998). Generally, people are hesitant to accept change if it was not their idea and they had no part in developing it. According to Nadler (1988) and others people resist or reject change because of the following:

Fear of the unknown.
Possibility of economic insecurity.
Threats to social relationships.
Failure to recognize the need for change.
Lack of confidence in the party promoting the change.
Lack of evidence of any benefit to be gained for themselves from the change.
Preference for things to remain comfortably the way they are; and
Fear that the change will affect them adversely.

For the performance approach to be implemented successfully and effectively, construction companies and other participants in construction-related activities will need to depart radically from their old way of doing things (Nadler and Tushman, 1989; 1990) until it becomes a corporate culture and part of the way business is done. Change may result in adjustments in the interconnection of any of the four components of people, task, technology, and structure. Such change will affect the culture of the organisation, transforming it in the process. Depending on the existing culture and the degree to which a change differs from that culture, an organisation might be more or less ready for such a change.

A model for determining the readiness of an organisation for change is offered by Sink and Morris (1995) as follows:

$$C = (a)\ (b)\ (d) > R$$

where

C = readiness for change;
a = level of dissatisfaction with the status quo;
b = clearly understood and desired future state;
d = practical first steps in the context of an overall strategy for actualising the desired future state; and
R = perceived cost or risk of changing.

The difference between what the organisation wants to achieve (variable b) and what presently exists (the status quo) creates a level of dissatisfaction (variable a). Once both of these variables are established, the first practical steps (variable d) and overall strategy for achieving the desired future state are decided. It should therefore become obvious that the degree by which these factors outweigh the perceived cost or risk of changing (variable R) will determine the readiness of the organisation for change (variable C). If the probability of achieving the future desired state is greater than the perceived cost or risk of changing, the more ready the organisation would be to change.

8.2   THE ROLE OF TOP MANAGEMENT

The importance of the role and commitment of management in supporting the safety and health effort in their organisations is well documented (Hinze, 1997; Levitt and Samelson, 1993).

Management's reaction to change determines [the] success [of change].
When upper management 'buys in' to the changes, it ensures success.
(Petersen, 1996:278)

Change, such as a paradigm shift from a prescriptive towards a performance approach, is difficult and almost impossible unless top management is totally committed to supporting and driving it. Management leadership, commitment and accountability are crucial (Statzer, 1999). Organisational change demands executive commitment and investment that is cognitive, emotional and financial (Diamond, 1998). According to Boles and Sunoo (1998), the largest barriers to managing change are lack of management visibility and support, employee resistance, and inadequate management skills.

Resistance to change is particularly relevant when the vision of management differs from the values and beliefs of the existing organisational culture. If the organisational culture fails to assimilate this vision and its implications, the desired change will never be accepted and will ultimately fail (Almaraz, 1998). Management is the key that allows safety performance improvements to occur in organisations (Freda, Arn and Gatlin-Watts, 1999; Hinze, 1997; Levitt and Samelson, 1993; Statzer, 1999). However, few managers acknowledge the need for a change in their management beliefs and values in order to support and nourish the new cultural reality (Almaraz, 1998; Boles and Sunoo, 1998) that the performance approach to construction worker safety represents. The importance of top management commitment and the issues of organisational culture cannot be underestimated. Improved safety and health performance within an organisation has to become a strategic choice. The extent to which top management chooses to support the programme of change will determine its ultimate success. It becomes apparent that the implementation of the performance approach to construction worker safety will be dependent on the capacity and willingness of management to introduce and support the changes necessary.

Another way in which behavior is strongly influenced is through modeling (learning by imitation). The research on modeling tells us that if we want to maximise approach (rather than avoidance) tendencies in workers, we [managers] must exhibit that behaviour ourselves.' (Petersen, 1996:266)

Managers and supervisors must strive to demonstrate safe work practices and make decisions that reflect their commitment to safety (Cook and McSween, 2000).

The influence of leaders on the performance of their organisations may be summed up as follows:

> *...organizational decision-makers, managers and professionals alike hope to ensure that their central values and beliefs influence the performance of their organizations by designing functional arrangements and hierarchies to facilitate and support those views.* (Ranson et al., 1980:199)

The values of individuals holding the top positions in organizations are the ones that are promoted and perpetuated throughout them (Hage and Dewar, 1973). Enz (1986:42) echoes this view when she claims

> *...clearly, top management is a critical group in examining values because of its control over organisational design and functioning. To understand the role of values in an organisational context requires close examination of the organisational leaders and how their beliefs operate to influence the activities within the firm.*

Major change is impossible unless the upper management of construction firms actively and demonstrably supports and understands the need for the changes they introduce (Freda, Arn and Gatlin-Watts, 1999). Not only is pressure to change required but also support in the form of time, financial resources, and decision-making authority.

8.3   RESEARCH DESIGN

Traditionally, research surveys have been characterised by the collection of data from large numbers of people to describe or explain the characteristics or opinions of a population using a representative sample (May, 1997). Ferber et al. (1980:3) agree that a survey is *a method of gathering information from a number of individuals, a 'sample', in order to learn something about the larger population from which the sample is drawn.*

Researchers have argued that there is a relationship between attitudes and behaviour by suggesting that the possession of a certain attitude necessarily means that a person will then behave in a particular way (May, 1997; Spector, 1981). Surveys have been used frequently as effective means to gain data on attitudes on key issues and causal relationships. For the most part, however, they can only show the strength of statistical association between variables.

In this particular study, the sample for the self-administered attitudinal questionnaire survey was drawn from a database of 843 construction organisations throughout the United States, compiled by the M.E. Rinker, Sr., School of Building Construction at the University of Florida. In order to select 100 organisations from the sampling frame, the probabilistic procedure of systematic random sampling was used. In terms of this practical procedure the researcher begins by making a random selection from the sampling frame, and then systematically samples every nth element (Salant and Dillman, 1994; May, 1997). Accordingly, the first construction organisation was randomly selected from the list. Since this sample would be a one-in-eight sample, every eighth (nth) organisation was systematically selected thereafter until the sample comprised of 100 organisations.

The number of completed questionnaires received was 67, representing an overall response rate of 68.4%, taking account of two questionnaires returned blank. The response was considered to be acceptable considering the nature of the study, the length of the questionnaire, and the time and budgetary constraints.

8.4   ANALYSIS OF FINDINGS

8.4.1   Upper management position

Most of the respondents (54.5%) held traditional upper or top management positions within their companies. Of these positions that were not directly related to safety and health involvement, 38.8% were CEOs, Presidents, Vice-presidents or General Managers of their firms. A further 14.9% were either Project or Contracts Managers. The spread of management positions of the sample is shown in Figure 8.1.

8.4.2   Duration of current employment in management position

Slightly more than half of the respondents (53%) had held their current position within the management structures of their firms for 5 years or less as is evident from the chart in Figure 8.2.

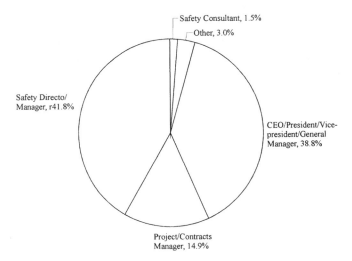

Figure 8.1 Management positions

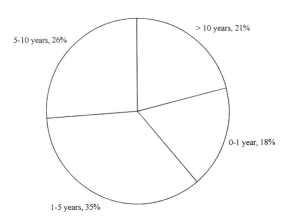

Figure 8.2 Length of employment in current position

8.4.3   Average number of employees

Of the sample, 42.4% firms employed between 0 and 100 workers; and 37.9% employed more than 250 workers. The distribution of workers employed by firms is shown in Figure 8.3.

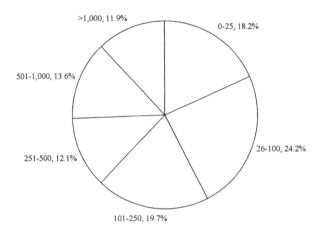

Figure 8.3 Average number of workers

8.4.4   Annual value of construction contracts

Most of the firms, namely 59.4%, had approximate annual construction contract values or turnover of less than US$100 million. The distribution is shown in Figure 8.4.

8.4.5   Contracting arrangements

The majority of firms in the sample (51.7%) engaged in general contracting, while 14.2% were subcontractors. Those that engaged in design-build contracting arrangements made up 11.5% of the sample.

8.4.6   Approach preference

If given a choice in a country where both the prescriptive and performance approaches to construction worker safety were acceptable and legitimate, 57.6% of the respondents indicated that they would prefer the performance approach. Of these, 42.1% made the choice because they perceived the approach to provide them with greater flexibility to dealing with safety issues, 23.7% because differing conditions required different approaches, 7.9% because they were responsible for selecting the most appropriate solution to the safety problem, and 7.9% because site conditions usually necessitated minor changes in approach from what is prescribed. The reasons for choosing either approach are listed in Table 8.1. These reasons compare very well with the features that characterise each approach.

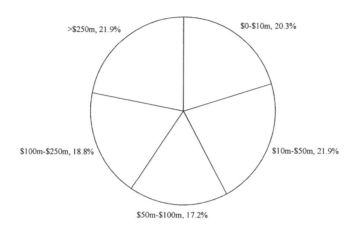

Figure 8.4 Annual construction contract revenue

8.4.7   Understanding of prescriptive and performance approaches

Most of the respondents (78.5%) felt that they understood the approaches well enough to be able to distinguish the differences between them.

8.4.8   Importance of issues with respect to an approach to construction safety management

Respondents were asked on a scale of 1 to 7 (1 = not important at all; 4 = neither important or unimportant; 7 = extremely important) to rate the importance of 5 issues with respect to construction safety management. The ranking of the means of the responses to each issue according to how important it was regarded is shown in Table 8.2. Respondents ranked the potential to improve safety performance on sites as being most important with respect to an approach to construction safety management. They regarded the ease with which the compliance requirements could be understood as next important.

Theo C. Haupt

Table 8.1   Reasons for Selecting Prescriptive or Performance Approach

Prescriptive	Performance	Total	Reasons for preference
	9	9	Differing conditions may require different approaches
	3	3	Minor changes allowed due to site conditions
16		16	More definitive and compliance can be measured objectively
6		6	Workers need specific instructions to avoid shortcuts
	16	16	Provides contractor with flexibility
	1	1	Easy for workers to understand requirements
	3	3	Responsibility of solution choice vests in contractor
	1	1	Allows for innovation and ingenuity
	1	1	Consistent structural strength better maintained
	1	1	Unit president concept resembles performance approach
1		1	Contractors caused safety issue in first place
1		1	Minimum prescriptive standards help subcontractor management
	1	1	Minimises liability exposure to general contractor
1		1	Eliminates subjective inspections
	1	1	Better working rapport with supervision
1		1	Lack of knowledge to use performance approach
	1	1	No strong preference
1		1	Contractor should be responsible for safety
27	38	65	

Table 8.2   Importance of Issues with Respect to an Approach to Construction Safety Management

Rank	Issue	N	Mean	Standard deviation
1	The potential to improve safety performance on sites	66	6.32	1.05
2	The ease of understanding compliance requirements	66	6.05	0.95
3	The ease of implementation of the approach	66	5.83	1.13
4	Support for innovation, new materials and technology	66	5.39	1.43
5	The cost effectiveness of approach	65	4.77	1.77

8.4.9   Sponsors of major change

With respect to who usually sponsors major change within their firms, respondents responded as follows:

- 53.6% top management;
- 16.1% middle management;
- 19.1% site management;
- 6.0% workers; and
- 5.2% supervisors (foremen).

8.4.10   Importance of change driving issues

Respondents who preferred the performance approach were asked on a scale of 1 to 7 (1 = not influential at all; 4 = neither influential or not influential; 7 = extremely influential) to rate the influence of 13 factors in driving change in their firms. These change-driving factors were ranked according to the means of their responses. The results are shown in Table 8.3.

Table 8.3   Influence of Factors in Driving Change

Issue	Sample Rank	CEO/President/Vice- president/MD/ General Manager Rank	Project/ Contracts Manager Rank	Safety Director/ Manager Rank
Improvement of financial performance	1	2	1	2
Improvement of your safety record	2	1	7	1
Generating of quality improvements	3	4	2	4
Complying with owner/client requirements	4	3	4	3
Exploitation of new market opportunities	5	8	5	8
Keeping up with competitors	6	6	3	6
Introduction of new technology	7	9	6	9
Meeting new insurance requirements	8	7	10	7
Responding to management initiatives	9	5	8	5
Meeting worker demands	10	10	9	10
Responding to third party claims	11	11	11	11
Occurrence of accidents	12	12	12	12
Staff turnover	13	13	13	13

The improvement of the safety record of the firm was the most influential factor in driving change according to CEOs and Safety Directors. Project Managers only ranked this factor 7th. To them, the improvement of the financial performance of their firms was the most influential change-driving factor. The next influential factors to them were the generation of quality improvements, and keeping up with competitors. The improvement of the financial performance of their firms, and complying with the requirements of owners and clients were the next influential factors to CEOs and Safety Directors. Factors such as meeting worker demands, responding to third party claims, occurrence of accidents, and staff turnover consistently ranked lowest according to all.

Theo C. Haupt

8.4.11 Importance of worker participation

Respondents were similarly asked to rate the importance of five issues to worker participation in change and change management. They regarded the receptiveness of supervisors and foremen to change as being most important in improving worker participation in change. Building credibility and trust was regarded as next important. The ranking according to the means of responses of issues affecting worker participation is listed in Table 8.4.

Table 8.4   Importance of worker participation in change and change management

Rank	Issue	N	Mean
1	The receptiveness of first-line supervisors and foremen	65	6.20
2	Building credibility and trust with workers	65	6.17
3	Enlisting the opinions of workers	66	5.74
4	Breaking down the resistance or workers to change	66	5.70
5	Willingness of workers to accept change	66	5.11

8.4.11 Importance of issues that affect the implementation of a new approach

Respondents were asked to rate the importance of ten issues that would affect the implementation in their firms of the performance approach. The ranking of the means of their responses is shown in Table 8.5. All management groups regarded open communication as being most important for the approach to succeed in their firms. CEOs and Safety Directors regarded top management support as being next important, followed by mutual trust between workers and management. Project Managers regarded the provision of adequate resources as being next important followed by effective coordination of construction activities. All the groups regarded incentives and rewards as being the least important issue. The continuous improvement of the safety performance of their firms did not rank as highly as was expected.

8.4.13 Importance of actions for the successful implementation of the performance approach

The introduction and support of appropriate training programmes was the most important action to be taken for the successful implementation of the performance approach according to CEOs and Safety Directors. Both groups regarded the demonstration of consistent and decisive leadership as next important. Project Managers regarded this action as most important, and the allocation of adequate financial, equipment and staff resources as next important. They also regarded the motivation of workers to implement changes for continuous improvement as important. The ranking according to importance of actions to be taken for the successful implementation of the performance approach is shown in Table 8.6.

Table 8.5 Importance of Issues that affect the Implementation of the Performance Approach

Issue	Sample	CEO/President/ Vice-president/ MD/ General Manager	Project/ Contracts Manager	Safety Director/Manager
	Rank	Rank	Rank	Rank
Top management support	1	2	4	2
Open communication	2	1	1	1
Mutual trust between workers and management	3	3	6	3
Effective coordination of construction activities	4	5	3	5
Continuous improvement of safety performance	5	7	8	7
Adequate resources	6	6	2	6
Workshops and training	7	4	9	4
Joint labour/management problem solving	8	9	5	9
Creativity	9	8	7	8
Incentives and rewards for supporting the change	10	10	10	10

8.5   CONCLUSIONS

The introduction and implementation of the performance approach to construction worker safety will require a major paradigm shift from prescriptive approaches encapsulated in most safety and health regulations around the world. Construction firms operating in these countries will have to embrace a new and different corporate culture and necessarily depart radically from their old way of dealing with worker safety and health. In particular contractors will have to learn to take responsibility proactively for finding solutions to the threats to the safety and health of their workers posed by the many occupational hazards found on their construction sites. For their part legislators will need to cope with an approach that is no longer based on enforcing compliance through punitive measures reactively. The successful implementation of the performance approach will be dependent on the capacity and willingness of the management of construction organisations to introduce and support the changes necessary. Without top management supporting and driving it, this change will be difficult and almost impossible. This management support needs to be cognitive, emotional, and financial, but also tangible and visible.

Theo C. Haupt

Table 8.6   Importance of Actions for the Successful Implementation of the Performance Approach

Issue	Sample Rank	CEO/President/ Vice-president/MD/ General Manager Rank	Project/ Contracts Manager Rank	Safety Director/ Manager Rank
Demonstration of consistent and decisive personal leadership	1	2	1	2
Introduction and support of appropriate training programmes	2	1	6	1
Allocation of adequate financial, equipment and staff resources	3	6	2	6
Encouragement of worker participation at all levels	4	4	8	4
Motivation of workers to implement changes for continuous improvement	5	5	3	5
Measuring and evaluating progress of the changes regularly introducing new plans of action if necessary	6	3	4	3
Changing the organisation's systems, policies and procedures to augment the changes	7	7	7	7
Rewarding workers for being innovative, and looking for new solutions	8	8	9	8
Amending the corporate vision and mission	9	10	11	10
Changing the organisational structure and hierarchy to make it more flexible and responsive to change	10	9	10	9
Comparing the performance of the company with competitors	11	11	5	11

The findings of this study suggest that where the management of construction firms has a good understanding of the performance approach, and given the opportunity to choose between the prescriptive and performance approaches they would prefer the latter. The study confirms that for the performance approach to succeed, open communication at all levels within the construction company, top management support and mutual trust between workers and management are imperatives. In order to improve worker participation in the process of implementation, construction supervisors and foremen must be receptive of the new approach. Further, top management must demonstrate consistent and decisive personal leadership that is accompanied by the introduction and support of appropriate training programmes.

While the study was conducted among contractors operating in the United States where the prevailing legislative framework under the Occupational Safety and Health Administration (OSHA) is a predominantly prescriptive one, its findings suggest that the performance approach might be an acceptable alternative approach to reducing the carnage caused by an industry that kills three to four construction workers each workday.

8.6   REFERENCES

Almaraz, J., 1998, Quality management and the process of change. *Journal of Organizational Change Management*, 1, **2**, pp. 129-152.

Bennis, W., 1993, *Beyond Bureaucracy: Essays on the Development and Evolution of Human Organization*, (San Francisco: Jossey-Bass Publishers).

Boles, M. and Sunoo, B.P., 1998, Three barriers to managing change. *Workforce*, 77, **1**, pp. 25-28.

Center to Protect Workers' Rights, 1993, *An Agenda for Change. Report of the National Conference on Ergonomics, Safety, and Health in Construction*, Washington, D.C., 18-22 July.

Coble, R. and Haupt, T.C., 1999, Safety and health legislation in Europe and United States: a comparison. In *Proceedings of International Conference of CIB Working Commission 99 and Task Group, Safety Coordination and Quality in Construction*, edited by A. Gottfried, L. Trani and L.A. Dias, 36, Milan, Italy, 22-23 June, pp. 159-164.

Coble, R.J. and Haupt, T.C., 2000, Performance vs. prescription based safety and health legislation: a comparison. *The American Professional Constructor*, 24, **2**, pp. 22-25.

Cook, S. and McSween, T.E., 2000, The role of supervisors in behavioral safety observations, professional safety. *Journal of the American Society of Safety Engineers*, 45, **10**, pp. 33-36.

Diamond, M.A., 1998, Organizational and personal vicissitudes of change. *Public Productivity and Management Review*, 21, **4**, pp. 478-483.

Enz, C.A., 1986, *Power and Shared Values in the Corporate Culture*, Ann Arbor, UMI Research Press.

Ferber, R., Sheatsley, P., Turner, A. and Waksberg, J., 1980, *What is a Survey?* Washington, D.C., American Statistical Association.

Freda, G., Arn, J.V., and Gatlin-Watts, R.W., 1999, Adapting to the speed of change. *Industrial Management*, Nov./Dec., pp. 31-34.

Hage, J. and Dewar, R., 1973, Elite versus organizational structure in predicting innovation. *Administrative Science Quarterly*, 18, pp. 279-290.

Haupt, T.C., 2001, *The Performance Approach to Construction Worker Safety and Health*, University of Florida, Unpublished Ph.D. dissertation.

Haupt, T.C. and Coble, R.J., 2000, International safety and health standards in construction. *The American Professional Constructor*, 24, **1**, pp. 31-36.

Hinze, J.W., 1997, *Construction Safety*, Prentice-Hall, Inc, New Jersey.

Levitt, R.E. and Samelson, N.M., 1993, *Construction Safety Management* (New York: John Wiley and Sons, Inc.).

Marshall, G., 1994, *Safety Engineering*, American Society of Safety Engineers.

May, T., 1997, *Social Research: Issues, Methods and Process*, Buckingham, Open University Press.

Nadler, D., 1988, Concepts of the management of organizational change. In *Readings in the Management of Innovation* edited by M. Tushman and W. Moore (Cambridge: Ballinger).

Nadler, D. and Tushman, M., 1989, Beyond the magic leader: Leadership and organizational change. In *The Management of Organizations*, edited by M. Tushman, C. O'Reilly and D. Nadler (New York: Harper and Row), pp. 533-546.

Nadler, D. and Tushman, M., 1990, Beyond the charismatic leader: Leadership and organizational change. *California Management Review*, 32, pp. 77-97.

Petersen, D., 1996, *Human Error Reduction and Safety Management*, (New York: Van Nostrand Reinhold).

Reason, J., 1998, Organizational controls and safety: the varieties of rule-related behavior. *Journal of Occupational and Organizational Psychology*, 71, 4, pp. 289-301.

Salant, P. and Dillman, D.A., 1994, *How to Design Your Own Survey* (New York: John Wiley and Sons, Inc.).

Sink, D.S. and Morris, W.T., 1995, *By What Method?* (Norcross: Industrial Engineering and Management Press).

Site Safe, 2000, *The Facts about Construction Injuries*, October 23, http://www.sitesafe.org.nz/facts.html.

Spector, P.E., 1981, *Research Designs* (Beverly Hills: Sage Publications).

Statzer, J.H., 1999, An integrated approach to business risk management. *Professional Safety* - American Society of Safety Engineers, August, pp. 30-32.

Weatherall, D., 1995, *Science and the Quiet Art: Medical Research and Patient Care* (Oxford: Oxford University Press).

Weston, S., 1998, The challenge of change. *Ivey Business Quarterly*, 63, 2, pp. 78-81.

Whetton, C., 2000, Loose change. *Hydrocarbon processing*, 79, 3, pp. 3.

CHAPTER 9

An Investigation into the Effectiveness of Partnering in Promoting Health and Safety Management on Construction Sites

Brian C. Heath

INTRODUCTION

To build is to be robbed: Johnson's edict still carries much weight in the modern age. The Latham Report (1994) paints a picture of distrust and conflict, not just between client and contractor but between the design and construction team and within the construction team itself.

The UK Government has attempted since the Latham Report was published in 1994 to change the ethos of the construction industry and make it less conflict orientated. Legislation followed on from Latham in the form of the Housing Grants, Construction and Regeneration Act 1996, and further initiatives in the form of The Egan Report and most recently the Movement for Innovation (M4i). The focus for these initiatives was the performance of the construction industry, principally its inability to satisfy customer expectations, the apparent lack of teamwork between the various contracting parties and the lack of post contract liability. In short, the emphasis was on value for money.

At the same time, initiated by European social legislation, the Construction (Design Management) Regulations (CDM) has provided a focus on health and safety in the design, construction and maintenance phases of construction projects allocating specific responsibilities to those named as function holders under the CDM Regulations.

More recently, publications by the Strategic Forum for construction such as Rethinking Construction 2002 (2002a), Rethinking the Construction Client (2002b) and Revitalising Health and Safety in the Construction Industry (HSE, 2002) have focused attention on health and safety and the role of the client. The client specifically is charged with having a major influence on the health and safety culture of the project being promoted.

The issue for consideration within this paper is whether both sets of initiatives can change the culture and focus of the construction industry and act in concert to promote health and safety on construction sites and improve the performance and image of the industry in this respect.

9.1   LATHAM AND EGAN: A CONSTRUCTION INDUSTRY PERSPECTIVE

Both Latham (1994) and Egan (1998) have concentrated on the concept of the team. Sir Michael Latham's Report 'Constructing the Team' was commissioned jointly by the Government of the day and the industry itself and represented an in-depth review of the procurement systems and contractual arrangements which permeated the industry. In doing this Latham considered the position of the main contractor as the focus for many of the long standing problems in the industry: principally, that it is conflict orientated and beset by poor working practices (Baden-Hellard, 1995, Critchlow, 1998). The problems highlighted by Latham in terms of poor level of service, poor end product and poor after sales service are linked to the main contractor and the transactional power afforded to him by his central position in the contractual framework (Berry 2000). It is of little surprise, then, that the aftermath of the Latham report was a series of legislative measures designed to limit this transactional power.

It could be argued, however, that the focus on the contractor as the cause of the ills of the industry ignored a more fundamental issue – the effect on the attitudes of the parties to the contract of the use of a procurement system which is based largely on securing the contract at the lowest price. The common view of the industry was that it was competitive on price but not on quality. Other researchers, Hatush and Skitmore (1997) and Lingard and Holmes (2001) support this view.

The Egan Report 'Rethinking Construction' was the report of the Construction Task Force chaired by Sir John Egan (Egan, 1998) commissioned by John Prescott to advise him on strategies for improving the performance of the construction industry. Like the Latham Report, Egan was concerned that the Construction Industry was not performing to the best of its abilities and was concerned with customer satisfaction and process and product development. It highlighted a number of drivers for change: committed leadership, a focus on the customer, integrated processes and teams, a quality driven agenda and commitment to people. It also stated that the industry must provide 'decent and safe working practices and improve management and supervisory skills at all levels' (Egan, 1998:7). Egan spoke also about commitment to people and the need to motivate operatives through a mixture of supervision and training in a 'no blame culture based on mutual interdependence and trust' and set a target to reduce the number of reportable accidents by 20% per year' (Egan, 1998:19). This target has been further defined in Revitalising Health and Safety in Construction (HSE, 2002:7) where the following goals are set:

- Cutting the fatal and major injury rate by 40% by 2004/5 and 65% by 2009/10.
- Reducing the number of working days lost per 100,000 workers from work related injury by 10% by 2004 and by 50% by 2010.
- Lowering the incidence rate of cases of work related ill health by 10% by 2004 and 50% by 2010.

Equally importantly, Egan made a number of points regarding the attitude of clients and the nature of the industry. Clients were seen as being overly concerned with cost, equating it to value, making selection decisions in relation to both contractors and designers based solely on price (Egan, 1998:10). The industry was seen as being fragmented with the vast majority of companies in the construction industry employing less than eight people (Egan, 1998:11). The picture is one of an industry dominated by small to medium sized enterprises (SMEs), a picture reinforced by Egan's comments regarding a "crisis in training" and the dearth of research and development. These two issues of client orientation and industry fragmentation are, it is suggested, central to the problems facing the construction industry in seeking to change attitudes and performance in respect of health and safety on construction sites.

Key to the achievement of the targets set was the strategy outlined by Egan of partnering. Inherent in this is the role of the client and the focus on the client as the lead player in implementing change. It is interesting to note that Rethinking the Construction Client (DTI, 2002:3) sets six key guidelines for rethinking construction as a client :

- Traditional processes of selection should be radically changed because they do not lead to best value.
- An integrated team which includes the client should be formed before design and maintained throughout delivery.
- Contracts should lead to mutual benefit for all parties and be based on a target an whole life cost approach.
- Suppliers should be selected by best value and not by lowest price: this can be achieved within EC and Government procurement guidelines.
- Performance measurement should be used to underpin continuous improvement within a collaborative working process.
- Culture and processes should be changed so that collaborative rather than confrontational working is achieved.

Although the focus is clearly on cost and fitness for purpose within the document, 'best value' is a recurrent theme within the document and is clearly defined as including 'Respect for People' issues central to which is the organisations' safety record.

9.2   LATHAM AND EGAN: A RESEARCH PERSPECTIVE

Latham and Egan focused upon partnering as a key strategy in promoting change. Partnering has been described in a number of ways. Typically, the CII (1991:iv) refer to it as 'long-term arrangements between companies to co-operate to an unusually high degree to achieve separate yet complementary objectives'; Day (1996) saw partnering as the essential element in helping conflict through the eradication of the traditional barriers between client and contractor. The emphasis in these and other definitions of partnering (CII 1991, NEDO 1991) is essentially financial: savings in cost, increases in productivity.

To be effective, therefore, partnering must produce tangible benefits for all parties (Critchlow 1998) and to do this requires, it is suggested, a radical reworking of the culture, attitudes and mind-set of all those engaged in the construction process (Marler 2000). Atkinson (1998) identified the importance of the individual in reducing errors on site and as a result improving health and safety and the quality of the construction process as a whole. The question remains, however, as to whether a solution which addresses business objectives is compatible with achieving a solution to health and safety issues which are essentially human relations orientated.

Berry (2000) in a case study based research project investigated the nature of conflict in the construction industry and came to a number of conclusions: that, given the nature and structure of the present procurement methodology, conflict in the construction process is unavoidable; that the dimensions of conflict identified are largely negative and that this negative conflict manifests itself in the form of stress and that negative stress leads to errors which themselves have repercussions which are time, cost and quality related; that management of the individual, though proved to be effective, is often neglected in the construction industry; that the contractor is in a central position to positively control conflict for his own and the client's benefit.

In arriving at these conclusions Berry undertook a study of six projects maintaining for each a site conflict diary in which events were initially recorded. From this the events were classified according to four operating levels. From this initial classification the conflict diary was re-written retrospectively and the events reclassified to identify both the causes and effects of conflict. A narrative was also provided and a flow chart of events was prepared to enable a clear picture of the chronology of events and to enable conflict and its financial and operational effects to be accurately traced back to the initial cause. The advantage of this approach was that it provided depth of analysis to the somewhat one-dimensional level provided by the quantitative analysis of the diary events alone.

From this combined analysis it was clear that the particular type of contract had little effect on emergent conflict within the case studies. More prevalent than inter-organisational conflict was intra-organisational conflict arising from both the client and the contractor. For the client this originates with his method of procurement of design and other consultants and continues with the on-going concentration on cost as being the dominant factor in the project. This attitude demonstrably influenced the behaviour of all parties to the contract and contributed to dysfunctional behaviour at critical points in the project life. For the contractor such behaviour exhibited itself *inter alia* through allocation of resources: failing to match resources to requirements through the inappropriate choice of site management personnel, inappropriate transfer of risk and responsibility at subcontract level.

The issues raised in this and other research (Djebarni, 1996; Loosemore 1999; Loosemore and Tan, 2000; Fraser, 2000, Bresnen and Marshall, 2000a) all support the points made by Latham and Egan in respect of the main drivers for change; committed leadership, a focus on the customer, integrated processes and teams, a quality driven agenda and commitment to people. All, however, are essentially critical of the focus of the procurement process which is essentially cost and, by extension, conflict orientated.

It can be further surmised that current methods of management within the construction industry appear to owe more to Taylor than to Locke. This is an important distinction when one considers that the drive in health and safety is towards personal responsibility and behaviour based safety management (Lingard and Rowlinson, 1998) and poses fundamental questions regarding the ability of the industry to change. An important issue is the attitude of client and contracting organisations towards partnering and the benefits it will provide.

Walker and Hampson (2003) in considering the benefits of alliancing as opposed to partnering provided a definition of the alliancing process which fits in quite easily with the demands made of the client in rethinking his approach to construction. Alliancing partners are selected on the basis of their ability to match set performance criteria prior to any consideration of price being made. Pre-selection on a basis other than price which centres on:

- Performance;
- Cost management;
- Quality;
- Resource management;
- Value management;
- Health and safety management;
- Environmental management;
- Ethical management.

The radical approach to procurement reflected in the alliancing system allows contractors to tender on a performance basis which is not dominated by price and which therefore allows health and safety to be considered in an open manner.

9.3   ATTITUDES TO PARTNERING: SURVEY AND RESULTS

Bresnen and Marshal (2000b) make the point that there is little empirical evidence of the effectiveness of partnering in practice. Indeed, there is little agreement on the form partnering should take: strategic or project based. For the most part the benefits are expressed in terms of profitability, productivity, cost and quality. The disadvantages are also couched in the same terms: lack of competition (Davies, 1995); additional partnering costs through workshops (Bennett and Jayes, 1995; Barlow, et al., 1997); lack of flexibility in the market place (Critchlow, 1998); the domination of existing adversarial attitudes (Berry, 2000). The negative comments apply equally to the contractor/subcontractor as well as the client/contractor interface raising concerns about the construction industry's ability to adopt supply chain management as part of its partnering strategy. The implications for the creation of a health and safety culture on site are, it is suggested, a cause for some concern. In an attempt to further investigate the issues raised above, an investigation was undertaken into the attitudes of parties to the construction process was undertaken to determine their attitudes towards partnering and the benefits its produces.

A total of 86 questionnaires were sent out: 40 to contractors, 46 to client organisations. In selecting the sample group a combination of purposive and systematic sampling was used. The questionnaire was designed to test a number of hypotheses: that partnering is infrequently used in the construction industry; that partnering has the potential to improve client/contractor relations; that partnering has the potential to improve product quality and reliability; that attitudes to partnering from the client side tend to focus on their own accountability; that attitudes from contractors tend to focus on the profit element; that partnering can convey to the parties a number of benefits.

The questionnaire was selected as the most appropriate method of data collection given the time limits imposed on the study and the intention, which was to provide empirical evidence of the use of partnering. The questionnaire content expanded on the hypotheses given and was comprised mainly of closed questions with only two open questions which asked the respondents to expand on their responses to questions relating to the impact of partnering on quality, reliability and supply chain management.

Of the 40 questionnaires sent out to contractors 35 were returned, a response rate of 87%. Of the 46 questionnaires sent out to clients 26 were returned, a response rate of 56%. A response rate overall of 71%. The breakdown of the respondents is as shown in Table 9.1 below

While Quantity Surveyors and Project Managers dominate the sample, the sample is, it is submitted, representative of organisational structures in client and contracting organisations (Berry, 2000). All respondents felt that they were aware of the general principles and objectives of partnering with 73% of the client group and 91% of the contractor group having had involvement with partnering in the past. For the most part partnering appeared to be client driven with only 20% of respondents reporting contractor driven partnering arrangements.

Eighty-one percent of the client group and 91% of the contractor group were aware of the Latham and Egan Reports though 75% and 51% of the groups respectively felt that it was too early to tell if the targets were being achieved. Confidence in the effect of the reports was marginally higher in the contractor group who were also more confident that the relationship between client and contractor had improved since the Latham Report: 60% compared with 38% in the client group and that the potential for conflict was lessened.

The vast majority of the sample group felt that partnering to succeed must be applied throughout the supply chain. This was one of the questions where the respondents were requested to expand on their choice of answer. The written responses echoed fairly closely the rationale for partnering provided by Egan and other researchers: cost reduction; elimination of waste; improvement in quality and value; fair apportionment of risk. This was supported in a further response where 77% of the client group and 72% of the contractor group felt that partnering was an aid towards quality enhancement and increased reliability. Again this was expanded upon in written comments which cited the following reasons: clearer communication; clearer risk apportionment; greater commonality of objectives. Overall, a high degree of synchronicity with the findings of Egan and the findings reported by Bresnen and Marshall (2000a) in their critical review of partnering in construction.

Table 9.1 The Breakdown of the Respondents

	Client %	Contractor%
Quantity Surveyor	38	29
Project Manager	27	28
Director	11	23
Client Manager	8	0
Engineer	8	2
Architect	4	0
Designer	4	0
Recruitment Manager	0	2
Head of Procurement	0	16

4

Given the level of compatibility with existing research this allows a number of conclusions to be drawn from the data gathered in the survey relating to issues of accountability, profit and specific effects. The results are shown in Table 9.2 below.

Despite the clear appreciation of the benefits of partnering there still remains a strong suggestion of cultural problems within the construction industry. There is clearly a reluctance to depart from the traditional methods of cost control through competitive tendering. This, combined with evidence that the contractor is still overly concerned with profitability and the client team with accountability, does not portray a picture of an industry likely to put health and safety first. This was reinforced to an extent by a further series of propositions put to the respondents the results of which are shown in Table 9.3 below.

Whilst it may be argued that providing a middle option enables the respondent to avoid a decision it can be proposed that any entries in the true or false columns carry further weight. This being said some interesting conclusions can be drawn from these responses.

On the positive side, there is a definite reinforcement of the utility of partnering in fostering communication and teamwork, this is supported by the small percentage of respondents who felt that there were no benefits in the form of either the avoidance or speedy resolution of disputes. Beyond this general position, however, the messages are considerably more divided and potentially negative.

Table 9.2 Survey Results (from Hughes 2001:61)

		Client	Contractor
Do you feel that a true partnering arrangement can never exist owing to the opposing interests of the two parties?	Yes	38	28
	No	62	72
Do you agree that partnering has been well received or disagree and feel that is there still a degree of apathy and cynicism?	Agree	31	23
	Disagree	69	77
Where partnering is used do you feel that clients place too much emphasis upon their own accountability at the expense of the true aims of partnering?	Yes	15	43
	No	27	8
	Sometimes	58	49
Would you say that contractors pay too much attention to the profit element that is generated rather than embracing the overall objectives of partnering?	Yes	77	34
	No	15	57
	Sometimes	8	9
Do you feel that partnering will one day replace the more traditional competitive tendering process?	Yes	15	14
	No	85	85

The contractor group were in general less enthusiastic about the financial impact of partnering; 32% did not feel that it increased turnover and profit and only 29% felt definitely that it increased productivity. In the same context 46% of the client group and 25% of the contractor group felt that partnering did not produce safer working practices and sites: only a minority, 31% and 26% respectively felt that there was an improvement. It is suggested that the contractor group responses reflect something of a split between attitudes to the client and attitudes to subcontractors and that whilst there may be better communication between contractor and client, the same does not necessarily hold true for the interface between contractor and subcontractor.

This suggestion is supported by statements made by respondents to the questionnaire where the relationship between contractor and subcontractor was still seen as confrontational despite the importance of the supply chain philosophy. This, it is further suggested, has fundamental implications for the implementation of effective health and safety management systems.

9.4 IMPLEMENTING HEALTH AND SAFETY MANAGEMENT

The overarching strategy for implementing health and safety is legislative; principally the Health and Safety at Work Act 1974 and the Construction (Design and Management) Regulations 1994 (HSE, 1994) (CDM) which became active in 1995. There is some debate, however, regarding the effectiveness of the CDM Regulations in improving the safety record of the construction industry. The HSE (1997) in investigating the effectiveness of the CDM Regulations found that both clients and contractors were generally supportive of the regulations and had reacted positively to them in terms of amending their own policies but that there were some concerns which existed relating to the increase in bureaucracy which the regulations imposed which had no apparent impact on efficiency.

Table 9.3   Benefits of Partnering (from Hughes 2001:63)

Partnering	Client			Contractor	
	True	P/T	False	P/T	False
Aids speedy resolution of contract disputes	38	58	4	54	6
Avoids costly overruns	15	69	16	69	2
Promotes better design criteria	23	46	31	43	17
Produces safer working practices and sites	31	23	46	49	25
Produces increased turnover and profit	27	65	8	57	32
Produces increased productivity	27	46	27	57	14
Apportions risk fairly	38	42	20	49	8
Helps promote teamwork	62	27	11	13	2
Merely provides a means to an end	38	31	31	40	31
Reduces exposure to litigation	15	53	31	51	12
Helps parties to understand each others objectives	65	27	8	24	2

The central role of the Pre-tender Health and Safety Plan has been considered by Tam, Fung and Chan (2001) in the context of the Hong Kong construction industry, which has a poor safety record comparative with the UK. They considered the effect of the introduction of the Supervision Plan (comparable with the Construction Health and Safety Plan) and recorded a positive response in terms of awareness on site of health and safety issues. The research findings were qualified, however, by comments regarding the strength of the change and the need to reinforce the message of safety against competing messages of cost and time. Tam and Fung (1998) had earlier established a correlation between the use of sub-contractors and the incidence of accidents on construction sites. These were related also to a lack of safety training and a lack of awareness of the operational and financial effects of accidents amongst small contractors.

Dainty, Briscoe and Millet (2001) considered the difficulties surrounding effective supply chain management in the construction industry where SMEs employing 24 or less workers comprise almost 98% of companies operating in the construction sector. Like Bresnen and Marshal, Dainty, Brisco and Millet found that partnering was restricted to the Client-Contractor interface with the SME section of the industry, who are largely subcontractors having little managerial input. In addition, they found that the SME section was itself layered with varying levels of managerial competence. Supply chains were in themselves fragmented with many SMEs working for a range of contracting organisations. It was clear from their investigations that relationships at the contractor-subcontractor interface were problematic and conflict orientated stemming from the transactional power of the contractor and the tendency of the contracting organisation to focus on costs rather than value. A view supported by Heath, Berry and Hills (1994) and Heath and Berry (1996).

In the same context, Bresnen and Marshall (2000b) investigating the utility of financial incentives as a basis for behaviour modification in the context of partnering, concluded that the financial incentives that form the basis of partnering agreements are of insufficient valence to effectively influence individual behaviour given the complexity of the social and organisational structures normally present on construction sites. The culture on site is not, it is suggested, one which promotes compliance as a normative response.

Griffith and Phillips (2001) in examining the influence of the CDM Regulations on the procurement and management of small building works commented upon the increase in managerial workload imposed by the Regulations, a workload which could not always be borne by the SME. They also found that this top down legislative control did not act to positively change the culture of the organisation and that the way forward to improved health and safety practice was through better training, clearer contractual accountability and improved workplace control.

This would seem to raise some doubts regarding the strategies for improving health and safety attitudes and training being considered by the HSE at a legislative level (HSE, 2002:20), with the proposition that construction workers should have as a legal requirement a verifiable knowledge of health and safety. The implications of such a scheme would, it is suggested be an acknowledgement that voluntary 'passport to safety' schemes do not work.

In a general sense it is accepted that the most effective implementation of health and safety takes place at the level of the individual (Duff, *et al.*, 1994). This viewpoint has been supported by Lingard and Rowlinson (1998), who linked the improvement of safety on site with motivation of the individual, and Lingard (2001) who associated site based first aid training with improved safety performance on site. The implication being that strategies for improvement must be centred on the individual, suggesting that a top-down approach to behaviour modification on site will be of limited effect. This results in something of an impasse for an industry dominated by top down management and populated by operatives who are traditionally resistant to change and often oblivious to the need to work safely

More recently, the HSE in their revision to the CDM Approved Code of Practice and Guidance, Managing Health and Safety in Construction (HSE, 2001), have reconsidered and redefined the role of the client in the management of health and safety in construction projects. Clients are required to ensure that the project is designed and organised so that 'people carrying out construction work, including upkeep, can do so safely and without risk to health.' Paragraph 15 is specific, the arrangements put in place by the client (or the client's agent) should ensure that:

- The project must allow sufficient time for design, planning, preparation and construction work to be carried out safely and without risk to health.
- Management of health and safety is carried out at all stages of the project by persons who are competent and adequately resourced.
- That all hazards associated with the project are identified and the residual risks assessed and communicated to all relevant parties.
- That the work undertaken is systematically and routinely monitored to ensure that it is carried out safely.

Paragraph 85 of the ACoP reinforces this: 'Clients must not leave it to contractors to discover hazards. Relevant information needs to be considered at the design/planning stage by the designer and those preparing the health and safety plan....The information needs to be sufficient to ensure that significant risks during construction can be anticipated, and avoided or properly controlled.'

The key to all of this is effective communication and continuity of communication through a systematic approach to the management of health and safety. In this context the client and the other duty holders need to clarify who will do what under the CDM Regulations. As Paragraph 66 of the ACoP points out, 'Many CDM duties require something to be done, but do not specify who must do it'. Clearly the HSE are looking for clients to take responsibility and it is perhaps significant that the Strategic Forum as part of their key recommendations in Accelerating Change (Egan, 2002) state that the HSE should consider publishing details of all companies, including clients, associated with sites where fatalities occur'.

9.5   CONCLUSIONS

There is evidence to suggest that the Government's legislative initiatives in respect of health and safety management in the construction industry have been beneficial in the sense that the CDM Regulations do have a positive effect on safety on site. It is not as easy to establish, however, that the effect is both wide ranging and long lasting - a change in culture. It appears that a number of barriers exist which act to prevent this. The structure of the construction industry, the financial orientation of the procurement process and the adversarial nature of contractual relationships within the industry all combine to limit the effect of health and safety initiatives in the workplace.

It is also the case that the need for further action to promote a clearer focus on health and safety has been recognised and that such action will come in the form of legislation and forced organisational change. It is clear also from the nature of the discussion documents referred to in this paper that the role of the client is seen as fundamental to this change as is a radical review of the procurement methods currently in use. The question is, is partnering such a radical change?

Partnering has been proposed as the strategic mechanism through which the culture of the industry will change. The research evidence presented in this paper suggests, however, that this is not the case. It is clear that partnering has some beneficial effects on the client-contractor interface but it is equally clear that this does not extend to the contractor-subcontractor interface. The result is an industry with a fragmented approach to the management of health and safety on site where the cost of health and safety is secondary to considerations of time, cost and quality.

Although the research would suggest that long-term improvements can only be achieved through strategies that encourage the individual to become more safety conscious and establish a minimum level for individual behaviour in this context, a top down strategy is, it is suggested, needed to provide the environment within which this change can occur. Despite the emphasis placed upon the pivotal role of the contractor in determining the culture of contractual relations, the key issue for health and safety lies in the pivotal role of the client in influencing the status of health and safety management as a key issue in the procurement process rather than its present position as being one item on a list dominated by the lowest cost as a criteria for selection. The approach taken in projects which promote alliancing and use as a basis for selection criteria other than price seem attractive and in line with the recommendations of the strategic documents beginning with and published since the Egan Report.

9.6   REFERENCES

Atkinson, A., 1998, Human error in the management of building projects. *Construction Management and Economics*, 16, pp. 339-349.

Baden-Hellard, R., 1995, *Project Partnering, Principles and Practice,* (London: Thomas Telford).

Barlow, J., Cohen, M., Jashapara, A. and Simpson, Y., 1997, *Towards Positive Partnering,* (Bristol: The Policy Press).

Bennett, J. and Jayes, S., 1998, *Trusting the Team: The Best Practice Guide to Partnering in Construction*, Centre for Strategic Studies in Construction/Reading Construction Forum, Reading.

Berry M.J., 2000, *The Incidence and Effect of Conflict arising as part of the Construction Process*, Ph.D. Thesis, University of Wales, Unpublished.

Bresnen, M. and Marshall, A., 2000a, Partnering in construction: a critical review of the issues, problems and dilemmas. *Construction Management and Economics*, 18, pp. 220-237.

Bresnen, M. and Marshall, A., 2000b, Motivation, commitment and the use of incentives in partnerships and alliances. *Construction Management and Economics*, 18, pp. 587-598.

CII, 1991, *In Search of Partnering Excellence*, Construction Industry Institute, Austin, Texas.

CII, 1994, *Benchmarking Implementation Results, Teambuilding and Project Planning*, Construction Industry Institute, Austin, Texas.

Critchlow, J., 1998, *Making Partnering Work in the Construction Industry*, Chandos, Oxford.

Dainty, A.R.J., Briscoe, G.H. and Millett, S.J., 2001, New perspectives on construction supply chain integration. *Supply Chain Management*, 6, **4**, pp. 163-173.

Davis, L., 1995, Two's Company. *Chartered Surveyor Monthly*, May, RICS, London.

Day, T., 1996, The key to success in construction contracts. *Chartered Surveyor Monthly*, October, RICS, London.

Djebarni, R., The impact of stress in site management effectiveness. *Construction Management and Economics*, 14, pp. 281-283.

Duff, A.R., Robertson, I.T., Phillips, R.A. and Cooper, M.D., 1994, Improving Safety by Modification of Behaviour. *Construction Management and Economics*, 12, **1**, pp. 67-78.

Egan, J., 1998, *Rethinking Construction, The Report of the Construction Industry Task Force*, DETR, London.

Egan, J., 2002, *Accelerating Change, A Report by the Strategic Forum for Construction*, Rethinking Construction, London.

Fraser, C., 2000, The influence of personal characteristics on effectiveness of construction site managers. *Construction Management and Economics*, 18, **1**, pp. 29-36.

Griffith, A. and Phillips, N., 2001, The Influence of the Construction (Design and Management) Regulations 1994 upon the Procurement and Management of small building works. *Construction Management and Economics*, 19, pp. 533-540.

HSE, 1994, *The Construction (Design and Management) Regulations) 1994* (London: HMSO)

HSE, 1997, *Evaluation of the Construction (Design and Management) Regulations 1994*, CRR158/1997, (London: HMSO).

HSE, 2001, *Managing Health and Safety in Construction HSG224* (Norwich: HMSO).

HSE, 2002, *Revitalising Health and Safety in Construction DDE20* (London: HMSO).

Hatush, Z. and Skitmore, M., 1997, Criteria for Contractor Selection. *Construction Management and Economics*, 15, **1**, pp. 19-38.

Heath, B.C., Berry, M.J. and Hills, B., 1994, The Nature and Origin of Conflict within the Construction Process. *International Council of Building Research, Studies and Documentation (CIB) Publication No. 171 Construction Conflict: Management and Resolution,* pp. 35-48. International Council for Building Research, Studies and Documentation.

Heath, B.C. and Berry, M.J., 1996, An examination of the issues arising on site from the use of standard and non-standard procurement methods and the implications of the same for project success. In *Proceedings of the CIB W92 Symposium,* pp. 200-212. International Council for Building Research, Studies and Documentation.

Hughes, G., 2001, *An Investigation into the Concept of Partnering within the Procurement Process*, B.Sc. Dissertation, University of Wales, Unpublished.

Latham, M., 1994, *Constructing the Team, Final Report on Joint Review of Procurement and Contractual Agreements in the UK Construction Industry* (London: HMSO).

Lingard, H., 2001, The effect of first aid training on objective safety behaviour in Australian small business construction firms. *Construction Management and Economics*, 19, pp. 611-618.

Lingard, H., 2001, The effect of first aid training on objective safety behaviour in Australian small business construction firms. *Construction Management and Economics*, 19, pp. 611-618.

Lingard, H. and Holmes, N., 2001, Understandings of occupational health and safety risk control in small business construction firms: Barriers to implementing technological controls. *Construction Management and Economics*, 19, pp. 217-226.

Lingard, H. and Rowlinson, S., 1998, Behaviour based safety management in Hong Kong's construction industry: the results of a field study. *Construction Management and Economics*, 16, pp. 481-488.

Loosemore, M., 1999, Responsibility, power and construction conflict. *Construction Management and Economics*, 17, pp. 699 -709.

Loosemore, M. and Tan, C.C., 2000, Occupational stereotypes in the construction industry. *Construction Management and Economics*, 18, pp. 559-566.

Marler, J., 2000, Construction Partners. *Local Government News,* April, 2000.

NEDO, 1991, *Partnering, Construction without Conflict* (London: HMSO).

Reading Construction Forum, 1995, Trusting the team: the best practice guide to partnering in construction, Centre for Strategic Studies in Construction, Reading, England.

Rethinking the Construction Clients, 2002, Department of Trade and Industry, UK available from www.rethinkingconstruction.org/rc/publications/reports.asp

Tam, C.M. and Fung, I.W.H., 1998, Effectiveness of safety management strategies on safety performance in Hong Kong. *Construction Management and Economics*, 16, **1** , pp. 49-55.

Tam, C.M., Fung, I.W.H. and Chan, A.P.C., 2001, Study of attitude changes in people after the implementation of a new safety management system: The supervision plan. *Construction Management and Economics*, 19, pp. 393-403.

The Strategic Forum for Construction, 2002a, *Rethinking Construction 2002*, Rethinking Construction, London.

The Strategic Forum for Construction, 2002b, *Rethinking the Construction Client*, Rethinking Construction, London.

Walker, D. and Hampson, K., 2003, eds., *Procurement Strategies: A Relationship-based Approach* (Oxford: Blackwell Science).

CHAPTER 10

Multi-layers Subcontracting Practice in the Hong Kong Construction Industry

Francis K.W. Wong and Lawrence S.L. So

INTRODUCTION

In 1998, the number of persons engaged in industrial activities constituted about 28% of Hong Kong's working population. However, the industrial sector recorded 43,034 loss-time industrial accidents in 1998 representing 68% of all worked-related loss-time injuries. Of these, 19,588 loss-time accidents or 45.5% happened in the construction industry. This accounted for an accident rate of 248 per thousand workers. The safety record of the construction industry was poor and much worse than other industries in Hong Kong. The reasons of the poor safety record may correlate with many factors such as complexity of the work or system, risk nature of works, management style, safety knowledge and commitment, and personal behaviour (Stranks, 1994). Surprisingly multi-layer subcontracting is unique to the Hong Kong construction industry and has been the most common practice being used with long history. When a principal contractor secured a project from a developer, usually it would break down the project activity into different trades and then sublet each category to individual subcontractors with the lowest bid (Lee, 1991). These subcontractors would normally further subcontract their work without the consent of their principal contractor to several smaller firms in order to minimise their overheads.

The problems of multi-layers subcontracting practice have long been an issue and a controversial subject in the industry. Recent studies (Linehan, 2000a and Tong, 2000) have all suggested having legislation for restricting the subcontracting practice in the construction industry for better safety performance. Although such restriction may help to improve the safety performance, it would be extremely difficult to implement in the workplace due to the traditional work practice. On the other hand, some people including Lee (1999a) and So (2000) had disagreed to restrict the multi-layers subcontracting practice. They considered that the subcontracting practice has great market values for the industry, which ought not be interfered by restriction in the form of legislation.

10.1   PREVIOUS STUDIES

10.1.1   Safety performance of the Hong Kong construction industry

It is generally acknowledged that poor safety performance of the Hong Kong construction industry is an unenviable fact. As shown in Table 10.1, the construction industry has a comparatively high accident rate than other major economic sectors such as manufacturing; transport, storage and communication; and others, between the periods 1998 to 2001. Also, Rowlinson (cited in Lo, 1997) identified that the construction industry accident rate was exceptionally high in Hong Kong, the fatality rate was ten times higher than UK, eight times than USA, four times than in Japan and two times than in Singapore. Thus construction should be one of the most important target areas of study for accident prevention in Hong Kong.

Table 10.1   Occupational Injuries in Major Economic Sectors in Hong Kong from 1998 to 2001

Major economic sectors	1998	1999	2000	2001
Financing, insurance, real estate and business services	3,500	3,900	4,300	4,500
Transport, storage and communication	5,600	5,700	5,600	5,100
Community, social and personal services	9,900	10,700	11,200	11,600
Whole and retail trades and restaurants and hotels	16,900	17,200	17,600	16,900
Construction	19,700	14,200	12,000	9,300
Manufacturing	7,700	6,800	7,000	6,000

Source: Adapted from Labour Department, 2001

10.1.2   Subcontracting practice of the Hong Kong construction industry

Recent researches (Wong, 1999 and Lee, 1996) have indicated that the high accident rate of the Hong Kong construction industry was related to the multi-layers subcontracting system. Lai (cited in Lee, 1991) found that the number of subcontractors in one construction site might be ranged from 17 to 54. The principal contractors' direct labour force in a project was small, and the subcontractors' workforce might actually carry out construction work without the knowledge of the principal contractor. Managing safety was a problem in terms of communication and monitoring. Lee (1999b) commented that multi-layer subcontracting is common and excessive in Hong Kong. The most extreme case quoted by Lee was subcontracting up to 15 layers. Rowlinson (1999) found in his study for the Hong Kong Housing Authority that average 84% of workers injured from 1995 to 1998 were subcontractors' workers. He further commented that subcontractors' workers tend to have less training and less awareness of safe working practice.

10.1.3   Problems of subcontractors

The structure of subcontractors is usually simple and small in size.   They had neither time nor inclination to keep abreast with legal requirements or technological developments in safety. Shaw (1998) found that small businesses faced with specific health and safety challenges, many firms lacked adequate resources and were often struggling to survive.   Furthermore, they lacked an understanding of their obligations and the health and safety issues of their processes.   Poon (1998) commented that the major cause of accidents was that subcontractors were rewarded according to work done. They were working under tremendous time constraints, which caused higher possibility of construction accidents. Capp (1973) identified that the common failure was arising from lack of direction and supervision of subcontractors' worker.

10.1.4   Advantages of subcontracting

Lee (1999a) however disagreed that subcontracting practice was the major cause of poor safety performance of the Hong Kong construction industry. He commented that the multi-layers subcontracting system was worth existing in the market. Wong (1997) found that subcontracting practice could be employed to cope with long-existing term demand uncertainty, allowing the firm to avoid the employment of a stable workforce and investments in fixed resources under conditions of the fluctuating demand, serving as an external buffering mechanism, absorbing uncertainties arising from availability of resources and operational conditions. Tse (2000) opined that multi-layer subcontracting was difficult to regulate due to the free market principle that operates in Hong Kong. We should determine the real causes of the problem, then to make improvements, and balance their advantages and disadvantages.

10.2   RESEARCH METHOD

In order to achieve quantitative and qualitative result, both questionnaire and interview survey methods were used for three types of target groups in the construction industry for this study (refer to Figure 10.1). Questionnaire survey was used for the contractor group and non-contractor group in order to collect quantifiable data to reflect the fact and norm of the industry. Non-contractor group here refers to the parties, which have close relation to the construction field, their works are actually affecting the operation of the industry such as the Government, developers, engineering consultants, trade unions and educational organisations.

Thereafter, face-to-face interview was carried out to a group of professionals in the construction industry who have not participated in the questionnaire survey in order to provide qualitative and objective explanation for the data collected from the questionnaire survey.

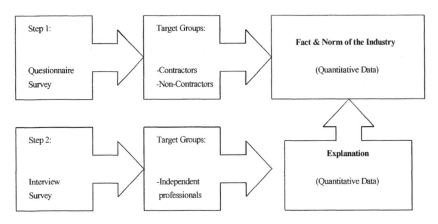

Figure 10.1 Research method

10.3   SURVEY RESULTS

Totally 55 out of 250 questionnaires were received, which represents a return rate of 22%. In addition, 15 professionals that included seven project managers, four consulting engineers and four safety professionals were interviewed. Data collected by questionnaire survey was treated by a statistical computer software SPSS to analyse the trend and correlation among accident records and different factors. While analysing the data provided by contractor and non-contractor groups, it is considered that both types of data have equal weighting for the research. The interview result is summarised for discussion purpose.

10.4   RESEARCH FINDINGS

10.4.1   Current status of the Hong Kong subcontracting practice

From the questionnaire survey, majorities of the respondents (45.5%) expressed that they would sublet 80-90% of their works to subcontractors, 22.7% had sublet 71-80%, 18.2% had sublet 91-100%, 9.1% had sublet 31-50% and 4.5% had sublet 51-70% of works. None of the respondents would carry out construction works that fully rely on their own effort, at least 30% of works would be subcontracted out. It reflected that considerable construction works were actually carried out by subcontractors. Though, the subcontracting practice was so common in the construction market, restricting of it might not be welcomed by the majority of construction contractors.

More than half of the respondents (63.6%) selected cost control as the major reason for subcontracting, 18.2% selected commercial purpose, 9.1% selected risk transfer and the remaining 9.1% selected technical support. Obviously, the major reason for contractors to subcontract works is because of cost control. It might become the greatest resistance for implementing the strategy of restricting the multi-layers subcontracting practice. An effective cost control system or method should be introduced to the industry before putting the proposed strategy into operation.

A contractor would normally employ over 100 number of subcontractors. In the current construction market, about 80-90% of construction workers are employees of subcontractors. If the subcontracting practice was restricted or regulated, it may cause a great impact to the society due to unemployment of subcontractors and their construction workers.

The survey also reflected a common industrial practice that the majority of contractors would allow their subcontractors further sublet works to other subcontractors. 95.5% of the respondents commented that they would allow their subcontractors further sublet works. Moreover, 77.3% of them have no requirement for subcontractors to gain permission or even inform them before subletting contracts. In addition, 90.9% of respondents expressed that they would not control or limit the subcontracting levels by their subcontractors. Only 9.1% of them would limit the subcontracting levels to one to two layers. A common trend was identified that the majority of contractors have not controlled the number of subletting levels. They would normally not consider their subcontractors' subletting status and therefore the practice of multi-layers subcontracting is fostered.

As shown in Table 10.2, the accident rates per thousand workers of the respondents were ranged from 33 to 200. These accident figures would be used for providing trend for analysis of factors, which were affecting the safety performance.

Table 10.2 Accident rate per thousand workers

Accident rates provided by the respondents

33	35	42	45	47	55	55
60	61	63	77	83	91	95
100	121	160	191	200		

10.4.2   Factors affecting safety performance

The regression analysis technique has been used to determine the relationship among the accident rates and other variables in the survey. The calculation result of Pearson's R-value reflects the trend of factors affecting safety performance.

10.4.2.1 Number of employees vs. accident rate

The number of employees in the organisation is one of the factors affecting the safety performance. The calculated R-value of the accident rate and the number of employees was at -0.52 (refer to Figure 10.2), which represented that there was negative correlation between them. That is to say a higher number of employees, a lower figure of the accident rate.

Accident rate is being defined as the number of accidents per 1,000 workers per annum. That is the total number of accidents divided by the number of workers employed and then times a constant 1,000.

R value = -0.52

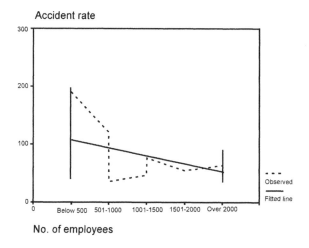

Figure 10.2 Scatter diagram for the accident rate and number of employees.

10.4.2.2 Percentage of subcontractors' employees vs. accident rate

The percentage of workers employed by subcontractors and the accident rate do have a positive correlation. The calculated R-value was at +0.41 (refer to Figure 10.3), which implied that more subcontractors' workers in an organisation, poorer would be the safety performance.

R value = +0.41

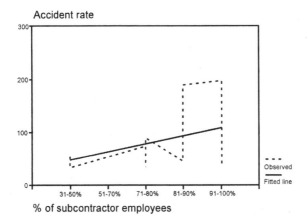

Figure 10.3 Scatter diagram for accident rate and percentage of subcontractors

10.4.2.3 Percentage of works sublet vs. accident rate

The scales of works being sublet and the accident rate do have positive correlation. The calculated R-value was at +0.42 (refer to Figure 10.4), which implied that more works being sublet to subcontractors, poorer safety performance may result.

R value = +0.42

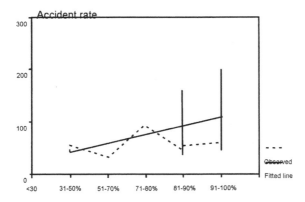

Figure 10.4 Scatter diagram for accident rate and percentage of works sublet

10.4.3   Perception of the industry

From the questionnaire survey, 69.1% of the respondents agreed that multi-layers subcontracting practice is one of the major causes of poor safety performance, 14.5% had no comment and 16.4% of the respondents disagreed with the statement. Majorities of the respondents at 61.8% agreed that the performance of subcontractors is difficult to control, 21.8% had no comment and 16.4% disagreed. Majorities of the respondents at 63.7% believed that reduction of subcontracting could help improve the safety performance of the construction industry, 20% had no comment and 16.3% disagreed with the statement.

However, multi-layer subcontracting has been widely adopted by the industry. Thus, industrial practitioners might not support restriction of the subcontracting practice. From the survey, 43.6% of the respondents believed that restriction of this practice is impracticable, 16.4% had no comment and 40% believed that it is practicable. If completely prohibiting the use of subcontracting practice is not possible, then the next question is restriction to how many layers of subcontracting is reasonable and practicable? 38.2% of respondents believed that restriction to 2 layers was most reasonable, 18.2% believed in both 1 and 3 layers, 7.3% believed in 4 layers, 1.8% believed in 5 layers and 16.4% had no comment.

If multi-layer subcontracting were regulated, 80% of the respondents expected that the local construction industry would have great impacts. Furthermore, Hong Kong is a place, which has been appreciated for its free market approach. 43.6% of the respondents expressed that legislation to restrict the use of subcontracting practice would damage the free market and reputation of Hong Kong.

Whether restricting multi-layer subcontracting is the best solution for improving the safety performance is arguable. However, the survey indicated that only 21.8% of the respondents agreed, 25.5% had no comment and 52.7% disagreed with it. Without the full support from contractors, implementation of the strategy to restrict the use of multi-layers subcontracting practice would be extremely difficult.

10.5   DISCUSSION

10.5.1   Problems caused by multi-layers subcontracting

Multi-layer subcontracting was considered as one of the major reasons for poor safety performance. The interviewees explained that subcontractors are normally small in size and simple in structure. They usually lack safety commitment because of limited budget, time and human resources. It resulted in insufficient provision of on-the-job safety training to employees, who have very limited knowledge to deal with safety matters. Tam and Fung (1998) found that there was a significant difference between the trained and un-trained employees in relation to accident rate. Safety training was an important tool in mitigating site accidents.

Subcontractors would rarely employ safety professionals as they had no interest in safety matters because most of them believed that safety should be the responsibility of the principal contractors. It could be explained by the fact that in Hong Kong, subcontractors are usually not accountable for serious accidents or for violation of safety regulations. Principal contractors would usually be responsible for workmen compensation and penalised for safety offences.

The short contract period and multi-layers subcontracting of works, which results in a subcontractor who is at the lowest subcontracting level, will not have realistic profit. Hence they have to resort to sub-standard works, which will increase the possibility of accident occurrence (Leung, 1997). High mobility of construction workers is also one of the major causes for poor safety performance. Simo (1995) found that subcontractors usually employed short-term workers. These short-term workers were strangers to the hazardous conditions on site and they would not take care of other workers because they do not know each other.

10.5.2   Assurance of subcontractors' performance and quality

It is considered that subcontractors' quality and performance were difficult to control. Subcontractors are operating as individual firms. They would have their own culture, structure, management style and business strategies. Most of the small size subcontractors have not been established formally and most of them have only a few staff or workers. They are not well organised and therefore communication between the principal contractor and subcontractors might have problems. Meanwhile, if the principal contractor provides too much effort on supervising subcontractors, the objective of minimising the use of limited resources would be lost.

Under the current subcontracting practice, the lowest bidder would get the contract and as a result the financial return is trivial but the risk is huge. Subcontractors have to shorten the completion time and resort to lower standards or to use unskilled or semi-skilled labour in order to save construction cost. As a result, the quality and performance are being affected accordingly.

Some interviewees opined that if the expected standards could be listed out clearly onto the subcontract documents, assurance for quality and performance would not be a problem. Furthermore, a suitable selection procedure for subcontractors such as pre-tender qualification is being considered as an effective tool to control subcontractors' quality and performance.

10.5.3   Restricting multi-layer subcontracting

Most respondents considered that restricting multi-layer subcontracting was impracticable. The interviewees suggested that even if the proposed strategy of restricting multi-layer subcontracting was adopted, there would be no effective and reliable method to monitor whether the contractors had sublet the work or how many layers of subcontract are being truly sublet. In addition, the expected financial impact and risk to principal contractors would be too great to be affordable if multi-layer subcontracting is restricted.

10.5.4   Expected impacts

Over 80% of the respondents expressed that impact to the industry was expected if multi-layer subcontracting is restricted. The industry will be adversely affected. If the subcontracting practice is drastically restricted or the subcontracting levels is highly regulated, principal contractors would have to employ extensive direct employees so that a large amount of routine turnover was required, and the operation cost would be greatly increased. Some contractors may be forced out of business. The society would also be affected because the construction cost would be increased and the increase in cost would eventually be transferred to the end-users.

10.5.5   Free market approach

Based on the questionnaire survey, it is considered that the free market approach adopted by Hong Kong would be damaged by restriction of the multi-layers subcontracting practice. Most small and medium size contractors could not survive. If multi-layer subcontracting is allowed to operate continuously, it could provide more tendering opportunities for small and medium size contractors and allow more subcontractors to join the competition.

10.6   PROPOSED ALTERNATIVE MEASURES TO IMPROVE SAFETY PERFORMANCE

10.6.1   Strengthen the control of subcontractors

As previously discussed, safety training is considered as a key factor affecting safety performance. It is recommended that sufficient safety training should be provided to employees at all levels. Subcontractors should be held accountable for safety and they should be encouraged to employ safety professionals in managing safety matters.

Careful control and selection of subcontractors is essential if pre-qualification process is adopted to assess subcontractor's quality and ability. The expected standards and safety requirements should be listed onto the subcontract documents as detailed as possible, and to correlate subcontractors' past safety performance with tendering opportunity.

It also suggests to replace the current practice of awarding the contract to the lowest bidder. The tender price should be assessed with specific safety criteria. Contracts should only be awarded to those contractors or subcontractors who have submitted tender at a reasonable price.

10.6.2   Legislation and enforcement

It is suggested a tightening the requirement of registered safety officer such that at least one registered safety officer should be employed for each construction site. The current legal requirements of one safety officer to be employed when the total workforce is at 100 or more should be tightened. Small construction sites usually have no safety officer to carry out safety supervision.

The current Pay for Safety Scheme operated by the Works Bureau is an effective tool to improve construction safety.   It is an incentive scheme to compensate the safety cost incurred by contractors. Under the conditions of contract, the contractors are entitled to pay on a monthly basis if they have completed the specified safety items as stated in the contract.   It is suggested that extending this incentive scheme to other Government and private contracts.

The enforcement of occupational health and safety regulations is suggested to hold an individual subcontractor or worker accountable for site safety. The enforcing department should prosecute those safety offenders directly.

It is recommended to implement a tradesman-licensing system. Special tradesmen must gain recognised licences before they are allowed to carry out high risk activities such as operation of plant, bend and fix steel bars, gas welding and working in confined spaces.   The licensing system could ensure that high-risk tradesmen have received sufficient safety training before they to work on site level.

Chan (2000) commented that there was very little social security to workers, and everyone puts in extra to ensure that they can feed their family.   As a result, workers either ignore or accept the danger, and in some cases they are being forced to take risks.   A better safety culture for the construction industry is required.   It is recommended that the Government should enhance the legal status and the negotiation ability of the trade unions and to educate workers to demand for the provision of safer plant, equipment and a safer working environment from their employers.

10.6.2   Technological changes

The construction industry should be encouraged to widely adopt the use of pre-cast construction techniques instead of using traditional labour intensive construction methods.   Chudley (1985) found that the traditional cast-in-situ concrete construction methods used to occupy large working areas, consume more labour forces and construction materials and produce large amount of construction wastes.

Wong (2000) opined that occupational health and safety issues were often the result of lacking safety as an element or inadequate safety consideration in the design of buildings and planning of building works. Remedial safety measures are often technically more difficult, less ideal, less acceptable to employees, and cost more money than well-planned safety measures, which have been considered at the design and planning stage.   Tang (2000) recommended introduction of the 'Construction Design Management' concept from Europe to the Hong Kong construction industry.   Also, to request architects to put consideration of design safety, construction safety and maintenance safety into account at the design stage. It is believed that this new management concept could help improve construction related occupational safety and health.

High mobility of construction workers is one of the major causes of poor safety performance.   These short-term workers are strangers to the hazardous conditions on site.   It is therefore recommended to employ as many long-term employees as possible.

Linehan (2000b) commented that much could be achieved in quality, costs and safety through an attempt to develop longer-term partnership between contractors and subcontractors.   Longer-term partnership enable both parties to work for good safety standards, to formulate effective means for working safely, and to develop a proper understanding of the reciprocal duties and responsibilities which exist on site.

10.7   CONCLUSION

This research, which was based on the perception of the survey respondents, has concluded that restricting multi-layer subcontracting might bring with it great impact which is not acceptable by the practitioners of the construction industry. Without their full support and cooperation, implementation of such idea would encounter tremendous difficulties that would make the regulation not applicable. Meanwhile, the subcontracting practice has its values and advantages for the construction industry, which should not be underestimated.   Linehan (2000b) commented that certain subcontracting by professional trades was unavoidable though excessive subcontracting should be minimised.   Thus, the proposed alternative measures, as well as the reduction of excessive levels of subcontracting are recommended for improving the safety performance of the Hong Kong construction industry.

10.8   REFERENCES

Capp, R.H., 1973, *An Engineer's Management to the Elements of Industrial Safety* (London: Contemprint Limited).

Chan, A., 2000, Tragedy in the use of Oxy-acetylene torches. *A Hong Kong Journal of Safety Bulletin*, 18, **4**, p. 8.

Chudley, R., 1985, *Construction Technology Volume IV* (Singapore: Longman Singapore Publishers).

Labour Department, 2001, Report of the Commissioner for Labour, Printing Department, HKSAR Government, Hong Kong.

Lee, H.K., 1991, *Safety Management - Hong Kong Experience*, Lorraine Lo Concept Design, Hong Kong.

Lee, H.K., 1996, *Construction safety in Hong Kong*, Lorraine Lo Concept Design & Project Management Ltd., Hong Kong.

Lee, S.S., 1999a, An article of interview by Labour Department. *A Hong Kong Journal of Construction Safety Newsletter*, Labour Department of HKSAR, 4, p. 4.

Lee, K.M., 1999b, An article of interview. *A Hong Kong Journal of Construction Safety Newsletter*, 4, Nov, p. 3.

Leung, P.C., 1997, A Review of the Report of Construction Safety by the Hong Kong Construction Association. *A Hong Kong Journal of Green Cross*, 7, **5**, p. 4.

Linehan, A.J, 2000a, Subcontracting in the construction industry. *A speech presented in Safety and Health Conference*, Hong Kong, 22-23 March.

Linehan, A.J., 2000b, The safety implications of subcontracting in the construction industry. *A Hong Kong Journal of Green Cross*, 10, 4, pp. 12-20.

Lo, A., 1997, A talk of safety experience. *A Hong Kong Journal of Safety Bulletin*, Occupational Safety and Health Association, 14, **5**, pp. 6-7.

Poon, T.C., 1998, Workers have no choice to work under an unsafe working environment. *A Hong Kong Journal of Safety Newsletter*, Labour Department, Hong Kong, 2, p. 3.

Rowlinson, S., 1999, *Accident Report for January 1999*, Hong Kong Housing Authority. http://hkusury2.hku/steve/housing/reports.1999/mainpage.htm, 27 June.

Shaw, M., 1998, Promotion of occupational safety and health for small and medium sized enterprises – Canadian experience. *A Hong Kong Journal of Green Cross*, Occupational Safety & Health Council, 8, pp. 40-45.

Simo, S., 1995, Serious occupational accidents in the construction industry. *A Journal of Construction Management and Economics*, 13, pp. 299-306.

So, K.C., 2000, The multi-layers subcontracting system. An article of interview from the Hong Kong newspaper of *Hong Kong Economic Post*, 12 May.

Stranks, J., 1994, *Health & Safety in Practice - Management System for Safety*, (U.K.: Bell & Bain Ltd).

Tam & Fung, 1998, Effectiveness of safety management strategies on safety performance in Hong Kong. *Journal of Construction Management and Economics*, 16, pp. 49-55.

Tang, W.S., 2000, Establish safe construction culture from work designs. *A Hong Kong Journal of Green Cross*, 10, **4**, p. 2.

Tong, Y.C., 2000, Scrapping of rotten subcontracting system urged. An article of interview by *South China Morning Post*, http://www.scmp.com/News/Hong Kong/Art.../FullText_asp_ArticleID-20000411015846713.as., 11 April.

Tse, L.L., 2000, Multi-layer subcontracting are difficult to regulate. An article of the Hong Kong newspaper of *Man Wei Post*, 26 May.

Wong, Y.Y., 1997, Safety Experience. *A Hong Kong Journal of Safety Bulletin*, Occupational Safety and Health Association, 15, **3**, pp. 12-13.

Wong, H.W., 1999, Health and safety management system. *A Hong Kong Journal of Green Cross*, Occupational Safety and Health Council, 9, **1**, pp. 20-21.

Wong, W.Y., 2000, Occupational safety and health at the workplace. *A Hong Kong Journal of Safety Bulletin*, 18, **5**, pp. 4-6.

CHAPTER 11

Integration of Safety Planning and the Communication of Risk Information within Existing Construction Project Structures

Iain Cameron and Roy Duff

INTRODUCTION - HEALTH AND SAFETY MANAGEMENT IN CONSTRUCTION

The performance standards traditionally associated with the delivery of construction projects are time, cost and quality. However, in recent think-tanks such as:

- The Construction Task Force
  (www.construction.detr.gov.uk/cis/rethink/index.htm);
- The Construction Best Practice Programme (www.cbpp.org.uk);
- The Key Performance Indicators Zone (www.kpizone.com), and
- The Movement for Innovation (www.m4i.org.uk).

have set additional targets, challenging construction to plan more thoroughly and improve performance in many areas, including a greatly increased attention to health and safety.

Today's thinking seriously challenges the old model of time/cost/quality trade-off, which suggested that an improvement in one must lead to deterioration in at least one of the others. It now extends the total quality management philosophy that 'quality is free' and embraces the premise that delivery in one area, safety, can actually lead to benefits in other areas, time and cost. The importance of effective construction planning and control in the communication and avoidance of health and safety risks cannot be overstated but the fundamental premise which underlies this research, is that this need not and should not be a separate exercise aimed solely at health and safety. Effective management will embrace all production objectives, as an integrated process, and deliver construction, which satisfies all these objectives, not one at the expense of the others.

11.1   HEALTH AND SAFETY TARGETS

The UK Government has recently set ambitious targets to reduce workplace accidents and ill health across all industries in *'Re-vitalising Health and Safety'*. However, construction's safety performance is worse than most industries, currently deteriorating and a major component of society's accident and ill health experience. Because of this the Health & Safety Commission and the Deputy Prime Minister hosted a Construction Summit entitled *'Turning Concern into Action'* on 27 February 2001. The summit communicated key 'Action Plans' to improve performance and meet new ambitious targets, set by CONIAC (Construction Industry Advisory Committee in the UK), for the construction industry. The Deputy Prime Minister has made it clear that he wants decisive action. This proposed research will, *inter alia*, develop key performance indicators which will assist HSE and industry to *'Work Well Together'* to realise some of these improvements. The targets currently being pursued can be summarised below.

'Re-vitalising Health and Safety' targets to be reached by the year 2010 for all industries include:

1. To reduce work days lost (per 100,000) by 30% due to accidents.
2. To reduce the incidence rate of fatal and major injury accidents by 10%.
3. To reduce cases of ill health by 20%.

'Turning Concern into Action' targets to be achieved by 2005 and 2010 by construction industries are:

1. To reduce the incidence rate of fatalities and major injuries by 40% (2005) and 66% (2010).
2. To reduce the incidence rate of cases of work-related ill-health by 20% (2005) and 50% (2010).
3. To reduce the number of working days lost per 100,000 workers from work-related injury and ill-health by 20% (2005) and by 50% by (2010).

11.2   COMMUNICATION AND INFORMATION FLOW

Effective health and safety planning to improve risk information, communication and control are fundamental parts of the UK's Health and Safety Executive (HSE) current construction focus - *'Working Well Together'* (wwt.uk.com). The flow of appropriate risk management information is also fundamental to much of UK health and safety legislation:

1. The Health and Safety at Work Act 1974.
2. The Management of Health and Safety at Work Regulations 1999.

3. The Construction (Design and Management) Regulations 1994.
4. The Construction (Health, Safety, Welfare) Regulations 1996.
5. The Health and Safety (Consultation with Employees) Regulations 1996.

However, the Construction (Design and Management) ACoP Consultative Document acknowledges that CDM compliance documents, written specifically for the purpose, are sometimes cumbersome and add little value to management processes generally. Because of this, the ACoP document indicates that stakeholders should treat CDM information provisions as part of the construction planning process. The ethos is: '*minimise bureaucracy and maximise performance*'.

To comply with all this legislation, and to establish effective health and safety planning and control, a large number of management procedures are required. The authors believe that health and safety is best managed within a contracting organisation's existing production management structure and procedures, as any attempt to overlay a safety specific structure is likely to meet resistance and organisational constraints.

11.3 INDUSTRIAL PROBLEM TO BE INVESTIGATED BY THE RESEARCH

Effective planning for health and safety is essential if projects are to be delivered on time, without cost overrun, and without accidents or damaging the health of site personnel. These are not easy objectives as construction sites are busy places where time pressures are always present and the work environment constantly changing.

The construction industry tends to be under resourced and under planned in relation to other industries and this promotes a crisis management approach to all kinds of production risk, a feature of construction culture which can impact on health and safety management. However, the industry frequently demonstrates that it can plan proactively, to manage production risk, if it is required to do so. Consider highly planned works, such as those requiring a temporary rail possession, which almost invariably run smoothly. This type of work is managed in a highly focused way and planned in great detail. This contrasts with routine work where recent collaborative EPSRC(IMI) research by Dundee University and UMIST (Dr. Duff) has demonstrated that productivity increases of at least 20% are readily available from a combination of more rigorous short-term planning, clear communication of goals, performance measurement and a participative approach to obtaining commitment at all levels. This approach is clearly cost-effective. There is no reason why health and safety performance should not benefit from the same approach and achieve at least as much improvement. Indeed, research for HSE (Duff *et al.*, 1993), adopting only the performance measurement and goal-setting principles, has shown that well over 20% reduction in unsafe operative behaviour is achievable (HSE Contract Research Report, CRR 229/1999); and more recent work has demonstrated comparable improvements in site management behaviour related to its health and safety responsibilities (Cameron, 1999).

The problem investigated by the programme of research outlined in this paper is how best to promote the effective integration of health and safety management into project planning and control in all construction activity - complex and routine - and thus achieve similar improvements. This will involve the development of methods that will ensure that health and safety management truly permeates construction sites by *'building-in'* safety planning and control as a core aspect of normal time and resource management, rather than attempting to run separate safety procedures as *'bolt-on'* extras.

In order to achieve a holistic approach to the management of construction, it is important to view health and safety planning as an integral aspect of production planning from the beginning. This should involve, for example, including safety risk assessment as part of the general management of construction risk; and including specific safety activities and milestones on the master construction programme. This allows safety activity to be monitored in the same way as production and embraces the *'what gets measured, gets done'* philosophy of management. For example, the authors' chose a 'Linear Responsibility Chart', detailing who does what, when and where (CIOB, 1991). This monitors health and safety responsibilities, along with other responsibilities, to avoid safety management being overlooked as a result of unclear priorities. Safety arrangements and effort then becomes controlled and visible as a central objective of production management.

This thinking demonstrates that strategic planning for Health and Safety begins at the very start of the project and a commitment to proactive safety planning can be evidenced by a review of the allocation of responsibilities for safety (Health and Safety Plan). This should be clear from a review of the project organisational structure chart which ought to detail who has responsibility for what (safety roles).

The following is an example of the extent of a comprehensive set of safety roles which form the foundation of the more advanced 'Linear Safety Responsibility Chart':

Site Manager
- Holds subcontractor pre start meetings
- Updates risk assessment control chart for all work activities
- Delivers safety induction talks
- Chairs safety committee
- Maintains records of statutory records
- Conducts safety audits with safety manager

Sub agent
- Fire Inspection and fire marshal and first aid
- Conducts Tool Box Talks
- Conducts works risk assessments

- Inspects and audits subcontractors and site for safety
- Deputy for safety induction talks
- Crane and lifting operations 'appointed person'
- Conducts works risk assessments
- Inspects and audits subcontractors and site for safety

Project Engineer
- Conducts excavation inspections
- Conducts RA for temporary works
- Temporary works co-ordinator who oversees safe installation and removal

General Foreman
- Conducts scaffold inspections
- Communicates visible commitment to safety
- Crane co-ordinator

Works Foreman
- Inputs into risk assessment
- Procures safety plant and PPE and inspects same
- Deputises for Site Manager and Sub Agent for safety acts
- Builds safety consideration into daily part of work instructions

Chargehands
- Inputs in to risk assessment
- Procures safety plant and PPE
- Inspects work equipment and PPE
- Identifies extra safety training needs of his squad

Tradesmen and Labourers
- Report unsafe acts and conditions
- Maintain their own PPE
- Co-operate with the main contractor

11.4   RESEARCH OBJECTIVES

1. To produce a theoretical model of construction management, planning and control, and practices for the integrated management of health and safety risks.
2. To improve the effectiveness and practicality of this model through consultation with experienced practitioners.
3. To investigate current construction planning, control, and supervisory methods and tools, and evaluate contractors' methods for the management of health and safety risk.
4. To identify gaps between current health and safety management practice and the model, and revise the model to benefit from any improvements suggested by this investigation.

5. To field test the revised model in order to further improve the validity of the safety planning model by taking account of these practical experiences.
6. To produce a guide to best planning and control practice in integrated construction management of health and safety risk, including a set of 'Key Integrated Safety Management Planning Procedures' and 'Key Safety Performance Indicators'.

11.5   RESEARCH METHOD EMPLOYED

11.5.1   Objective 1

The first objective will be achieved by investigating available literature on good practice in construction management, in general, and legislation and other literature on health and safety management, in particular. This will cover site investigation, site set-up and mobilisation, site layout planning, contractor design (including temporary works), method statements, risk assessment, programming and allocation of resources, short-term planning and control, construction site supervision, the management of design variations throughout construction, and commissioning and client handover. A draft, theoretical model of 'best-practice' procedures, information flows and planning tools will be produced, showing source, destination, content, timing and frequency of communications and the roles and responsibilities of typical participants in the construction management process. This will show the integration of health and safety management with the general planning and site management processes.

11.5.2   Objective 2

The draft model will be tested for completeness, perceived value, practicality and the potential for integration into real construction sites, by consultation with a range of experienced practitioners; and modified accordingly. This interaction will use a selection of interview, Delphi and focus group techniques, as appropriate and depending on opportunity and availability of participants.

11.5.3   Objective 3

The next objective is to investigate how far actual practice diverges from the modified model. This will involve the collection and mapping of data on a variety of live case-study construction sites by: interviews with key project actors; passive observation of site induction meetings, planning meetings, tool-box talks, etc.; and, review of site management documentation, such as site investigations, risk assessments, construction programmes, method statements, etc. A gap analysis of health and safety content and quality will then be carried out, against criteria derived from the model. Instances of deviation from the model, particularly instances of failure to comply with accepted good practice or legal requirements, or instances of dealing with health and safety as a peripheral issue, will be recorded and reasons sought. Instances of alternative or better ways of achieving the model objectives will be recorded for inclusion in the model.

11.5.4   Objective 4

This will involve developing strategies for the introduction of any previously untried health and safety management procedures in the model (effective 'best-practice' procedures discovered during observation of current practice may not require any further field testing). This will be done with the assistance of industrial collaborators, and recording practitioner feedback.   Strategies will focus particularly on ways of integrating health and safety management, such as health and safety plans and risk assessments, into equivalent mainstream construction management procedures.   The drivers for change will include improved overall communication flows, methods of reducing the duplication of safety effort and ways of promoting teamwork and consultation etc., to the benefit of the whole construction management process. The model mechanisms are clearly not yet determined but current thinking suggests that they will include: structured meeting agendas; safety management performance indicators; goal setting and $360°$ feedback; risk assessment workshops; integration of accident/dangerous occurrence reporting with other feedback into short-term planning procedures; collection and feedback of employee contributions into the whole short-term planning process, including health and safety; assessment of employee experience and competence across all factors of performance, including health and safety; re-engineering site management supervisory practices, such as induction and tool-box training, to integrate risk awareness with production related information.

11.5.5   Objective 5

This will involve the production of a set of documented Key Integrated Safety Management Procedures which can be used by HSE and industry to support a strategy for integrating health and safety management within existing construction management systems. These aids will also be used as the basis for a set of Key Performance Indicators which can help monitor health and safety activity within construction planning and control.

11.6   CONCLUSIONS

This chapter has presented a planned programme of study which is currently being negotiated with the UK's Health and Safety Executive (HSE). The initial findings of the study and early collaborator consultation have suggested that health and safety is not seen as an integral aspect of work planning. Instead, risk information and avoidance strategies are conducted at a later date and risk assessments and control plans are viewed as separate activities sometimes 'the domain of the specialist'. For example, construction barcharts seldom make reference to key safety items (e.g. completion of risk assessment, approval of method statements, training events, etc.). This suggests that safety planning is being conducted too late in the process in order to truly permeate site planning and supervisory practices. This chapter recommends that more research is required to find effective and efficient ways of (easily) embedding safety planning and risk communication within the day-to-day planning tasks of the construction site management team.

11.7   REFERENCES

Cameron, I., 1999, *A Goal Setting Approach to the Practice of Safety Management in the Construction Industry: Three Case Studies*, PhD Thesis, University of Manchester Institute of Science and Technology (UMIST), Department of Building Engineering.

CIOB, 1991, *Planning and Programming in Construction - A Guide to Good Practice*, Chartered Institute of Building Publications (Ascot: Englemere).

Duff, A.R., Robertson, I.T., Cooper, M.D. and Phillips, R.A., 1993, *Improving Safety on Construction Sites by Changing Personnel Behaviour*, Contract Research Report 51/1993, Health & Safety Executive, ISBN 011 882 1482.

HSE, 1997, *Evaluation of the CDM Regulations 94*, The Consultancy Company. HSE Contract Research Report 158 (HSE Books).

HSE, 2000, *Re-vitalising Health & Safety: StrategySstatement*, June (HSE/DETR).

HSE, 2000, *Proposals for Revising the Approved Code of Practice on Managing Construction for Health and Safety*, CDM Consultative Document, Aug (HSE).

HSE, 2000, *Securing Health Together*, MISC225, Aug (HSE/DETR. Books).

HSE, 2000, *Re-vitalising Health and Safety in the Construction Industry: Developing an Agenda for Action.* In Working Well together Conference, London, June.

HSE, 2000, *Tackling Health Risks in Construction: Developing an Agenda for* (HSE Books).

HSE, 2000, *Health and Safety Benchmarking: Improving Together*, HSE *Action*, Working Well together Conference, London, Oct.

11.8   INTERNET REFERENCES

HSE's, Working Well Together campaign, *wwt.uk.com.*
Egan's Report, *www.construction.detr.gov.uk/cis/rethink/index.htm.*
Key Performance Indicators Zone, *www.kpizone.com.*
Movement for Innovation (M4I), *www.m4i.org.uk.*
Construction Best Practice Group (KPI's), *www.cgpp.org.uk*

CHAPTER 12

The Use of Project Management Techniques in the Management of Safety and Health in Construction Projects

Malcolm Murray

INTRODUCTION

In many countries around the world the construction industry continues to be one of the sectors with the highest indices of accidents leading to deaths or temporary or permanent disabilities among members of the workforce.

In developed countries, governments attempt to address the problem by implementing legislation which punishes infracting companies or persons. In the European Union, for example, fines may be applied (New Civil Engineer International, 2000a). In the USA infractions may lead to imprisonment as well as to fines (Engineering News Record, 2001a). A complementary approach in the USA is for Government to exhort the construction industry to improve its safety record (Engineering News Record, 2001a).

In developing countries, appropriate legislation quite often exists but it is frequently not applied in the case of infraction of laws, probably because construction workers are often migrant, low-paid, are not afforded union-protection and are held in low esteem by the society in which they live. Another factor which influences attitudes to safety and health is cost: theoretically, an unskilled worker from a developing country who earns US$0.50 an hour and who dies in an accident at age 25 would be assessed for future loss of future earnings of US$40,000, a US worker at least 40 times that amount. In many countries, however, a wage of US$0.50 an hour would be considered generous - the World Bank estimates that more than 1.2 billion people worldwide live on less than US$1 a day. In practice therefore, unskilled workers' families may receive ludicrous amounts on the workers' death - in India, for example, the amount would be three months' wages - about US$400. (Koehn and Reddy, 1999)

At the same time, in the corporate world, in the author's experience issues of safety and health are rarely really seriously addressed by owners, engineers and contractors during the construction planning and execution processes. The general approach seems to be a top-down one where head-office based inspectors visit and monitor sites or provide guidelines. Two examples of this are the approaches taken by AMEC (AMEC, 2000) and Skanska (Särkilahti, 2001).

In University construction management courses, safety and health issues, while being often dealt with adequately, are not addressed as being one of the core management topics. Nunnally (1998), for example, provides an overview of the USA's OSHA (Occupational Safety and Health Act) and describes risks that need to be avoided when operating different types of equipment. The writer is not aware of any serious approach to safety and health matters being taken in other built environment university courses, such as civil engineering, architecture and quantity surveying. Integration of safety and health management into construction management and civil engineering courses would lead to construction site managers dealing with the matters in a routine manner, precluding the need for a top-down, head-office approach.

The climate within which international corporations, and international construction companies, carry out their business, however, is changing. Globalisation is leading to the dissemination of first-world attitudes and legislation to developing countries. Recent corporate scandals are leading to business leaders being made more accountable for their financial responsibilities and the responsibilities of corporations are being extended to cover environmental and social issues. In the UK, Latham and Egan (DETR, 1999) have recommended that the UK construction industry become more productive, innovative and safer. The outcome of all of these shifts in attitude is that international construction companies, wherever they perform throughout the world, will tend to be held responsible for a wide range of their actions, these including their performance in the areas of safety and health, as well as in the areas of the environment and social impact.

To safely manage safety and health matters throughout a company's organisation these areas, as well as others such as social impact and environmental impact, which will not be dealt with here, will need to be integrated into the company's contract, divisional and corporate business plans. One way to do this would be to use the Project Management disciplines, specifically the Project Management Body of Knowledge (PMBOK), suitably adapted to the specific requirements of the modern international construction industry.

Following an overview of the problem of accidents, approaches to solving the problem, and the construction process, the project management body of knowledge will be reviewed, its adaptation to the needs of the international construction industry described and examples of its application given.

This chapter is accordingly organised following the sub-headings listed below:

- The problems of accidents.
- Approaches to solving the problem.
- The construction planning process.
- The project management body of knowledge (PMBOK).

- Adaptation of PMBOK (C-PMBOK) to the needs of the international construction industry.
- The application of the construction-PMBOK to the international construction industry.
- Organisation for implementation of C-PMBOK.
- Conclusions and recommendations.

This chapter will focus on accident-prevention management. A similar approach could be taken to health-implementation management.

12.1   THE PROBLEMS OF ACCIDENTS

Accidents on construction sites cause a series of problems, moral, productive and financial.

The moral problem is that workers die or become permanently or temporarily disabled; their families suffer the consequences of this.

In terms of productivity, deaths lead to a drop in worker morale, with a consequent drop in productivity which may last for days or weeks. In the case of a death or deaths, the site may be embargoed while the safe and health authorities, or the insurance company, carry out investigations, or while key equipment is replaced. Productivity may also drop due to the need to rework damaged structures or materials.

Costs would go up as a result of rework, insurance premiums would probably increase and productivity drops would evidently lead to cost increases.

According to the US Bureau of Labor Statistics, the construction industry was responsible for 1154 deaths in 2000. This figure represents nearly 20% of all workplace fatalities in the USA during that year (Reid, 2001). Furthermore, construction-related deaths have increased from under 900 in 1992 to a peak of 1200 in 1999, dropping to 1154 in 2000. Non-fatal accidents rates are similarly depressing and health problems are of concern too.

Figure 12.1 compares the rates of fatal injury and rate of injury leading to an absence of work of over three days for a number of European countries and the USA in 1996 (Atkinson, 2001). The figure also shows rates of fatal injury in the construction industry for five European countries.

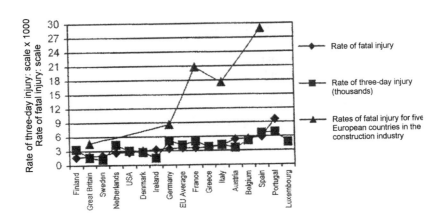

Figure 12.1 Workplace injuries in Europe and the USA 1996

Figure 12.2 shows the rise of construction-related deaths in the UK from 1995/96 to 2000/01 (Mathiason, 2001) and Figure 12.3 shows similar absolute figures for construction-related deaths in the USA between 1992 and 2000.

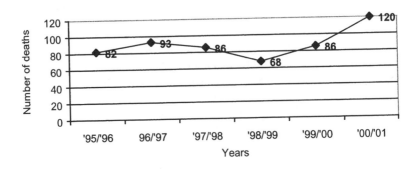

Figure 12.2 Construction-related deaths in the UK 1995/96 to 2000/01

According to Krueger (2000), from 1992 to 1998, in the USA, the number of work-related injuries and illnesses has fallen 25%, to 6.7 per 100 from 8.9. This improvement translates to an annual saving to the US economy of US$125 billion. Again, according to Nunnally (2000), although composing only about 5% of the US workforce, it has been reported that construction accounts for 20% of work fatalities and 12% of disabling injuries. The total annual cost, direct and indirect, of construction accidents has been estimated to exceed US$17 billion.

Meanwhile in the UK, the Health and Safety Executive (2000) believe that accidents and ill health among the workforce cost the British economy up to

£18 billion a year. According to the HSE, over £180 million could be saved in work related illness costs in the construction industry alone.

The economic implications of accidents have been examined by Casals *et al.* (2001), Weber (2001), Haupt, Hinze and Coble (2001), Tang (2002), Smallwood (1999), Ngai and Tang (1999).

In the next section solutions to these problems will be examined.

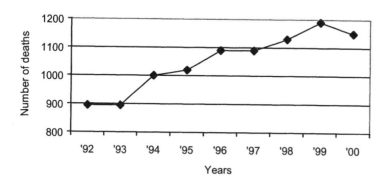

Figure 12.3 Construction-related deaths in the USA 1992 to 2000

Source: ENR/Bureau of Labor Statistics

12.2   APPROACHES TO SOLVING THE PROBLEM

In the USA the Occupation Safety and Health Administration (OSHA) was set up to administer the Occupational and Safety Health Act of 1970. OSHA penalties may be applied following administrative or criminal proceedings and may entail fines of up to US$10,000 and imprisonment of from 6 months to life. Heavier penalties may be imposed in the civil and criminal courts.

In the UK, European Union legislation was incorporated into British law in the early 1990's. Smith (1996) has examined the legal and cost implications of this legislation.

More recently, the British government has set targets to cut work-related deaths, accidents and illness, as mentioned earlier (Health and Safety Executive, 2000).

The Government and the Health and Safety Commission (HSC) have established a ten-point strategy supported by a 44-point action plan. Strategy highlights include:

- The health and safety system must promote a better working environment as well as preventing harm.
- A focus on occupational health as a priority.
- The need to motivate all employers, particularly small firms, to improve their health and safety performance.
- The need for government to lead by example.
- The importance of education at all levels for improving Health and Safety.
- The role of effective design in preventing risk.

The action plan includes:

- Tougher penalties, including fines of up to £20,000.
- New, innovative penalties, such as the suspension of managers without pay.
- A director's code of practice, which will make a named person responsible for health and safety matters within every company.
- New help for small businesses.
- Abolition of Crown Immunity for public servants who infract health and safety legislation.
- Exploration with the insurance industry of the implementation of incentives to reward good health and safety performers.

In the UK, and to a lesser extent in the USA, the onus will increasingly be on industry managers to implement effective health and safety policies.

12.3   THE CONSTRUCTION PLANNING PROCESS

12.3.1   Introduction

The construction process is in a state of flux, influenced by wide-ranging changes that have, and are, occurring in the international construction industry.

Fifty years ago, a developer would engage an architect or engineer to carry out a design, tender documents would be prepared, tenders called for, and a contractor appointed. The contractor would carry out the contract under the engineer or architect's supervision. The engineer/architect designed and the contractor built – there was little integration between the design and the construction team.

Today, with globalisation and the emergence of large multinational corporations through a process of growth, mergers and acquisitions, the international construction industry is dominated by large multi-disciplinary companies from the US, Europe, Japan and, increasingly, China, with turnovers of around US$10 billion a year (Murray and Appiah-Baiden, 2000), (Murray and Mavrokefalos, 2000). These companies often carry out Engineering, Procurement and Construction activities (EPC) and often raise finance to carry out Build Operate and Transfer (BOT) contracts and others of a similar nature. Since the operation of a facility, such as a motorway, would typically take place over a time-span of twenty or thirty years the organisation responsible for a BOT contract would evidently look closely at the lifetime operating and maintenance costs, as well as the design of the facility itself.

In the building industry, similar, holistic approaches to construction costs have emerged where owners and their design team examine not only initial costs but the lifetime running costs, including those involved with managing the building's facilities.

A partnering approach, where the efforts of the client, the design team and the contractor are integrated, is being adopted on a number of projects.

In the following, for the sake of simplicity, the different processes involved in a typical construction project will be described for a project being carried out according to traditional procedures.

12.3.2   Conception

The first process would be initiation where the project owner defines the scope of his requirements. This may be accompanied by a preliminary feasibility study of the project.

12.3.3   Design

The owner's requirements are then translated into a design which is then worked up into a detailed design by the professional design team.

12.3.4   Call for tenders

Once the owner has decided to go ahead with the construction project, after carrying out further feasibility studies, he then instructs his representative to prepare tender documents. These, among other things, may include specifications which give guidance to the tendering contractors on the expected quality of the finished product. Sometimes the Bill of Quantities may include items to be priced dealing with environment or health and safety matters: the contractor may be required to establish on site a first-aid station, for example. In some countries, such as Brazil, local legislation requires that sites employing more than 200 workers would require a medical doctor and a safety engineer in attendance; in the author's experience rarely does the tender documentation emphasise the need to comply with this type of legislation or, indeed, safety and health issues in general; if they are mentioned it is in a token fashion. This is probably because owners and designers do not usually perceive construction safety to be an area with which they should concern themselves, the contractor being considered to be responsible for the construction process.

12.3.5   Tender preparation

During the tender preparation process contractors usually focus on planning and scheduling (or programming) to achieve a competitive price, given that contract awards are still frequently based on lowest-price criteria. Material and labour costs are common to all tenderers and tender advantage can usually only be gained by creative planning and programming.   The contractor will usually examine the specifications with care to ensure that its tender includes sufficient money to produce a product to the required quality within a defined timeframe. Sums will be included in the preliminary and general sections of the budget to cover such safety and health requirements as a first-aid station and safety equipment for the workers. The primary concern of the contractor, however, is generally to deal with the 'golden triangle': the time, cost and quality inputs required to deal with the defined scope, as shown in Figure 12.4.

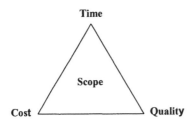

Figure 12.4 The 'golden triangle' of construction performance

12.3.6   Construction planning

On award of the contract the tender plan is usually detailed and developed into a construction plan, or Programme of Action, which is used as a guide during actual construction.

12.3.7   Construction execution

It is often only during the construction execution period that safety and health issues are really addressed, usually by a process of inspection. There may be a process of induction of workers to safety matters and procedures when they join the project.

The Project Management Body of Knowledge (PMBOK), which has been progressively developed since the 1970s, includes the basic initiation and planning processes described above and has also developed complementary processes which will be described in the next section.

12.4 THE PROJECT MANAGEMENT BODY OF KNOWLEDGE (PMBOK)

12.4.1 Introduction

Project management techniques were developed during the late 1950's and the 1960s when the space race got underway. Since the 1970's the growth of the use of the techniques has been phenomenal and today they are used in a wide number of industries, including the space, aeronautical and software sectors as well as the construction industry. In the USA, project management knowledge has been collected by the Project Management Institute into a so-called Project Management Body of Knowledge (PMBOK) which regulates the processes and areas of knowledge. The latest edition of PMBOK was issued in 2000. It is important to remember when applying it that PMBOK is a generic framework and needs to be adapted to specific industries on a case-by-case basis.

12.4.2 Processes

The 2000 edition of the Project Management Body of Knowledge (PMBOK, 2000) describes the five groups of processes which are followed during the carrying out of a project; these comprise the initiation, planning, execution, control and closing process groups. Figure 12.5 provides an indication of the groups and of the linkages between them.

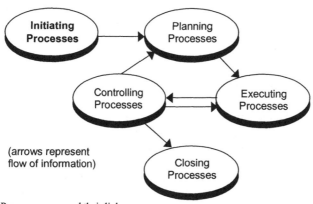

Figure 12.5 Process groups and their links
Source: PMBOK (2000)

12.4.3   Areas of knowledge

The carrying out of the processes involves the application of so-called areas of knowledge which comprise the management of project: integration, scope, time, cost, quality, human resources, procurement, communications and risk. The areas of knowledge are given equal weighting by PMBOK, as it has been developed as a guide and is not industry-specific.   In the writer's experience, however, in the construction industry the following priority ranking is often given to the areas, as mentioned earlier when describing the 'golden triangle':

1.   scope, time and cost
2.   procurement
3.   human resources
4.   quality
5.   integration and communication
6.   risk

The reason for this is probably because construction managers mainly have an engineering background and are more concerned with numerical, measurable issues than with human-related matters. It will not escape notice that, with the exception of risk management, the above listing shows a gradation of priority from 'hard' to 'soft' management topics.

In PMBOK (2000), compliance with health and safety regulations is listed under project human resources management, together with other topics concerned with dealing with people. This HR area comprises the following major processes: organisational planning, staff acquisition and team development. Project quality management, however, is considered to be an area of knowledge in its own right, consisting of quality planning, assurance and control.   Similarly, project risk management is also considered to be another knowledge area and consists of the following major processes: risk management planning, risk identification, qualitative analysis, quantitative analysis, response planning and monitoring and control.

12.4.4   A generic framework

Finally, it should be remembered once again that the Project Management Body of Knowledge is only a guide, or generic framework, which can and should be adapted for use in a variety of project scenarios, which include the development of products such as software, the implementation of information and communication technology (ICT) in industry, and the carrying out of projects in the aerospace, aeronautical and construction industries.   Evidently the application of PMBOK should be tailored to meet the particular needs and peculiarities of each industry. Its application in the construction industry will be discussed in the next section.

12.5   ADAPTATION OF PMBOK TO THE NEEDS OF THE INTERNATIONAL CONSTRUCTION INDUSTRY

12.5.1   The International Construction Industry

As is known the construction industry differs from others in a number of ways, although it does share the characteristic of being project-based with an increasing number of areas of industry and commerce.

The larger companies of the construction industry work internationally and may work in an ever-changing number of countries which are often of the type characterised as 'developing', possessing diverse cultures, populations with different levels of education and safety and health legislation of varying quality. (Murray, 2001)

The traditional way of carrying out business in developing countries was for the contractor to mobilise, carry out the contract and, unless they obtained further work, to demobilise. Responsible contractors complied with local legislation but there was no overwhelming pressure for them to do so.

However, society now increasingly expects that businesses act in an ethical manner, towards their shareholders and towards the society in which they operate. This means that contractors, whether they are operating in their home country or abroad, can and should be expected to manage environmental and social issues in a responsible manner. The social issues, of course, include the management of the safety and health of their workforce. Environmental issues include the responsible management of the contractor's operations to prevent pollution of water, ground and air in the region in which they are working, as well as to ensure the protection of flora and fauna.

Finally, it is expected that the contractor interacts with local communities in such a way that these benefit from the presence of the intruders (by being given, for example, employment preference and training) and are not adversely affected by the presence of the contractor, culturally or socially. An example of negative impact would be the spread of HIV/Aids by the contractor's migrant workers.

An example of how companies working internationally are being held to their social responsibilities is the ruling handed down by the House of Lords in July 2000 holding that a group of South African asbestosis and mesothelioma sufferers could sue Cape plc in England. Cape plc is a British company which had previously worked in South Africa through a subsidiary and it was sued for compensation as a result of its workers' exposure to asbestos and related products arising from Cape plc's earlier operations in South Africa (Stein, 2001).

Companies will also be held to their safety responsibilities and their financial consequences according to New Civil Engineer International (2000b).

It is becoming increasingly clear that, as globalisation develops, construction companies will be obliged to carry out their business following first-world criteria, independently of in which part of the developed or developing world they are working.

12.5.2   Phasing of the construction process

As mentioned earlier, the modern construction process is a much more integrated affair than in former times. The different steps to be taken in the evolution of a scheme from inception to demolition can be represented by phases. As discussed by Murray, Nkado and Lai (2001) the life cycle of a typical construction project can consist of the following phases:

- inception
- preliminary design
- feasibility
- detailed design
- preparation of tender documentation
- tendering and contractor choice
- construction
- operation
- demolition and recycling.

Each phase could be treated as an independent project, comprising linked processes as shown in Figure 12.6 and driven by areas of knowledge as described in the following section.

12.5.3   Extension of areas of knowledge

It would make sense to have the areas of knowledge of the PMBOK, when applied to the construction industry, extended to include items dealing with the additional areas described earlier. A revised list of knowledge items would therefore consist of the management, throughout the duration of the project, of:

- integration
- scope of work
- time, i.e. planning and scheduling (or programming)
- costs and budgets
- quality
- human resources
- procurement
- safety
- health
- community impacts
- environmental impacts
- risk
- communication.

Table 12.1 Mapping of Project Management processes to the Process Groups and Knowledge Areas, revised for use by international construction contractors.

Process groups / Knowledge area	Initiating	Planning	Executing	Controlling	Closing
Project integration management		Project plan development	Project plan execution	Integrated change control	
Project scope management	Initiation	- Scope planning - Scope definition		- Scope verification - Scope change control	
Project time management		- Activity definition - Activity sequencing - Activity duration - Estimating - Schedule - Development		- Schedule control	
Project cost management		- Resource planning - Cost estimating - Cost budgeting		- Cost control	
Project quality management		- Quality planning	- Quality assurance	- Quality control	
Project human resource management		- Organisational planning - Staff acquisition	- Team Development		
Project procurement management		- Procurement planning - Solicitation planning	- Solicitation - Source selection - Contract administration		- Contract closure
Project safety management		- Safety management planning - Safety risk identification - Safety risk response	-.Safety Plan execution	- Safety risk monitoring and control	
Project health management		- Health management planning - Health risk identification - Health risk response	- Health plan execution	- Health risk monitoring and control	
Project community management		- Community management planning - Community impact identification - Community impact response	- Community management plan execution	- Community impact monitoring and control	
Project environment management		- Environment management planning - Environment impact Identification - Environment impact response	- Environment management plan execution	- Environment impact monitoring and control	
Project risk management		- Risk management planning - Risk identification - Qualitative risk analysis - Quantitative risk analysis - Risk response planning		- Risk monitoring and control	
Project communications management		- Communication planning	- Information distribution	- Performance reporting	- Administrative closure

Source: Adapted from PMBOK (2000)

Table 12.1 shows the mapping of the project management processes to the process groups and knowledge areas, adapted for use by international contractors. The added areas of knowledge are shaded.

The management of the environment and community issues would be carried out by using techniques similar to those used for Environmental Impact Assessments which involve identification of impacts and their subsequent mitigation and control.

Similarly, safety and health issues would be managed by identifying the areas of risk, planning for risk mitigation and then controlling operations to make sure that the safety and health plans are executed correctly. Areas of risk would be identified by examining safety and health indicators which are discussed in the next section.

This revised map would be applied to each of the phases which comprise the construction project life cycle as shown in Figure 12.6.

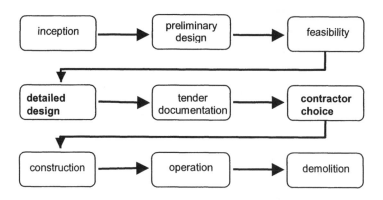

Figure 12.6 The construction project life cycle

12.6   THE APPLICATION OF THE CONSTRUCTION C-PMBOK (C-PMBOK) TO THE INTERNATIONAL CONSTRUCTION INDUSTRY

12.6.1   Introduction

In the following sections, examples will be given of the application of C-PMBOK (Construction-PMBOK) to projects carried out internationally, emphasising how safety and health would be dealt with in an integrated manner.

When managing projects, or phases of a project life cycle, certain areas of knowledge are prioritised, depending on the type of project being carried out. Following disastrous fires on off-shore deep-water extraction platforms in the North Sea and off the Brazilian coast, any offshore oil platform project would be flagged for rigorous fire risk examination as part of a safety review (the North Sea Piper Alpha disaster, for example, which occurred in 1988, caused 167 deaths). A nuclear power station project would have its quality management prioritised (as part of a safety management process). A five star hotel project would also have quality as a priority. A project with an important deadline, such as the Millennium Dome, would require focus on time management issues. While some projects are evidently high risk as regards safety or health issues, it is not always apparent when these areas should be prioritised. The use of indicators in these cases would be useful.

Following a description of the use of safety and health indicators on projects, the project life cycle will be examined and examples will be given of how safety and health issues can be managed or mitigated during the different stages of the project life.

12.6.2 Safety and health indicators

It is well known that there are areas of operation in the construction industry where safety risks or health risks can be inherently high. In the area of safety, for example, projects involving trenching or the use of tower cranes should immediately lead to preventative safety planning and operation monitoring to guarantee that accidents do not happen. These mitigative activities should include, in the case of tower cranes, examination of erection procedures, operations procedures, training of operators and signallers and monitoring of operations by a qualified and experienced construction engineer (who may, by the way, not necessarily be the safety engineer).

In the health management area, winning of a contract in sub-Saharan Africa would immediately alert the responsible contractor to the need to plan for, monitor and control HIV/Aids, TB and malarial infections (Murray and Appiah-Baiden, 2002).

Using published information, lists of safety and health indicators can be drawn up. In the case of safety, operations or activities would be classified on a scale of 1 to 5, one indicating a fairly safe operation, five a dangerous operation. In the case of health, countries or regions of countries would be classified using a similar scale: the level of health danger would be indicated, and the source of that danger identified. Operations of danger to health would be similarly classified. For example, the existence, or possible existence, of asbestos in a building being renovated would merit an indicator of 5.

Engineering News Record has published the top ten causes of deaths in the US construction industry in 1999 and from 1991 to 1998 (Engineering News Record, 2001c). Table 12.2 lists the causes; in 1999 23.8% were fall-related and 15.3% were due to being run-over by equipment or a vehicle. Comparing the total fall figures for 1999 and 1991-98, the figure of 23.8% for 1999 compares unfavourably with the 1991-98 figure of 21.7%.

Table 12.2 Top 10 causes of death in 1999 and 1991-1998 in the US construction industry

	INCIDENT TYPE	1991-98 [% of total]	1999 [% of total]
1	Fall from/through roof	11.2	10.6
2	Non-operator run over by equipment	7.8	9.2
3	Fall from structure	7.6	8.1
4	Run over by highway vehicle	3.8	6.1
5	Operator crushed by equipment	5.2	5.8
6	Lifting operation accident	5.4	5.4
7	Equipment contacts power source	8.4	5.2
8	Fall from scaffold	2.9	5.1
9	Collapse of structure	4.3	5.0
10	Shock from equipment/tools	4.1	4.7

Source: ENR/OSHA

Table 12.3, from the same source, lists numbers of construction fatalities, again in the USA which occurred from 1995 to 1999 and in 2000. In 2000, 58% of fatal accidents happened to people working for speciality trades contractors, which usually tend to be smaller organisations.

Table 12.3 Construction fatalities

	1995-1999 Average	1999	2000	2000 %
Total Construction Fatalities	1,115	1,191	1,154	100
General Building Contractors	190	183	175	15
Heavy Contractors	260	280	284	25
Speciality Trades Contractors	652	710	672	58
Unidentified	13	18	23	2

Source: ENR/U.S. Dept. of Labor, Bureau of Labor Statistics

Transport engineers use techniques similar to that involving the use of safety indicators; one system is to assign a danger index to accident black-spots which is based on the number of crashes resulting in claims by insurance company policyholders (Engineering News Record, 2001d).

Following the identification, and the indication of the level of safety and health risks, a risk evaluation exercise would be carried out where monetary values would be assigned to the costs and benefits of not carrying out, or carrying out, risk mitigation activities. Chapter 11 of PMBOK (2000) deals with risk management in general and outlines suitable procedures to follow.

12.6.3   Safety and health management throughout the project life cycle

One of the themes of this chapter is that safety and health should be managed in the same way as other areas of knowledge such as time, cost and quality and should also be managed during each of the phases of the project life cycle. The project life cycle has been illustrated in Figure 12.6.

The life-cycle project process is iterative, with feedback of knowledge being given to previous phases for reworking. For example, if a project is found not to be financially feasible, feedback would be given to the preliminary design team, the preliminary design would be adapted to the budgetary limits and the cycle would then continue.

Information and knowledge should flow through the cycle to maintain continuity of the initial inception objectives. This can be done by setting up a project web site, as has been described by Murray and Lai (2001). An example of how safety and health management processes can be applied during the preliminary design phase will now be given.

12.6.4   The preliminary design and safety and health management

Safety and health management are essentially areas of knowledge where risks are identified, quantified and responded to by the development of a plan of action.

The detailed processes to be followed are similar to those used in the risk management of the project as a whole, namely:

- planning
- identification of risks
- qualitative analysis
- quantitative analysis
- response.

Ideally, project safety and health risks should be identified at the design stage of the project cycle as it will be possible to design the risks out of the project and/or identify them so that the subsequent project phases are fully aware of them and able to address them. This is an example of how knowledge should flow through the phases, either via the project web site or, more traditionally, via the tender and subsequent paper-based documentation.

During the preliminary design stage, a short-listed group of contractors may be involved in the preparation of the design concept to help eliminate risk by helping to develop a design which is constructible and safe and risk free.

Indicators would be used to highlight risks and the risks may be quantified by estimating the potential number of accidents which would statistically be expected to occur under a certain risk circumstance. The feasibility cost equation should now consider the life cycle cost instead of the more usual initial cost and would comprise the costs of:

- the professional team
- construction
- financing
- safety
  - insurances (workmen's compensation; contractors' all-risk, plant and equipment and civil risk)
  - legal (e.g. negligence)
  - re-work
- health
- operation over the project's lifetime
- demolition.

The usual cost/benefit analysis would be carried out using present values of future costs. The probability of risks becoming a reality would be assessed using probability theory.

As mentioned earlier it is now possible to design projects so as to minimise safety and health risks. Some examples of project innovation being used to mitigate safety risk will now be examined.

As has been described earlier, in the USA 23.8 % of the accidents which happened in 1999 were due to falls of one type or another. To avoid fall-type accidents in the UK some contractors now pre-assemble the whole roof of a traditional house and crane it into place.

Again in the USA, 58% of accident fatalities occur in specialised contractor organisations. There is preliminary evidence that small contractors in the USA are less regulated safety-wise than larger, more established contractors and that they also employ unskilled immigrant labour. It is believed that similar conditions may prevail in the UK among smaller contractors. Recent housing projects in the UK involve the prefabrication of housing units under factory conditions, where trained, skilled labour tends to be used, transporting them to site and assembling them by crane. This evidently reduces the risk of falls and the risk of accidents to untrained workers (Bagenholm, Yates, McAllister, 2002).

Another initiative consists of building office and apartment blocks using cast in-situ walls and pre-cast flooring. The walls are successively cast up to windowsill height, which provides a solid safety barrier for workers, again preventing falls.

Another example of changing the design of a facility to enable its assembly to be carried out more safely would be the pre-assembly of the steel-trussed roofs of large stadia on the ground and their jacking into place. During the construction of a roof at a Manchester soccer ground in 2001 two workers died from falls. (Mathiason, 2001)

Finally, as trenches are notoriously apt to collapse without prior warning, leading this type of accident to appear regularly high in the accident statistics, traditional trenching may in time be replaced by the use of thrust-boring or directional-drilling techniques.

Coble and Blatter (1999) have described the advantages of integrating the design-construction process and Coble and Haupt (1999) have examined partnering as a project-delivery system capable of improving construction safety. Hinze, Coble and Elliot (1999) have discussed the integration of construction worker protection into design and MacKenzie, Gibb and Bauchlagen (1999) have examined the communication of health and safety requirements in the design phase of projects. Finally, Casals, *et al.* (2001) recommended the integration of safety planning into the design phase of projects.

12.6.5   Safety and health management in phases of a project subsequent to the design phase

The philosophy of analysing the risks to workers safety and health (and to the projects' safety) would be carried through subsequent phases of the project life cycle. To provide further examples:

In the contractor-choice stage the owner may reject a contractor offering a low bid because it has a poor safety record and if it were appointed could, because of accidents, lead to delays in project completion. Completion of a recently concluded stadium in the USA was, for example, delayed for a period of six months by the collapse of a large crane.

During the construction phase the operation of cranes may be critically examined and rigorous rigging procedures prepared – cranes have a low factor of safety against both toppling and structural member failure and publications such as Engineering News Record have featured over recent years numerous examples of accidents involving large cranes and tower-cranes.

To address another type of risk, the owner may require the professional team to carry out risk management of the possibility of legionnaires' disease occurring or mould forming in a building during the operations phase of the project life cycle.

Finally, while construction processes are often carefully planned, demolition planning is usually left to the demolition contractor. With increased emphasis being placed on the recycling of materials, the project owner may require the project to be designed and planned rationally and safely disassembled at the end of the project life cycle to enable future income to be generated, by the recycling of materials.

12.6.6   Risk management

The carrying out of safety (and health) risk management exercises, even when they involve construction processes, does not obviate the need to carry out an overall project risk evaluation exercise. This would be carried out, for example, when a project is carried out in a country notorious for guerrilla activity or expatriate kidnapping, such as Colombia. This would be an example of a so-called country risk. Another example of a generic risk would be the widespread existence of land mines in a country which had recently been at war. Any marine work or heavy lifting operations would require close attention to risk as would tunnelling using NATM (New Austrian Tunnelling Method) which requires stringent safety monitoring and close designer-constructor integration.

Other areas of potential concern include the design and construction of bridges over trafficked waterways, where the collision of boats and barges with piers is not uncommon and the designs and operations of long vehicular tunnels - a number in Europe have experienced devastating fires in recent years.

12.7   ORGANISATION FOR IMPLEMENTATION OF C-PMBOK

To be successful, any safety and health management activity must have a person, or persons, responsible for its implementation and accountable for the process.

Traditionally, safety issues are the responsibility of a 'safety officer' who is usually allocated to a staff function. Health topics are often the responsibility of medical staff or even a 'safety and health officer', even though the knowledge required to manage the two topics is not common to the two areas.

In the writer's view, responsibility and accountability for the two areas should be invested in the line managers of the project. These would be supported by specialised staff where necessary.

To illustrate the concept, Figure 12.7 shows an organisation chart typical of a dam project constructed in a developing country.

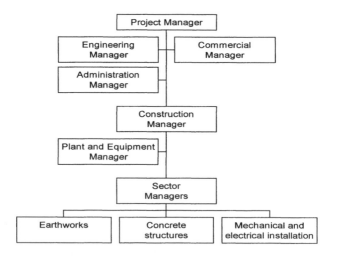

Figure 12.7 Organisation chart for a typical dam construction project.

As an example, the manager responsible for the construction of the concrete structures would be responsible for all safety issues in his/her sector, as well as for health matters related to the sector activity such as dust and heat exposure. The administrative manager, on the other hand, would support the line manager in his safety and health-related responsibilities by having safety equipment on site when needed, by organising induction courses, and by maintaining clean and healthy canteens, ablutions and accommodation. All costs related to health and safety matters related to the concrete structures sector would be debited to that sectors' cost plan as would be the cost of workers put on sick-leave following accidents or ill-health. In this way, accountability would be guaranteed.

Concerning health, the site sector manager would ensure that workers used dust masks, for example, while he or she would be supported in managing the health of the sector's workers by medical staff, subordinated to the administrative manager, who would screen the workers for diseases, such as TB, and carry out orientation on health practices, such as basic hygiene, STD and HIV-infection prevention and so on.

12.8   CONCLUSIONS AND RECOMMENDATIONS

Given the continuing high rate of accidents in the construction industry and the high risk of workers contracting ailments and diseases, especially in developing countries, it is recommended that safety and health management should be accorded the same level of importance as given to the more usually considered areas of time, cost and quality management. Indicators can be developed to alert construction managers to the existence of dangerous or unhealthy operations on a project-by-project basis. Risk management procedures could then be applied to the identified risky operations or unhealthy situations. A similar approach could be taken to the management of the social impact or environmental impact of construction operations on an existing society or environment. All of the above would be included in the owner or contractor's Plan of Action or Business Plan. The presence, or otherwise, of a competent Business Plan, developed at contract level, would help health and safety inspectors and others in evaluating the degree of attention and care given by site management to safety, health and other issues.

Further research needs to be carried out in quantifying safety and health indicators which would classify unsafe construction activities and unhealthy operations in projects carried out in developed and developing countries respectively.

12.9   REFERENCES

AMEC, 2000, *Safety, Health and Environment*, www.amec.com.

Atkinson, A.R., 2001, Corporative killing and the role of senior managers in accident prevention. In *Proceedings of the International Conference on Costs and Benefits Related to Quality and Safety and Health in Construction*, Barcelona, Spain, pp. 1-10.

Bagenholm, C., Yates, A. and McAllister, I., 2002, *Prefabricated Housing in the UK*, BRE Information Paper IP 16/01, Part 3.

Casals, M., Forcada, N., Peñaranda, F. and Roca, X., 2001, Economic Implications derived from accidents in construction. In *Proceedings of the International Conference on Costs and Benefits Related to Quality and Safety and Health in Construction*, Barcelona, Spain, 2001.

Coble, R.J. and Blatter, Jr. R.L., 1999, Concerns with safety in design-build process, American Society of Civil Engineers. *Journal of Architectural Engineering*, June, pp. 44-48.

Coble, R.J. and Haupt, T.C., 1999, Improving construction safety through partnering. In *Proceedings of CIB W55 and W65 Symposium on Customer Satisfaction*, Cape Town, South Africa.

DETR, 1999, *Rethinking Construction*, The Egan Report.

Engineering News Record, 2001a, Contractors face criminal charges in workers' deaths. *Engineering News Record*, USA, August 27, p. 7.

Engineering News Record, 2001b, Summit tackles work zone safety. *Engineering News Record*, USA, July 23, p. 10.

Engineering News Record, 2001c, Construction leads in deaths despite lower fatality rate. *Engineering News Record*, USA, August 27, p. 12.

Engineering News Record, 2001d, State farm (insurance company) pinpoints nation's most dangerous intersections. *Engineering News Record*, USA, July 9, p. 20.

Haupt, T.C., Hinze, J. and Coble, R.J., 2001, The loss of workers' compensation premiums: the case of Florida. In *Proceedings of the International Conference on Costs and Benefits Related to Quality and Safety and Health in Construction*, Barcelona, Spain.

Health and Safety Executive, 2000, *Press release CO24:00*, 7th June, Government sets targets to cut work-related deaths, accidents and illness, www.hse.gov.uk/press/c00024.htm.

Hinze, J., Coble, R.J. and Elliot, B.R., 1999, Integrating construction worker protection into project design. In *Proceedings of the Conference on Implementation of Safety and Health on Construction Sites* (Rotterdam: Balkema).

Josepth, A.J., 1999, Safety costs money and can save money. In *Proceedings of the Conference on Implementation of Safety and Health on Construction Sites* (Rotterdam: Balkema).

Koehn, E. and Reddy, S., 1999, Safety and construction in India. In *Proceedings of the Conference on Implementation of Safety and Health on Construction Sites*, (Rotterdam: Balkema).

Krueger, A.B. (Princeton University), 2000, Economic Scene; fewer workplace injuries and illnesses are adding to economic strength. *The New York Times*, Business/Financial section, 14th Sept.

Mathiason, N., 2001, Building death toll rises, *The Observer*, UK, 27 May, Work, p. 20.

MacKenzie, J., Gibb, A.G.F. and Bauchlagen., N.M., 1999, Communication of health and safety in the design phase. In *Proceedings of the Conference on Implementation of Safety and Health on Construction Sites* (Rotterdam: Balkema)

Murray, M., 2001, The influence of cultural differences on the performance of international contractors. *Proceeding of the Seventeenth Annual Conference of ARCOM, Association of Researchers in Construction Management*, University of Salford, UK, 1, pp. 101-110.

Murray, M. and Appiah-Baiden, J., 2000, Difficulties facing contractors from developing countries: problems and solutions. In *Proceedings of the 2nd International Conference of the CIB Task Group 29 (TG29) on Construction in Developing Countries*, Gabarone, Botswana, pp. 277-286.

Murray, M. and Appiah-Baiden, J., 2002, The influence of the HIV/Aids pandemic on the construction industry in sub-Saharan Africa. In *CIB W65 Symposium on Construction Innovation and Global Competitiveness*, USA, September.

Murray, M. and Lai, A., 2001, The management of construction projects using web sites. In *2nd Worldwide ECCE Symposium, Symposium Report: Information and Communication Technology (ICT) in the Practice of Building and Civil Engineering*, Espoo, Finland, pp. 163-168.

Murray, M. and Mavrokefalos, D., 2000, Opportunities for consultants from developing countries in the global village. In *Proceedings of the 2nd International Conference of the CIB Task Group 29 (TG29) on Construction in Developing Countries*, Gabarone, Botswana, pp. 306-315.

Murray, M., Nkado, R. and Lai, A., 2001, The integrated use of information and communication technology in the construction industry. In *Proceedings of the CIB W78 International Conference, IT in Construction in Africa*, Mpumalanga, South Africa, pp. 39/1-13.

New Civil Engineer International, 2000a, Heathrow Express report slates poor risk management, *New Civil Engineer International*, London, August, p. 5.

New Civil Engineer International, 2000b, Construction backs calls for EU-wide fine enforcement, *New Civil Engineer International*, London, March, p. 5.

Ngai, K.L. and Tang, S.L., 1999, Social costs of construction accidents in Hong Kong. In *Proceedings of the Conference on Implementation of Safety and Health on Construction Sites* (Rotterdam: Balkema).

Nunnally, S.W., 1998, *Construction Methods and Management*, 4th ed (Prentice Hall).

Nunnally, S.W., 2000, Safety and environmental health. *Managing Construction Equipment*, 2nd ed (Prentice Hall).

PMBOK 2000, *A Guide to the Project Management Body of Knowledge*, 2000 ed., Project Management Institute, USA.

Reid, J., 2001, High-risk industry baits kids, construction safety. *Engineering News Record*, Oct, 29, p. 75.

Särkilahti, P., 2001, Skanska OY's Integrated Management System. *Proceedings of the International Conference on Costs and Benefits Related to Quality and Safety and Health in Construction*, Barcelona, Spain.

Smallwood, J.J., 1999, The costs of accidents in the South African construction industry. In *Proceedings of the Conference on Implementation of Safety and Health on Construction Sites* (Rotterdam: Balkema).

Smith, J.M.C.K., 1996, The legal and cost implications of recent European Union Legislation relating to health and safety. In *Proceedings of the 1996 COBRA Conference*, RICS, UK.

Stein, R., 2001, Cape plc to cough up and pay. *M.Sc. Environmental Law Course, Part B*. University of the Witwatersrand, South Africa.

Tang, S.L., 2002, Financial and social costs of construction accidents 2002. In *Proceedings of the CIB W99 Conference on the Implementation of Safety and Health on Construction Sites*, Hong Kong, pp. 213-222.

Weber, C., 2001, Costs and benefits related to quality and safety and health in construction. In *Proceedings of the International Conference on Costs and Benefits Related to Quality and Safety and Health in Construction*, Barcelona, Spain.

CHAPTER 13

Integrating Safety into Production Planning and Control: An Empirical Study in the Refurbishment of an Industrial Building

**Saurin, T.A., Formoso, C.T., Guimarães, L.B.M.
and Soares, A.C.**

INTRODUCTION

In spite of the high costs of work accidents (Hinze, 1991), many construction companies in Brazil adopt as their only safety management strategy being in compliance with safety mandatory regulations. In Brazil, the main regulation related to construction industry is the NR-18 standard (Work Conditions and Environment in the Construction Industry). However, compliance with these standards might not be sufficient to guarantee excellence in safety performance, since they cover only minimal acceptable preventive actions. This statement is also applicable to regulations on safety and health management systems, such as OHSAS 18001 (Occupational Health and Safety Management Systems). Like quality systems standards, OHSAS 18001 also does not establish performance targets, but is concerned with the compliance to some managerial routines.

Safety planning appears as a core requirement at OHSAS 18001 as well as at NR-18. On one hand, OHSAS 18001 safety planning requirements consist of a risk management cycle, involving the continuous risk identification, evaluation and control. On the other hand, NR-18 (like similar regulations in many other countries) requires the elaboration of a safety and health plan, named PCMAT (Plan of Conditions and Work Conditions in the Construction Industry), which has a minimal mandatory scope. Since NR-18 was established in 1995, most companies produce the PCMAT only to avoid fines from governmental inspectors and do not effectively use it as a mechanism for managing site safety. Its main shortcomings are presented below:

- PCMAT implementation is usually regarded as an extra activity to managers, since it is not integrated to routine production management activities. NR-18 does not require its integration to other plans, except for site layout planning.

● Safety experts (usually external from the company) produce it and normally there is no involvement of production managers, representatives of subcontractors or workers.

● PCMAT does not usually take into account the uncertainty of construction projects. A fairly detailed plan is produced at the beginning of the construction stage and it is not usually updated during the production stage. The lack of update is caused by another common problem: the absence of formal control.

● PCMAT emphasises physical protection, normally neglecting the necessary managerial actions (for instance, implementing proactive performance indicators) for achieving a safe work environment; and

● PCMAT does not induce risk elimination through preventive measures at the design stage.

Such shortcomings in both conception and implementation of mandatory plans indicate that it is necessary to improve safety planning and control (SPC) methods, beyond what is required by the regulations. Some studies that have investigated causes of accidents and good practices to avoid them support this statement. Suraji and Duff (2000), for example, based on the analysis of approximately five hundred accidents in the UK, found that planning and control failures contributed to 45% of the accidents. Consistent with these findings, a study carried out at the Construction Industry Institute (Liska et al., 1993) found that, among several preventive actions that had been used by the industry, detailed safety planning was the most effective one towards achieving the zero accident target. However, that study did not provide details on how safety planning worked in the companies that were investigated.

The integration of safety into planning has been studied by a number of authors such as Ciribini and Rigamonti (1999) and Kartam (1997). However, those research studies have been limited to the introduction of safety measures into construction plans, using CPM or line of balance planning techniques. This approach tends to have not much impact, since it has been accepted that planning should not be limited to the application of techniques for generating plans. By contrast, planning should be treated as a broader managerial process composed by several stages, including data collection, implementation of corrective actions, and information diffusion (Laufer and Tucker, 1987). These stages, as well as some of the main requirements for effective production planning and control, such as hierarchical decision-making, cooperation, continuity and systemic view (Laufer et al., 1994), are also requirements for safety management. This indicates that there are several similarities between those two processes.

Thus, there seems to be an opportunity for improving SPC methods, based on concepts and principles that have been successfully used in production planning and control (Ballard, 2000; Hopp and Spearman, 1996; Laufer et al, 1994; Laufer and Tucker, 1987). This chapter reports on a model for SPC, integrated to the production planning and control process. The model was tested in an empirical study, which is reported in the following sections.

13.1   CONTEXT OF THE EMPIRICAL STUDY

13.1.1   Description of the site

The empirical study took place in the refurbishment of a steel mill building, in the Metropolitan Region of Porto Alegre, South of Brazil. The construction project was carried out by a small sized construction company, involving six main processes: (a) demolition and construction of the roof, (b) substitution of the windows, (c) demolition of external walls, (d) construction of brick walls, (e) wall plastering and painting, and (f) structural repairs in columns and beams. The peak number of workers in the site was around fifty.

The construction stage took around six months. The implementation of the model took place during the first four months of the project. As in most industrial refurbishment projects, the steel mill production was not interrupted to allow the construction project to be undertaken. In this situation, the steel mill health and safety risks also affected construction workers. The most important risks in the existing mill were: (a) noise level (110 dBA), specially near blast furnaces while these were working, (b) widespread dust, (c) heavy vehicle traffic, and (d) blast furnaces chemical reactions, which sometimes expelled some melt iron around them. Besides, the main steel mill building was relatively old, and had not been designed for performing maintenance and repairs easily. Such constraints made this a high risk project, especially in terms of safety issues. In order to comply with contractual requirements, the construction company assigned a safety specialist to work full time on the site.

13.1.2   Existing production planning and control process

The construction company had a fairly well-structured production planning and control system, based on the Last Planner Method (Ballard, 2000). There were four hierarchical planning and control levels: one-week or one-day short-term commitment planning, three-week look-ahead planning, and long-term planning, that was concerned with the whole construction stage.

At the short-term level, work packages were assigned to different gangs, and an initial negotiation concerned the release of working areas took place in weekly meetings. Due to the variability of the work environment, caused by the interference between the steel mill production and the construction process, weekly plans needed to be reevaluated in daily meetings. In such meetings, the final definition of working areas for each crew was made, and, based on that, the client provided work permits. The Percentage of Plans Completed (PPC) indicator (Ballard, 2000) was collected on both weekly and daily basis.

Regarding the look-ahead planning level, its main function was to support the removal of constraints related to work packages. A three-week plan was produced weekly, containing a list of constraints (e.g. space, materials, labour and equipment), and the deadline for its removal. Finally, the master plan, including the whole construction project was updated on a monthly basis. Based on the long-term plan, an initial resource schedule was produced this was updated through the look-ahead constraints analysis. The site layout was roughly defined before starting construction.

An action-research strategy was adopted in this project. The aim of the study was to devise and test the SPC model through its implementation in a real project. The construction company was particularly interested in successfully implementing SPC, since many of their clients have strict safety requirements. The implementation of the model in the construction project was co-ordinated by the contractor's quality manager. The stages of the empirical study are presented in following sections.

13.1.3   Integration to long-term planning

Long-term safety planning was carried out based on the construction stages established in the long-term production plans as a basis. For each construction stage (i.e. demolition and construction of the roof, replacement of the windows, etc.) a plan was produced using the technique of preliminary hazard analysis (PHA). PHA is a widely used tool in safety planning (Kolluru et al., 1996) dealing with three out of the four major stages of risk management: risk identification, risk evaluation and risk response. Risk monitoring is the fourth one. These four stages together correspond to the risk management cycle (Baker et al, 1999) that should be repeated throughout the project.

Although one safety plan has been made for each stage of the construction, this is not necessarily a general rule. If an activity takes place in different construction stages, it might be better to produce a specific plan for it. For instance, if welding performed several stages, it is easier to produce a specific safety plan for this activity instead of including its risks into several PHA. Besides the plans were directly related to productive activities, also two additional types of safety plans were produced: a plan for the risks in temporary facilities (lodges and bathrooms), and a plan for the risks in the circulation areas of the site the majority of the risks in the latest category were related to the steel mill operation.

Safety plans were produced by a working team, which was formed by the site manager, the safety specialist, a representative of the research team, subcontractors, and representatives of the steel mill management. The participation of different stakeholders was important, since they could provide different insights about risks, allowing safety plans to be more realistic and effective. The main steps for producing the safety plans are presented below:

(a) establish the necessary steps to carry out the activity: both conversion (for e.g. place bricks on the wall) and flow (for e.g. moving or storage of materials) should be considered, as suggested by Koskela (2000);

(b) identify the risks: this must include risks of any nature, i.e. health risks, ergonomic risks, environmental risks, etc. in each step of the process. This is a critical task, since if a risk is not identified it cannot be controlled. The effort to identify risks can be more successful if this activity uses supporting tools such as check-lists and brainstorming, as well as technical literature or plans from past projects (Baker *et al.*, 1999). In order to establish a common language for all plans, it is also helpful to adopt a risk classification (e.g. caught in, stuck by, etc.) at this stage. In addition, different stakeholders should be involved, as mentioned above, since they are all valuable sources of information;

(c) define how each risk will be controlled: considering that safety control will be based on what has been written down in the plans, it is important not to establish a control if there are no resources to apply them or if they are not considered to be necessary. Although the aim should be to eliminate all risks, such objective will be rarely possible and residual risks will always remain. The solution is to put such residual risks within an acceptable level. Managers are the ones who must decide what is acceptable or not (obviously, following regulations as minimal requirements). In this study, an informal risk evaluation procedure for establishing the magnitude of the safety measures was adopted. A formal risk evaluation procedure was considered not to be cost-effective at this stage, due to the subjectivity involved in this activity (Tah, 1997). A formal risk evaluation was only used to re-evaluate controls after near misses or accidents, as explained later in this paper.

Considering that construction is a dynamic environment, procedures were established to update safety plans when necessary. Such updates took place through the integration of safety planning and look-ahead and short-term production planning. At these planning levels, it is easier to anticipate risks and to evaluate the impact of the production methods on safety than at the long-term planning level.

13.1.4   Integration to look-ahead planning

Safety constraints were systematically included in the look-ahead constraints analysis, which took place on a three-week basis. At this level of planning, only the production manager, the safety specialist and a member of the researcher team were involved. Four types of safety constraints were identified: training, collective protections, design of safety facilities and space in the steel mill building. Personal Protective Equipment (PPE) was not considered as a constraint since this was readily available from the site warehouse.

Considering five weeks of constraints analysis, safety constraints represented, on average, 41% of all constraints. Safety related space constraints corresponded to 20%. In this way, safety constraints were made more visible in advance, avoiding stoppages of construction processes.

13.1.5   Integration to short-term planning

At this level, safety measures were discussed in both weekly and daily planning meetings. Such meetings were effective for discussing safety issues, since most key people involved in production management were involved. Production methods were discussed and detailed, providing an additional opportunity to evaluate their impact on safety. Also, the meetings were useful to report near misses or any other safety or production problem. In addition, safety and production performance indicators were presented and discussed.

If a new risk was identified or risk control measures were changed, this information was used to update the safety plans and retrain workers. Such changes were documented in a specific form (Table 13.1), and copies were distributed to the steel mill management and subcontractors, as attachments to the plans.

Table 13.1   Form to document changes in safety planning, for future updates

Data	Nº PHA	Risk	Control
01/03/29	PHA 06	Break energy cables during windows demolition *(the cables are inside the building)*	Remove windows from inside to outside

Short-term planning also provided an opportunity to apply one of the core techniques of the Last Planner method, shielding production. Such technique recommends that a work package must only be assigned if five quality requirements have been fulfilled: definition, soundness, sequence, size and learning (Ballard, 2000). In this study, safety was considered to be an additional requirement.

13.1.6   Safety control

The main performance indicator used to evaluate safety effectiveness is fairly similar to PPC (Percentage of Plans Completed). It is called PSW (Percentage of Safe Work Packages), indicating the percentage of work packages that were carried out 100% safe. A work package is considered 100% safe when all preventive measures planned have been implemented and when no accident, near miss or other unforeseen safety event has happened. The formulae used to calculate PSW is presented below:

$$PSW = \frac{\sum \text{number of work packages 100\% safe}}{\sum \text{total number of work packages}}$$

PSW emphasizes preventive measures, since it measures the degree of safety planning effectiveness, which may influence the probability of accidents to happen. The original reasons for not followings the plans are identified and classified according to a checklist of problems. In spite of its importance, calculating PSW is relatively time consuming, since some problems can only be identified by intensive observation of the site activities, if possible on a daily basis. Regarding PSW target, it is believed that it should be higher than the one pursued in PPC, since the lack of safety can lead to an accident (or even to a delay), while a delay in the schedule does not have such a serious consequence. Besides PSW, other performance indicators were also collected. Some of them were related to the impact of the lack of safety, such as the number of accidents and delays provoked by work stops caused by the lack of safety.

However, in this research project emphasis was given to controls that had a preventive character. In this respect, two proactive indicators were proposed: (a) the degree of compliance to NR-18, and (b) a training indicator, calculated through the ratio between man-hours of training and total man-hours. The documentation and investigation of all near misses was another important preventive measure. Such events were reported by safety specialists and by PSW observers. Ideally, workers should also report such events this could be achieved by training them to report near misses or unsafe conditions. Table 13.2 presents an example of a near miss investigation report.

Table 13.2 Example of near miss investigation report

N°	Description	Immediate causes	Corrective actions	Severity *	Probability **	Zone
3	Mortar recipient fell down from scaffold, due to a rolling bridge becoming stuck during the night	The recipient should be removed after the work has been finished	After work, all materials must be taken off scaffolds	Low	Remote	Green

* Severity if the accident takes place
** Probability of the accident to take place after the corrective actions have been implemented.

A review of safety controls took place after each near miss or accident. Such review was made through a subjective risk evaluation, adopting the risk zones presented in Table 13.3. Due to the difficult in obtaining reliable data to calculate probability and severity accurately, authors such as Tah (1997) support the use of subjective criteria to estimate these variables. Although this is not a key element of the model, it helps to perform a systematic analysis of near misses and accidents, and establish priorities in terms of corrective actions.

Saurin, Formoso, Guimarães and Soares

Table 13.3 Re-evaluation of controls after accidents or near misses

PROBABILITY	SEVERITY				
	Very high	High	Moderated	Low	Minor
Extremely remote			9		
Remote	6.7			3	5
Unlikely	1		10	4	2.8
Likely					
Frequent					

Risk not acceptable, controls must be improved
Residual risk is in an acceptable level
Residual risk is not meaningful

- The numbers in the cells correspond to the number of the accident or near miss

In order to evaluate the safety performance of the construction, a monthly meeting was carried out involving a director, the quality manager, the production manager and all the safety staff of the company. Such meetings were based on reports which presented performance indicator results.

13.1.7 Workers participation

Workers should take an active role in the SPC process, since they are its main customer. In order to get workers involved, the SPC model proposes a cycle of risk identification and control based on workers perceptions, as illustrated in Figure 13.1.

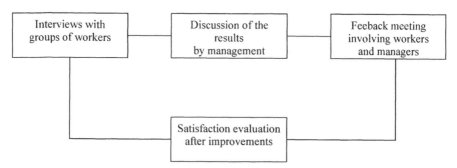

Figure 13.1 Risk identification and control cycle based on workers perceptions

The first stage involves interviews with small groups of workers. The interviews are divided in two stages: (a) an open section, in which workers are encouraged to talk about their work not only about the tasks performed by them, but mainly about both good and bad aspects of their work, and (b) an induced section in which workers are asked to talk about specific issues. In the latter section a check-list is used, including the following topics: material handling, awkward postures, PPE, workload, relationship with colleagues and managers, food, tools, the most difficult tasks, knowledge on the environmental risks, emergence procedures and temporary facilities. When a problem is reported, workers are asked to suggest ways to solve it.

The interviews have a role to identify new risks as well as to evaluate the effectiveness of existing controls. Regarding risk identification it must be pointed out that, through the interviews, workers are prone to point out risks that are related to organisational issues, such as job enrichment, rhythm or workload. According to the macro-ergonomic approach (Hendrick, 2001), such variables have a strong influence on safety, health and productivity performance.

The second stage consists of discussing the results of the interviews in a meeting involving project managers (including a company director). In this meeting the first draft of an action plan aiming to solve some of the problems reported by the workers is established. The third stage consists of a meeting involving both workers and management. The action plan is presented by the management and discussed with the workers. A justification is presented by the management if any of the demands by the workers have not been dealt with. The meeting is also another opportunity to report both new problems and suggestions and to solve communication gaps between managers and operatives. Finally, the fourth stage aims to evaluate workers' satisfaction after the improvements have (or have not) taken place. This evaluation is carried out in a group interview, in which new risks may be identified and controls are re-evaluated. No strict interval between interviews has been proposed in this study. However, new interviews should be carried out when new teams come into the site or after some substantial improvements have taken place.

13.2   SAFETY PERFORMANCE INDICATORS

Figure 13.2 presents the number of near misses and losses observed on site. In addition to accidents, there were three situations in which tasks did not take place due to safety failures. One of these situations illustrates the major importance of safety constraints. Crane maintenance personnel could not come into the steel mill when that machine broke down because they had not attended the safety training programme provided by the client. Since this training was provided only on Mondays, the maintenance personnel had to wait for a few days before coming into the steel mill.

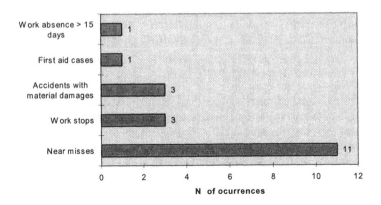

Figure 13.2 Losses and near misses due to lack of safety

The PSW indicator was collected in thirty-two working days, between 2nd of March and 1st of June. This sample corresponded to 40.5 % of the total number of working days in that period. Such data was collected by two researchers. The decision was made not to assign this task to the safety specialist in the research project, since his observation could be biased. Figure 13.3 compares the evolution of PSW and PPC in the empirical study.

PSW was on average 75% (S.D. = 16.5 %), while PPC was on average 65% (S.D. = 33.8%). Sometimes PPC was higher than PSW, while in other days PSW was higher than PPC. In one study carried out by Hinze and Parker (1978) safety and productivity were not in conflict, but appeared to be dependent on each other. These authors stated that superior production performance cannot be achieved in the absence of good safety performance. According to those findings, PSW and PPC should have a positive correlation. However, no statistical correlation was found between these two indicators (p-value = 0.7).

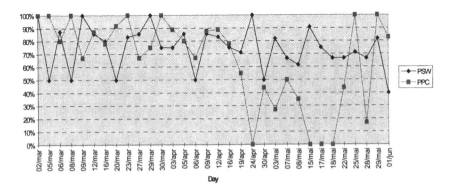

Figure 13.3 PSW and PPC results

Similarly to the Last Planner Method, the reasons for not following the plans were analysed. Figure 13.4 indicates that failures in collective protections planning were the main problem. Many problems have been included in this broad category, such as the lack of isolation under scaffoldings and the lack of supports on ladders. However, the main problem within this category (32% of the problems) was failure in the arrangement of cables to tie body harnesses. Often, such cables were not properly installed in both length and directions needed to allow workers to move easily move on the roof.

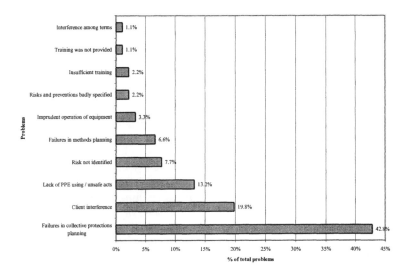

Figure 13.4 Reasons for not following safety planning

The category client interference involved situations in which the steel mill production did not allow the complete installation of the physical protections or when construction activities took place near a dangerous area inside the building. Although work permits were provided daily, it did not mean that steel mill activities were completely stopped in the construction area. Regarding unsafe acts, it must be pointed out that four workers were removed from the site by client representatives, because they were found working without body harnesses. However, this procedure was not fully effective since they did not carry out a detailed investigation to check why those workers were not using the harnesses.

Four evaluation meetings took place during the study. In the first two meetings, some participants adopted a reactive posture. Such posture can be exemplified by a statement of the safety specialist in one of the monthly meetings: 'accidents sound like incompetence of the safety specialist'. They often tried to blame client interference or unsafe acts as the source of safety failures. In fact, a similar attitude is often found on PPC analyses, when, for instance, rain or other uncontrollable factor are blamed for production failures. Photographs of the problems, pressure exerted by directors and repeated explanations that the objective of safety control is not to find 'who made the mistake', contributed to change this posture. However, the reactive postures avoid that some corrective actions were better implemented, so some problems remained almost the same during all the study.

13.3   SUMMARY OF WORKERS' PERCEPTIONS

The participatory cycle took place twice during the empirical study. The second round of interviews happened forty days after the first one. At this time, a new team had came into the site (painters), as well as others bricklayers had been hired. In addition, some of the suggestions given at the first round had already been implemented. The interviews involved groups of eight workers, on average, and all reports were recorded. The results of the first round are presented in Table 13.4.

Several of the complaints indicated that some risk controls were not being efficient. For instance, complaints 6 and 7 pointed out that other models of safety belts should be provided. Others complaints revealed some risks that had not been identified in safety planning. For instance, in problem 9 the subcontractors of structural repairs complained that they were assembling their own scaffoldings, although they had no previous experience in this task. In reality, the root cause for this problem was that the responsibility for scaffold assembling was not clearly defined in the contracts. Following the cycle proposed, an action plan was produced by the management. Table 13.5 illustrates one of the decisions included in the action plan.

Such action plan was presented and discussed in the feedback meeting involving workers and managers. The meeting took place in the site. The first participatory cycle ended with the second round of interviews, in which workers were asked to comment on the measures that had been carried out. Based on their evaluation, a new table was produced, indicating whether the complaint was totally solved, partially solved or not solved. A new improvement cycle is then started. Table 13.6 illustrates this evaluation.

Twenty-one new complaints were also reported in the second interview. This increase might be explained for two main reasons: painting team came into the site and workers were more confident to express their opinions. Also, it is normal that new demands are continuously appearing, due to the dynamic nature of construction sites.

Table 13.4   Problems according to workers (first round)

	Problems	MC*	SC1*	SC2*
1.	Widespread dust		x	
2.	Excessive noise		x	
3.	Heat on the roof, especially near blast furnaces			x
4.	Lodges are small and badly divided	x	x	x
5.	Lodges are untidy and dirty	x	x	
6.	Body harnesses do not fit properly		x	
7.	Body harnesses with two lanyards are necessary to work on scaffoldings		x	
8.	Rubber gloves have poor quality	x		
9.	Assembly of scaffoldings without previous experience		x	
10.	Lack of knowledge on risk areas in the steel mill (where am I not allowed to walk?)		x	x
11.	Access to the steel mill bathroom is risky		x	x
12.	The initial training provided by the steel mill to their subcontractors does not highlight the risks in the steel mill		x	
13.	Safety specialist spends too much time operating the crane. He should walk more around the site		x	
14.	There are not enough ladders to leave the roof in case of emergence or rain	x		x
	Equipment for horizontal transportation are not adequate	x		
15.	Horizontal transportation distances are too large (layout problems)	x		
17.	Some tools are not well maintained	x		
18.	Bad quality of food	x		

Notes:
* MC: main contractor (bricklaying, change of windows and general support);
  SC1: subcontractor 1 (structural repairs); SC2: subcontractor 2 (change of roof).

Table 13.5   Example of action plan to attend workers' demands

Problems	Action	Responsible	Deadline
Lodges are untidy and dirty	Assign a labourer to carry out a daily cleaning	Safety specialist	03/22

Table 13.6   Example of workers' opinions on the solving of their complaints

	Solved?	New suggestions or comments
Lodges are untidy and dirty	Partially	The worker assigned to clean lodges is often required to carry out other tasks

13.4   CONCLUSIONS

This chapter has presented a safety planning and control model (SPC) which was tested in an empirical study in the refurbishment of an industrial building. The technique of preliminary hazard analysis (PHA) was used for producing basic safety plans, which were continuously updated through the integration of safety into look-ahead and short-term level of planning. It was found that several concepts and methods of production planning and control (such as constraints analysis, shielding production or analysis of causes for not following the plans) can be easily extended to safety management. Although this integration makes it easier to manage both safety and production, its effectiveness depends on the existence of a well-structured planning system.

Two elements of the model must be emphasised. The first is the percentage of safe work packages (PSW), a new indicator to evaluate safety performance. By collecting PSW, it is possible to carry out an objective and detailed inspection on the site, using safety plans as references to the control. The second is the participatory cycle, a method to identify and control risks based on workers perceptions.

13.5   REFERENCES

Ballard, G., 2000, *The Last Planner System of Production Control*, PhD Thesis, School of Civil Engineering, The University of Birmingham.

Baker, S., Ponniah, D. and Smith, S., 1999, Risk response techniques employed currently for major projects. *Construction Management and Economics*, 17, **2**, pp. 205-213.Ciribini, A. and Rigamonti, G., 1999, Time/space chart drawings techniques for the safety management. In *Proceedings of International Conference of CIB Working Commission W99*, Haiwaii, pp. 25-32 (Rotterdam: A.A. Balkema).

Hendrick, H., 2001, An introduction to work system design. Santa Monica. *Macroergonomics*, Human Factors and Ergonomics Society.

Hinze, J., 1991, *Indirect Costs of Construction Accidents* (Austin: The Construction Industry Institute).

Hinze, J. and Parker, H., 1978, Safety: productivity and job pressures. *Journal of the Construction Division*, 104, **1**, pp. 27-34.

Hopp, W. and Spearman, M., 1996, Foundations of manufacturing management. *Factory Physics*, (Boston: McGraw-Hill), 668pp.

Kartam, N., 1997, Integrating safety and health performance into construction CPM. *Journal of Construction Engineering and Management*, June, 123, **2**, pp. 121-126

Kolluru, R.V., (ed.), 1996, *Risk Assessment and Management Handbook* (McGraw Hill).

Koskela, L., 2000, *An Exploration towards a Production Theory and Its Application to Construction*, VTT, Technical Research Centre of Finland, 258 pp.

Laufer, A. *et al.*, 1994, The multiplicity concept in construction project planning. *Construction Management and Economics*, London, 12, **1**, pp. 53-65.

Laufer, A. and Tucker, R.L., 1987, Is construction planning really doing its job? A critical examination of focus, role and process. *Construction Management and Economics*, London, 5, pp. 243-266.

Liska, R.W., (ed.), 1993, *Zero Accident Techniques* (Austin: The Construction Industry Institute).

Suraji, A. and Duff, R., 2000, Construction management actions: a stimulation of construction accident causation. In *Proceedings of Association of Researchers in Construction Management*, 16, Glasgow, www.eci-online.org.

Tah, J., 1997, Towards a qualitative risk assessment framework for construction projects. In *Proceedings of IPMA Symposium on Project Management*, Helsinki, pp. 265-274 (London: E&FN SPON).

13.6   ACKNOWLEDGEMENTS

The authors are thankful to Brazilian agencies CAPES (Coordenação de Aperfeiçoamento de Pessoal de Nível Superior) and FAPERGS (Fundação de Amparo à Pesquisa do Estado do Rio Grande do Sul) for the funding provided to this research project.

CHAPTER 14

Management of Construction Site Safety during Projects: The Use of '5s' System

Yang, X., Zhang, L.Y. and Zhang, J.

14.1   BACKGROUND

In the UK the Health and Safety Commission (HSC) produces annual reports on accident statistics throughout the industry. The surveys show that construction is one of the most dangerous industries with 581 fatal accidents between 1986 and 1992. There is no doubt that maximum effort is required to significantly reduce these statistics.   One particularly worrying observation is that almost identical situations continue to cause death and injury. The construction industry seems unable to learn from its past mistakes. There is a tendency to blame external factors for the poor safety record. Factors (The Establishment and Implement of Assurance System for Safety Producing in Construction Site, 1999) such as:

- The transient nature of the industry.
- The complete disregard for safety of many of its employees.
- The need to use a partially completed permanent structure, or a regularly changing temporary platform, to access works at a higher level.
- The constantly changing hazards as the project is constructed.
- In general every corporation has its own safety policy. A corporate safety policy will include.
- Policy statement: states what the organisation will do.
- Operation of the policy: explains how the organisation will ensure that the policy is adhered to.
- Organisation: states who is responsible for safety at different levels of the organization; explains how the policy affects departments, sections or projects within the organisation.
- Communication: explains how the policy will be communicated throughout the organisation, and how senior management will be briefed on its implementation.

There is a danger that, once produced, the safety policy is filed out of sight. The safety policy must be a working document, used regularly to help monitor site practice. If practice begins to ignore or deviate from the policy, the main contractor must act immediately.

14.2   THE MAIN CONTRACTOR'S ROLE IN MAKING PROJECTS SAFER

In 1988, the UK's HSC published a guide to managing health and safety. The guide described (*Health and Safety Commission Annual Report*, 1992) the roles and responsibilities of the various parties under the following types of construction contract; management contracting; design and manage; construction management.

The main contractor's responsibilities for managing health and safety (Project Management Institute (1996) include:

- to help the design team identify information on major health and safety matters which will be passed on to the package contractors;
- to contribute construction and health and safety expertise to the design team. This should lead to design features which are easier and safer to construct;
- to identify high-risk activities during the pre-contract planning stage and to formulate site-wide method statements for dealing with them. These will be included in the contract documents for the package contractors. High-risk activities will differ from contract to contract, but are likely to include: confined spaces and excavation; demolition including asbestos removal; cranes and hoists; scofflding and temporary works; multi-storey frames; cladding and roofing; hot-works (welding/buming/cutting, etc.); substances hazardous to health;
- to identify essential, separately priceable health and safety items (for example, access scaffolding, edge protection and welfare facilities) which could be included in contracts with package contractors;
- to shortlist package contractors who will be invited to tender, taking account of their skills in managing health and safety;
- to produce a site safety policy that includes rules and conditions, procedures, guidance notes and codes of practice. The policy should incorporate client requirements, where appropriate, and be included in the contract documents for the package contractors;
- to set up the site organisation for the management of health and safety taking into account the following factors:

    (1)   overall programme for the project;
    (2)   planned procedures;

(3)   arrangements for coordination, liaison and communication;

(4)   safety representatives' functions and arrangements for joint consultation;

(5)   arrangements for monitoring site health and safety;

(6)   arrangements for training, instruction and information;

(7)   policy on the use of common facilities, plant and equipment;

(8)   arrangements for record-keeping and statutory examination;

(9)   extern liaison;

(10)   responsibilities of package contractors;

(11)   responsibilities of individuals.

- to ensure that package contractors are briefed about anticipated construction method, site/design factors, relevant hazards, precautions, general site safety rules and conditions, and are clear about divisions of responsibility. Similarly the package contractors should inform the major contractor, and interfacing package contractors, about possible hazards arising from their own activities;

- to ensure that package contractors have made plans to work safely, have priced their bids accordingly and have the necessary resources. Each package contractor should produce a contract-specific safety policy;

- to ensure that package contractors produce detailed method statements for high risk activities, to monitor the package contractor's performance against the method statements and take action where necessary. It is good practice to consider safety as the first item on the agenda of the regular package progress meetings;

- to manage health and safety on site by coordinating activities, ensuring that planned procedures are implemented and monitoring performance so that revised arrangements can be made as necessary. The major contractor should ensure that he does not become remote from day-to-day problems on site;

- to consider the creation of a joint safety committee operating on a site-wide basis and involving representatives of management and operatives from all package contractors;

- to convene regular, site-wide coordination meetings, attended by both the major contractor's staff and each package contractor's site management. Safety is one of the key aspects of coordination;

- to make site-wide arrangements for emergencies, fire prevention, safe access, lighting, etc.

However, the advice given is applicable to all major construction projects, especially those where the major contractor has some influence over detailed design and employs package contractors to construct some, or all, of the project.

14.3   '5S' SYSTEM

14.3.1   The source of '5S'

'5S' is a series of management activities canonised in Japan. It includes: tidy (seiri), place (seiton), clean (seiso), clear (seiketsu), attain (shitsuke). The five items begin with 's' in Roman pronunciation of Japanese, so they are called '5S'. '5S' is so simple and clear that employees can understand it as they look at this. The purpose of '5S' is to create a clean, comfortable, suitable and tidy environment for the employees.

14.3.2   The three sections of '5S'

The activities of '5S' are implemented one by one and can be divided into three sections:

1. Firstly creating a structured process for project. '5S' changes the way of people's behaviour, so it is very important to train each person to be responsible.
2. Secondly, creating a clean environment, that is cleaning every nook and cranny, getting rid of dust to make the site looks brand-new and spotless. This is a revolution of the mind.
3. Lastly creating a visual approach to management of the project. Observe with the eyes to discover where the abnormality is, to help each man finish his work and to avoid making mistakes. This is the standardisation of '5S'.

*   Tidy (seiri)
    1. Definition: to distinguish whether the object is necessary or not and get rid of the unnecessary.
    2. Purpose: to make space for other objects.
    3. Essentials: to examine all work areas including the visibile and invisible, establish the criterion to justify the necessary and unnecessary, get rid of the unnecessary objects, investigate the utility frequency of the necessary and determine everyday usage, formulate the way of dealing with the castoff, and examine all things every day.
*   Place (seiton)
    1. Definition: to place the necessary according to the rule to allow easy identification.
    2. Purpose: orderliness and do not waste time to look for something.

3. Essentials: arrange the flow and determine the area of placing, rules about placement, and identify the objects there (the emphasis on visual management).
4. Emphasis: an orderly state so that anyone can find the objects in need at once; make it explicit that something should be somewhere; as for location and placed objects, all measures should be taken to use them immediately.

- Clean (seiso)
  1. Definition: to get rid of the dust and dirt in the concourse and prevent pollution occurring.
  2. Purpose: to eliminate the 'dust' to keep the concourse clean and bright.
  3. Essentials: to set up the area for cleaning (indoor and outdoor), execute and routinely sweep and clean up the dust, investigate the source of pollution and stop it, establish the benchmark of cleaning and look at it as a criterion.
  4. Emphasis: make the concourse free of dust and pollution. After the tidying and placing, the objects can be made available immediately but can they be used normally? The first purpose of cleaning is so that the equipment can be used normally.

- Clear (seiketsu)
  1. Definition: to institute and standardise the activities of the first 3S, fulfil and maintain production.
  2. Purpose: to keep the production system moving and show where the 'abnormality' is.
  3. Essentials: to fulfil the first 3S; formulate the benchmark of visual management and colour management; formulate the way of auditing; formulate the system of rewards and punishment and execute it seriously; keep in mind the '5S'; the senior director should usually take the initiative to go on a tour of inspection.

- Attainment (shitsuke)
  1. Definition: everybody performs in accordance with the rules and adopts good practice.
  2. Purpose: to change people's attitudes and develop the custom of serious work.
  3. Essentials: to keep on promoting the first 4S; formulate the related rule and regulation and abide by them; educate and train; promote all kinds of spirit activities (morning meeting, smile sports, etc.).
  4. Emphasis: attainment is not only the last result but also the ultimate purpose aspired by director in the enterprise. If every employee adopts good practice and abides by the rule, the work order can be fulfiled, the workers can be united, and all activities can be fulfilled.

- '5S' and safety

The safety committee should meet regularly and issue safety directives because safety is the most important element of the project. If there is no exhortation every day safety will be neglected (*Total Project Management of Construction Safety, Health and Environment, 1993*). A project which fulfils 5S completely (Shaoxiong Shun, 2000) will be well run, so it is usually said that '5S' is the mother of the safety'.

14.3.3   The virtue of fulfilling the 5S

The achievements of '5S' are significant in preventing accidents; be good to tidy the site, steady and raise the qualification; make the flow (Morris and Hough, 1987) orderly, faults discovered easily and efficiency improved; the work space can be increased and used effectively; reduce the damage of the things; in favour of health and sanitation; in favour of forming comfortable air; make the concourse bright and clean; conducive to preventing fire.

The emphasis of work safely

(1) work clothes
    • keep clean usually;
    • if the fastener fall off or damage;
    • if the heel of work shoes disrupted, bootlace loose;
    • put on the safety helmet in the appointed work;
(2) the things for safeguard (glasses, respirator, earplug, and safety helmet etc.)
    • put on glasses and earplug;
    • put on respirator correctly, not chin;
    • filter things;
    • get store at any time;
    • determine the spot for storage and keep correctly.
(3) passage
    • place anything in the passage;
    • if there is any accident and hazard;
    • if sprinkle the oil or water;
    • if there is any dangerous situation that the wire and tube cross the passage;
    • determine the height and width of the entrance.
(4) the floor of the concourse
    • if there is any unnecessary appliance;
    • assure wire and tube safety;
    • keep the sweep appliance and store in directed place;

- do not put anything nearby the switch, hydrant, fireproof equipments, and safety gate;
- if any problem about tidiness and lighting.

(5) organic acid and tinder (Adapter from *Managing Health and Safety in Construction: Management Contracting*, 1998)

- mark the content and functionary of the things at the custodial place, and tidy them by class;
- prevent evaporation by the lid of vessels, avoid overfalling and leaking;
- ensure the exchange and discharge equipment run normally in the work concourse;
- examine the keeping warehouse termly, prepare the hydrant and sign 'fire forbidden';
- keep the scrap cotton yarn in the fireproof vessel;
- if oil is spilled it is cleaned up immediately;
- survey if there is any fire or fumes nearby;
- do not execute the work with fire (belt etc.) near the tinder.

The emphasis of visiting and examining at concourse

The safety guideline (Cooper, 1993) should be fulfilled and strengthen everyone' attitude to safety:

(1) to make the safety sign board, the safety record, and calendar;
(2) plan the safety advertisement and selves' purpose;
(3) sign the items that can almost become accident and constitute the examination team.

Master the affairs to resolve hazard problems and include:

(1) to illuminate what degree has the action reached;
(2) ensure to use the safety clamp practically and determine the spot for placement obviously;
(3) examine every team, patrol and record the items that may be accidents.

Confirm if there is any dangerous state includes: to take the measures (Major Project Association Beyond 2000, 1992) that settle the equipments' safety cover, interlock and transducer etc., consummate the safety installation of equipment, especially notice the temperature of drier, manipulator and auto running equipment.

14.4   CONCLUSIONS

As the main contractor, it is necessary that much attention should be paid to the construction site safety. Though by now 5S system has not been applied in the field of construction, in other fields it shows very strong functionality, especially in Japan it is usually used by many directors of enterprise. Through the discussion in this paper, 5S system must become one of the most effective means to resolve the problem of site safety.

14.5   REFERENCES

Managing Health and Safety in Construction: Management Contracting, 1988, Construction Industry Advisory Committee (London: HMSO).

Cooper, K.G., 1993, The rework cycle: Benchmarks for the project manager. *Project Management Journal*, 24, **1**, pp. 100-111.

Health and Safety Commission Annual Report (UK), 1991/1992, 1992 (London: HMSO).

Levitt, A.M. and Parker, H.W., 1976, Reducing construction accidents: Top management's role. *Journal of the Construction Division, ASCE*, 102, **3**, September.

Major Project Association Beyond 2000, 1992, *A Source Book for Major Projects*. Major Projects Association, Oxford, UK.

Morris, P.W.G. and Hough, G.H., 1987, *The Anatomy of Major Projects*, UK (Chichester: John Wiley).

Project Management Institute, 1996, *A Guide to the Project Management Book of Knowledge*, Project Management Institute, Upper Darby, PA, US.

Shaoxiong Shun, *How to Push 5S*, 2000, (Xiamen: The Press in Xiamen University.

The Establishment and Implement of Assurance System for Safety Producing in Construction Site, 1999, *The Superviser Seminar of Project Construction in Shanghai* (Beijing: The Press of Construction Industry in China).

Total Project Management of Construction Safety, Health and Environment, 1993, ISBN 0-7277-1923-8 (London: Thomas Telford Services Ltd).

Section 3

Health Issues

CHAPTER 15

Health: The Poor Relation in Health and Safety?

Steve Rowlinson

This chapter reviews the current situation with regards to health in the construction industry and highlights significant failings in terms of health provision. The magnitude of the problem is explored and key issues identified. The following sections provide a brief overview of some of the health issues arising in the construction industry.

Occupational health hazards are common to all industries and can be particularly acute in the construction industry. However, we rarely see discussions in the press and programmes on TV relating to occupational health. Occupational accidents are commonly reported and their seriousness is obvious but the losses due to occupational health problems are at least as great as those stemming from occupational accidents. Common health issues include decompression sickness, tendonitis of hand or forearm, inflammation of skin, inflammation of mucous membrane, gas poisoning, silicosis, asbestos - related diseases and contact dermatitis. Every year many thousands of construction workers suffer from work-related ill health. Gibb, Haslam and Gyi (2001) report that in 1995, the UK's 1995 Self-reported Work-related Illness Survey found an estimated 134,000 construction-related workers reported a health problem caused by their work, resulting in an estimated 1.2 million days lost in a work-force of 1.5 million. The construction sector prevalence rate for self-reported work-related illness was 6.8% compared with an all-industry average of 4.7%. In particular for construction, asbestos related disease resulted in about 700 deaths annually; there were 96,000 musculoskeletal disorder cases; 15,000 respiratory disease cases; 6,000 cases of skin disease and 5,000 noise induced hearing loss cases. Recent research also indicates that construction workers are particularly at risk from hand-arm vibration although this was not evident in the 1995 figures. Many of these conditions have a long latency period between exposure and onset of symptoms, leading to difficulties attributing the root causes of the ill health condition.

The UK Government estimates that asbestos-related diseases will kill up to 10,000 people each year by the year 2020 (Croner, 1998) and 36,000 people will suffer from 'vibration white finger' with over a million considered to be at risk (HSE, 1998). UK and US construction workers have the third highest rate of skin disorders (Burkhart, *et al.*, 1993) and cement is believed to cause up to 40% of the UK's occupational dermatitis (Beck, 1990). The problem is just as bad elsewhere in Europe. For example, in 1993, the 2000 largest Portuguese enterprises lost more than 7.7 million working days as a result of illness, representing 4.5% of all working days for these companies (Gründemann and van Vuuren, 1998). 20% of sickness absence among 15 to 66 year-olds in Denmark is caused by the working environment (Gründemann & van Vuuren, 1998). Dermatitis is becoming the main reason for absence from work in Germany (Berger, 1998).

15.1   RISKS FROM CONSTRUCTION OPERATIONS

Traditionally, we classify health hazards as being chemical, biological and physical. However, we should also consider the effects of the weather, particularly in a place such as Hong Kong which has temperatures in excess of 35 degrees Centigrade in the summer and humidity approaching 99%. Physical and mental stresses are also additional health hazards and should be recognised as such along with other environmental hazards apart from the weather. It is not always obvious as to how a health hazard should be classified. King and Hudson (1985) discuss the hazard presented by fine sawdust in the air, which can cause nasal cancer, and ask the question should this be classed as a chemical or a biological hazard. They then go on to discuss the synergistic effect of different hazards and use the example of a mixture of benzene and carbon tetrachloride vapour. When mixed these chemicals are much more dangerous than when exposure is to each individually as carbon tetrachloride reduces the capacity of the liver to convert benzene into excretable products. Most health hazards take time to impact upon those who are exposed to them but some hazards have an immediate effect, such as being close to a loud explosion which can cause immediate loss of hearing or tinitis. Exposure to carbon monoxide is a common and highly dangerous form of health hazard which can have an almost immediate effect.

Construction workers are exposed to health hazards from the materials that they use on a daily basis. The most obvious hazard for most construction workers is cement which contains chromium and other constituents which are responsible for dermatitis. The continued use of hand-held equipment is another regular exposure hazard for construction workers which can lead to serious numbness developing in the joints and extremities, Raynaud's disease. These issues are dealt with in more detail later in this chapter.

15.2   HEALTH MANAGEMENT

The issue of managing health in the construction industry has been poorly dealt with to date. Few construction companies commit the level of resources to health management that they do to safety management. When comparing the levels of commitment to major manufacturing companies then the construction industry's performance can be described as very poor. In order to deal with health issues construction companies need to consult at least, or preferably employ, occupational health professionals. Typical professionals required are:

- Occupational health physician
- Occupational hygienist
- Other specialists including:
  - Epidemiologist
  - Dermatologist
  - Ergonomist
  - Industrial psychologist
  - Occupational nurse.

15.2.1   Occupational health professionals

It is obvious from this list that most small construction companies would not have the resources to employ such a range of staff; they find it difficult to employ an adequate number of construction professionals let alone the occupational health professionals. On this basis it would be sensible for such organisations, and many larger organisations also, to retain a number of consultants in these areas and work together to retain these on a group basis where a number of firms call on the services of these professionals on a regular or irregular basis.

- Occupational health physician
  The role which the occupational physician can play is in dealing with issues such as:
  - Examination of people on first entering employment.
  - Regular examination of workers at work.
  - xamination of workers exposed to more extreme than normal hazards.
  - The design of health programmes and campaigns.
  - The investigation of industrial disease incidents.
- The occupational hygienist
  The occupational hygienist is concerned with recognising, evaluating and controlling those environmental factors which may cause sickness or

impair health or cause discomfort and inefficiency amongst the workforce and which come about in the workplace. Occupational hygienists usually have a qualification in a discipline such as engineering or chemistry and have been further trained, by way of an MSc or other course, within the field of industrial or occupational hygiene.

The occupational hygienist is expected to be able to recognise and evaluate hazards in the workplace and to devise methods to eliminate, control or reduce these hazards. Typical hazards that have to be dealt with are:

- Chemical hazards such as liquids, vapours, gases, etc.
- Physical hazards such as ionising radiation, noise, vibration, pressure, etc.
- Biological hazards such as moulds, fungi, bacteria, viruses, etc.
- Ergonomic hazards which relate to the workers, body position, repetitive tasks, work place design, etc.

Hence, it is important that the occupational hygienist has an understanding of both the construction process and its operations and the whole range of hazards which may manifest themselves during these operations.

- Epidemiologist
  The epidemiologist's task is to quantify the hazards that a worker is exposed to and determine the causes of any physical ailments which emanate from these exposures.
- Dermatologist
  The dermatologist is concerned with treating and preventing skin diseases such as allergic reactions, dermatitis and other skin disorders associated with continuous exposure to allergens or hazardous substances.
- Ergonomist
  The ergonomist deals with the relationship between the worker and the workplace. The aim of ergonomics is to match tools and working positions and controls to the physical capabilities and characteristics of the worker.
- Industrial psychologist
  The industrial psychologist is concerned with human relations at work and with the mental well-being of individual workers and managers. The role of the industrial psychologist runs the whole gamut from staff selection, job analysis and motivation to job enrichment and counselling.

15.2.2   Health education in construction

• Craft education
  With the demise of apprenticeship schemes many construction workers
  now have little or no craft education. The situation is exacerbated by the
  fact that many processes are now highly automated or based on precast
  construction technology. Given these circumstances it is not surprising
  then that many construction workers are ill prepared for the hazards
  which occur during the building process. As a consequence, it is
  necessary for workers to be periodically trained and re-trained so that
  they are aware of the technologies that they use and of the risks that are
  associated with these technologies.

• Health education
  Health education cannot be divorced completely from craft education
  and training. Many construction workers with poor craft backgrounds put
  themselves into dangerous situations because of the lack of craft
  knowledge in education and training. As a consequence, it is not possible
  to instruct workers in terms of safety education without referring to the
  physical circumstances and skills needed to undertake the various tasks.
  Hence, the incidence of occupational ill health as well as occupational
  accidents is highly dependent on educating and training workers in both
  skills and safety awareness. These two issues cannot be separated.

15.3   SPECIFIC HEALTH ISSUES

15.3.1   Skin cancer

Research in the construction industry has focused little on exposure to adverse weather
conditions. Excessive exposure to the sun and its ultraviolet radiation is an important
risk fast factor in the development of skin cancer. Estimates indicate that the incidence
of skin cancer is doubling every decade. Outdoor workers, such as construction
workers, are potential victims of skin cancers, non-melanoma skin cancers, and yet
this is an area where a health intervention can easily be undertaken. Non-melanoma
skin cancers are associated with the long-term exposure to sunlight and the body parts
most commonly affected are, of course, the face hands legs and torso. These skin
cancers take two forms: basal cell cancer accounting for 70% of all lesions with
squelmost cell carsinomas accounting for the majority of the rest. The risk factor from
exposure to the sun depends on skin type and skin types have been classified on a
scale of 1 to 6. Type 1 is fair skin that never tans and always burns whereas type 6 is

genetically black skin. Those with the skin types 1 and 2 will burn readily after 20 to 40 minutes exposure to strong sunshine. It has been found that construction workers are exposed to synergistic effects when it comes to cancer exposure. Site materials such as Polycyclical hydrocarbons, which are found in oils and tars, not only act as carcinogens themselves but also increase photosensitisation of the skin.

In order to combat the risk of exposure to sunlight and so the risk of cancer development there are a number of possible actions which can be taken. Obviously, the use of shaded areas is important. The use of personal protective clothing is important and the covering up of the torso, arms, legs and neck is an essential precaution. Sunscreen with a protection factor of 15+ should be used. However, this is only a final line of defence. The Australians refer to skin protection by 'slip, slop, slap' - slip on a shirt, slop on sun screen and slap on a hat.

Given the nature of the construction worker health education, interventions are particularly important in dealing with this issue. If workers do not recognise the hazards of exposure which come from the work they will not make any attempt to prevent the exposure. The transient nature of construction workers is one of the main problems in developing health promotion education strategies. A large proportion of construction workers are unlikely to participate in such initiatives.

15.3.2   Work method induced illness

15.3.2.1 Carpal tunnel syndrome

There is evidence from the United States that many hand and wrist injuries in the construction injury industry are in fact attributable to carpel tunnel syndrome (CTS). Carpel tunnel syndrome is a neurological disorder that results from compression of the median nerve of the wrist which in turn leads to tingling sensations and a numbness in the fingers, gradual weakening of the grip of the hand and eventually permanent disability. The disorder is associated with a highly repetitive work tasks and the excessive use of force in the hands and wrists. A study in the United States looked at the occurrence of CTS among construction apprentices. The study indicated that more than 10% of young workers had CTS. They noted that the majority of those suffering from CTS had not sought medical attention. One of the problems with this situation was that workers felt that the disorder was 'just part of their job' and were also afraid of employer reprisal should they report the syndrome. It is essential with CTS that the problem is recognised early so that treatment can lead to recovery and prevention of permanent disability. Also, workers should have their tasks modified to reduce exposure to the movements that will aggravate CTS.

15.3.2.2 Raynaud's disease

Mechanical vibration is a hazard to which many workers are exposed on construction sites. The vibration may be a whole body vibration, such as when a worker is seated in the cab of a vehicle, or much more focused vibrations such as hand and arm vibration which comes from holding power tools for long periods of time. The impact of such vibrations takes some period of time to develop but typically drivers of diesel engined vehicles may suffer back problems and excessive vibration can damage the internal organs, such as the lungs. The issue for drivers is really a function of design of the cab and of the seats on which the driver works. Modern technology has gone a long way to improving the situation for drivers.

Hand and arm vibration is a different matter however. Vibrating, rotary and reciprocating hand-held tools, such as a pneumatic drills and concrete vibrators, cause vibration induced injuries. Examples of these are:

Raynaud's disease, also known as a white fingers or dead fingers, which results in the loss of circulation to the extremities, the fingers, and can in severe cases lead to the need for amputation. This phenomenon can also be associated with cold weather.

Osteoarthniritis, or arthritis of the joints, particularly the elbows.

Hardening of the palms of the hands which are normally soft tissues.

15.3.2.3 Decalcification of the bones

The problems associated with vibration from tools can be dealt with by limiting work periods and using specially designed tools which minimise vibration. Shock absorbers and isolation of handles is one way of reducing exposure. King and Hudson point out that gloves are of little use in reducing hand vibration but they are helpful in dealing with problems such as Raynaud's disease where the problem is exacerbated by cold hands and reduced circulation of the blood.

15.3.2.4 Silicosis

Silicosis is a serious problem in the construction industry and it is caused by crystalline silica dust entering the lungs. Concrete and other masonry which is sawn, cut or hammered leads to the development of high concentrations of crystallised silica in the air. These dust concentrations need to be controlled if exposure is to be limited to acceptable levels. Silicosis is basically permanent lung damage and it is a disabling illness. In Hong Kong the practice of constructing hand-dug caissons lead to high incidences of silicosis and pneumoconiosis amongst the workers. This practice has now been banned in all but the most exceptional of circumstances.

Simple measures can be undertaken to reduce the risk of exposure to silica in the air. Simply providing water to saw blades and breaking equipment will help to reduce the exposure through reducing dust concentrations in the air. Saws can be fitted with extractors or vented to keep the dust away from the worker. In addition, respiratory apparatus can be provided but, as with all personal protective equipment, this should not be the primary means of preventing exposure to the airborne silica. Exhaust ventilation, wet (water flow) methods and enclosed systems are the more sensible and effective approaches to reducing exposure to silica.

15.3.2.5 Asbestos

Asbestos has been used for many years in the building and construction industries. Blue and brown asbestos are no longer used in most countries but white asbestos is used and it can still create problems. Asbestos is used in roofing, cladding, thermal and acoustic installation and fire resistance areas. Asbestos boards can release considerable quantities of fibres. One of the problems with asbestos is that it can be damaged or adapted during its lifetime and it is during these periods when intense exposures may take place. Hence, it is important that asbestos is identified in any working situation and appropriate methods planned and implemented to ensure that airborne particles are not present in a significantly high concentration. Asbestos exposure has been associated with mesophilioma, lung cancer and pulmonary fibrosis or asbestosis. The problem associated with asbestos fibres in the lungs is one in which the fibre's durability and how long it remains in the lungs is of crucial importance, as is the size and shape of the fibres.

15.3.3   Stress

Lingard and Sublet, in Chapter 16, discuss job stress which is commonly defined in terms of role demands originating in the work environment. An increasing body of research suggests that a major reaction produced by job stressors is 'burnout' (see Lee and Ashforth 1993 and Cordes and Dougherty 1993 for a review of the early research). Since Freudenberger (1974) coined the term 'burnout' to describe a state of chronic emotional fatigue, this phenomenon has been the focus of much research interest. The most widely accepted definition of burnout conceptualises the phenomenon as 'a syndrome of emotional exhaustion, depersonalisation and reduced personal accomplishment that can occur among people who do 'people work' of some kind' (Maslach, Jackson and Leiter, 1996). This reflects the fact that early research work was conducted in the 'caring' professions. It is now recognized that burnout can and does occur in a wide range of occupations and is not unique to the human service sector. They report that the first component in Maslach's burnout model is emotional exhaustion. This describes feelings of depleted emotional resources and a lack of energy. In this state, employees feel unable to 'give of themselves' at a psychological level. The second component, depersonalisation, is characterised by a cynical attitude and an exaggerated distancing from clients. The last component of burnout, diminished personal accomplishment, refers to a situation in which employees tend to evaluate themselves negatively and become dissatisfied with their accomplishments at work.

They go on to point out that evidence suggests that burnout is associated with negative outcomes for both individuals and organisations. For example, at an individual level, burnout has been associated with the experience of psychological distress, anxiety, depression and reduced self-esteem. Burnout has also been positively correlated with unhealthy behaviours, such as substance abuse (Maslach, Schaufeli and Leiter, 2001). In addition to its effect on mental health, the association between burnout and physical health has also come under scrutiny and there is a growing understanding that burnout is associated with physical health problems, such as coronary heart disease (Appels and Schouten, 1991, Tennant, 1996). When burnout occurs as a response to job stress, it should be regarded as an occupational health issue warranting organizational intervention rather than a matter for the individual. This claim is supported by the fact that occupational stress cases represent Australia's most rapidly increasing category of employees' compensation claims (Schaufeli and Enzmann, 1998).

Failure to address this issue is also likely to have negative impact on organizational effectiveness because burnout has also been associated with absenteeism, turnover, reduced productivity or effectiveness and lower levels of satisfaction and organizational commitment (Maslach, Schaufeli and Leiter, 2001). Furthermore, some research suggests that burnout is contagious, spreading to affect colleagues of those who experience it and even resulting in a negative spillover into home life (Cordes and Dougherty, 1993, Bacharach, Bamberger and Conley, 1991, Westman, Etzion and Danon, in press). In an industry that already struggles to attract and retain a skilled workforce, a failure to protect the well-being of employees is likely to reduce an organisation's efficiency and ability to compete.

15.4   CONSTRUCTION HEALTH PROGRAMMES

Gibb, Haslam and Gyi discuss what they describe as Construction's Ill-Health Problem and introduce the system APaCHE - a partnership for construction health. They report that WHO, the World Health Organisation, defines health as a 'state of complete physical, mental and social well-being, and not merely the absence of disease or infirmity' (WHO, 1946). The authors have developed this argument to apply to organisations as well as individuals. They postulate that healthy organisations operate to meet their objectives efficiently, providing healthy, fulfilling employment for their workers.   In promoting healthy individuals and organisations, prevention is considerably more effective than cure. To be effective, prevention requires a holistic, whole-system, complete life-cycle approach recognising and addressing the complex influences arising from individual and collective attitudes and actions.

To date significant effort has been directed towards the more immediate, high profile (and perhaps more easily solvable) problem of safety rather than the more elusive health matters. Health and safety is often used by experts and non-experts alike to represent only safety. Most health and safety managers, supervisors or inspectors have little more that a very rudimentary knowledge of occupational health issues. However, ill-health continues to kill and disable many construction workers world-wide. In fact, the delay in the effects becoming evident is one of the key reasons why the subject should be taken more seriously.

The European construction sector is aware of these ill health issues, but they have difficulty managing them because of the highly mobile workforce, lack of continuity of employment, predominance of self-employed workers, and short-term, rapidly changing workplaces.   There are also other, more general, life-style-health issues for construction workers, often linked with the peripatetic aspects of their employment. Furthermore, it has been estimated that 80% of construction workers have not even registered with their local doctor.

O'Neal-Coble (Sink, *et al.*, 2003) reports that in recent years indoor air quality (IAQ) problems leading to the so-called 'sick building syndrome' have fed considerable litigation. Personal injury claims, which utilise remedies under insurance disability and workers' compensation laws, or the Occupational Safety and Hazard Act, as well as property damage, constructive eviction, and design and construction defect claims may relate to IAQ problems.

The World Health Organisation estimates that nearly 30% of all new and remodelled buildings worldwide may be afflicted with indoor air quality problems (US Environmental Protection Agency, 1991).   This and studies performed by the US Environmental Protection Agency rank IAQ concerns among the top environmental risks to public health. Not surprisingly, commentators expect IAQ litigation to proliferate (Cross, 1996). Perhaps 8% of commercial buildings in the US fall short of compliance with engineering standards for acceptable indoor air according to the National Energy Management Institute.

This chapter, Indoor Air Quality Problems in Buildings in the United States, outlines the various types of IAQ problems related to buildings, and addresses some of the practical and legal problems stemming from IAQ issues.

15.5 REFERENCES

Appels, A. and Schouten, E., 1991, Burnout as a Risk Factor for Coronary heart disease. *Behavioral Medicine*, 17, pp. 53-59.

Bacharach, S.B., Bamberger, P. and Conley, S., 1991, Work-home conflict among nurses and engineers: mediating the impact of role stress on burnout and satisfaction at work. *Journal of Organizational Behavior*, 12, pp. 39-53.

Beck, M.H., 1990, Letter to editor. *British Medical Journal*, 301: 932.

Burkhart, G. *et al.*, 1993, Job tasks, potential exposures and health risks of labourers employed in the construction industry. *American Journal of Industrial Medicine*, 24: pp. 413-425.

Cordes, C.L. and Dougherty, T.W., 1993, A review and an integration of research on job burnout. *Academy of Management Review*, 18, pp. 621-656.

Croner, 1998, TUC urges asbestos switch. *Construction Safety Briefing*, Croner Publications Ltd., Iss 35, August, p.3.

Cross, Frank B., 1990, Legal Responses to Indoor Air Pollution, 169. In *The Fungusamoungus: Sick Building Survival Guide*, edited by G. Rober, 1996, 8 St. Thomas L. Rev. 511, 513.

Freudenberger, H.J., 1974, Staff burnout. *Journal of Social Issues*, 30, pp. 159-165.

Gibb, A.G.F., Haslam, R.A. and Gyi, D.E., 2001, Cost and benefit implications and strategy for addressing occupational health in construction. In *Conseil Internationale de Batiment, Working Commission W99*, Barcelona, October, ISBN 84-7653-790-5, pp. 29-38.

Gründemann, R.W.M. and van Vuuren, C.V., 1998, Preventing absenteeism at the workplace: A European portfolio of case studies, *European Foundation for the Improvement of Living & Working Conditions*, Dublin, Ireland.

HSE, 1998, Ignore workers health and profits will suffer warns chairman at launch of new HSE health campaign, *Health and Safety Executive Press Release*, 22 May, E114: 98.

King, R.W. and Hudson, R., 1985, *Construction Hazards and Safety Handbook* (London: Butterworths).

Lee, R.T. and Ashforth, B.E., 1993, A longitudinal study of burnout among supervisors and managers: comparisons between the Leiter and Maslach (1988) and Golombiewski *et al.* (1986) models. *Organinisational Behavior and Human Decision Processes*, 54, pp. 369-398.

Maslach, C., Jackson, S.E. and Leiter, M.P., 1996, *Maslach Burnout Inventory Manual*, Palo Alto (CA: Consulting Psychologists Press), 3rd edition.

Maslach, C., Schaufeli, W.B. and Leiter, M.P., 2001, Job burnout. *Annual Review of Psychology*, 52, pp. 397-422.

Schaufeli, W. and Enzmann, D., 1998, *The Burnout Companion to Study and Practice: A Critical Analysis*. London; Philadelphia (PA: Taylor & Francis).

Sink, Charles M., Gibson, C. Allen, Oles, Douglas S., Gwyn, Allen Holt and O'Neal-Coble, Leslie, 2003, *Construction Damages and Remedies*, ABA Publishing.

Tennant, C., 1996, Experimental stress and cardiac function. *Journal of Psychosomatic Research*, 40, pp. 569-583.

U.S. Environmental Protection Agency, 1991, *Air & Radiation Publication*, No. ANR-445-W, Indoor Air Facts No. 4 (revised): Sick Building Syndrome 1, April.

Westman, M., Etzion, D. and Danon, E., Job insecurity and crossover of burnout in married couples. *Journal of Organizational Behavior* (in press).

WHO (World Health Organization), 1946, *Official Records*, 2, p. 100, Preamble to the Constitution.

CHAPTER 16

Job, Family and Individual Characteristics Associated with Professional 'Burnout' in the Australian Construction Industry

Helen Lingard and Anna Sublet

WHAT IS BURNOUT?

Job stress is commonly defined in terms of role demands originating in the work environment (Cooper and Marshall, 1977; Carayon, 1992). An increasing body of research suggests that a major reaction produced by job stressors is burnout (see Lee and Ashforth (1993) and Cordes and Dougherty (1993) for a review of the early research). Since Freudenberger (1974) coined the term 'burnout' to describe a state of chronic emotional fatigue, this phenomenon has been the focus of much research interest. The most widely accepted definition of burnout conceptualises the phenomenon as 'a syndrome of emotional exhaustion, depersonalisation and reduced personal accomplishment that can occur among people who do 'people work' of some kind' (Maslach, Jackson and Leiter, 1996). This reflects the fact that early research work was conducted in the 'caring' professions. It is now recognised that burnout can and does occur in a wide range of occupations and is not unique to the human service sector. For example, Dolan (1995) studied burnout among senior executives, Bacharach, Bamberger and Conley (1991) undertook a comparative study of burnout in samples of engineers and nurses and Cordes, Dougherty and Blum (1997) investigated burnout among human resource managers and professionals in corporate settings. In response to the recognition that burnout is not unique to human service workers, Maslach and her colleagues (1996) have developed a new version of their burnout inventory (the MBI) designed to measure burnout outside the human services sector (the MBI-GS). This instrument has been found to possess good factorial validity and satisfactory internal consistency reliability across a range of different occupational groups and countries (Schutte *et al.*, 2000).

The first component in Maslach's burnout model is emotional exhaustion. This describes feelings of depleted emotional resources and a lack of energy. In this state, employees feel unable to 'give of themselves' at a psychological level. The second component, depersonalisation, is characterised by a cynical attitude and an exaggerated distancing from one's work. The last component of burnout, diminished personal accomplishment, refers to a situation in which employees tend to evaluate themselves negatively and become dissatisfied with their accomplishments at work.

16.1 WHY SHOULD COMPANIES BE CONCERNED ABOUT EMPLOYEE BURNOUT?

Research evidence suggests that burnout is associated with negative outcomes for both individuals and organisations. For example, at an individual level, burnout has been associated with the experience of psychological distress, anxiety, depression and reduced self-esteem. Burnout has also been positively correlated with unhealthy behaviours, such as substance abuse (Maslach, Schaufeli and Leiter, 2001). In addition to its effect on mental health, the association between burnout and physical health has also come under scrutiny and there is a growing understanding that burnout is associated with physical health problems, such as coronary heart disease (Appels and Schouten, 1991; Tennant; 1996). This suggests that if burnout occurs as a response to job stress, it should be regarded as an occupational health issue warranting organisational intervention rather than a matter for the individual. The need for organisational intervention is also supported by the fact that occupational stress-related illnesses result in substantial losses to both companies and the economy as a whole. For example, Australian research has found that, while the number of claims for stress-related illness represents only a small proportion of the total number of employees' compensation claims, the costs of these claims represents a major proportion of all compensation costs (Cotton, 1995).

Failure to prevent or treat burnout is also likely to have negative impact on organisational effectiveness. For example, research indicates that burnout is associated with absenteeism, turnover, reduced productivity or effectiveness and lower levels of satisfaction and organisational commitment (Maslach, Schaufeli and Leiter, 2001). Furthermore, some research suggests that burnout is contagious, spreading to affect colleagues of those who experience it and even resulting in a negative spillover into home life (Burke and Greenglass, 2001; Cordes and Dougherty; 1993, Westman, Etzion and Danon, in press). In an industry that already struggles to attract and retain a skilled workforce, a failure to protect the well-being of employees is likely to reduce the Australian construction industry's efficiency and threaten its long-term competitiveness.

16.2 SOURCES OF BURNOUT

16.2.1 Work stressors

Researchers have identified many job demands (stressors) to be associated with employees' experience of burnout (Lee and Ashforth, 1996; Gmelch and Gates;

1998). Maslach, Schaufeli and Leiter (2001) broadly classify these stressors into the following six areas of working life: workload; control; reward; community; fairness; and values. Work overload arising from having too many tasks to carry out in the time available or being faced with the wrong kind of work is associated with the exhaustion component of burnout. An inability to control one's work environment, for example being unable to determine resource issues or work-pace, has been associated with the personal accomplishment component of burnout. The lack of appropriate rewards, either in the form of remuneration or career development has also been associated with feelings of inefficacy. Maslach, Schaufeli and Leiter (2001) also suggest that people who do not enjoy positive relationships with co-workers, and therefore enjoy little social support in their work, become disengaged from their work. Finally, where there is a conflict between the individual's values or personal aspirations and those of the organisation, or where there are conflicting organisational goals, such as safety and cost, the risk of burnout is high. Maslach, Schaufeli and Leiter (2001) suggest that it is possible that the weighting of the sources of burnout might reflect individual differences, some people placing more importance on one aspect of their work than others. Furthermore, there may be a trade-off effect. For example, some employees may be willing to tolerate a high workload if they are satisfied with their pay. It is possible that these trade-offs are related to demographic characteristics, such as parental status.

16.2.2   Ecological models of burnout

While the early theories of burnout focused exclusively on work-related sources, recent work adopts a broader theoretical framework in which individual and situational factors are integrated (Maslach, Schaufeli and Leiter , 2001). For example, Caroll and White (1982) present an 'ecological' model of burnout, in which the phenomenon is understood in terms of the complex interaction between the person and his/her work and non-work environments. Recent research has sought to understand this interaction. For example, Dolan (1995) investigated the relationship between job/organisational demands, personal characteristics and burnout and found support for an ecological model of burnout. Thus, the present study considered individual personality and demographic characteristics and aspects of respondents' non-work life in addition to the known work-related stressors.

16.2.3   Work-family interface

As noted above, burnout studies that focus only on sources of burnout in the work environment are conceptually limited by their adoption of a 'segmentation' viewpoint. They examine the relationship between stressors in the work environment and affective work outcomes, such as burnout, satisfaction or intention to turnover, within a framework, which assumes the work and non-work realms are independent of one another (Bacharach, Bamberger and Conley, 1991). However, if conflicting roles and expectations in the work context are associated with burnout, it is not unreasonable to suppose that conflict between one's work and non-work roles may also be a contributory factor. Indeed, research on the work-family interface has highlighted the negative impact of work-family conflict, conceptualised as an inability to simultaneously satisfy demands arising from work and family roles (O'Neil and Greenberger, 1994). Bacharach, Bamberger and Conley (1991) also found work-home conflict to be directly related to burnout. Furthermore, early research documented the 'spillover' effect, whereby negative experiences in one domain (e.g. work or non-work) have a damaging impact on experiences in the other domain (Barnett, 1994; Eckenrode and Gore, 1990). The role of this spillover effect in employees' experience of burnout warrants further investigation and Westman, Etzion and Danon (in press) recently found that burnout could be passed from husbands to their wives. In the light of the growing body of evidence, it seems that human resource management models that separate the work and family spheres are no longer appropriate (Kamerman and Kahn, 1981).

16.2.4   Individual characteristics

Consistent with the ecological model of burnout, a multitude of individual factors have been investigated as predictors of burnout. Demographic characteristics of interest include age, gender, marital status and level of educational achievement. Of these variables, age has been most consistently related to burnout. Research suggests that burnout is higher among younger employees than among employees over 30 or 40 years of age (Maslach, Schaufeli and Leiter, 2001). Although caution must be exercised when interpreting these results because the cross-sectional nature of most burnout studies means that this may be partly explained by the 'survival effect' whereby employees who are susceptible to burnout ultimately leave their jobs leaving the survivors who exhibit lower levels of burnout. Gender has not proved to be a strong predictor of burnout and, while some studies show higher levels among women, others find no significant differences. Despite this, Maslach, Schaufeli and Leiter (2001) note that studies consistently find that men report higher levels of cynicism than women. Unmarried employees, especially men, are more prone to burnout than those who are married and single people experience higher levels of burnout than divorcees (Maslach, Schaufeli and Leiter, 2001). Research also suggests that people who have attained a higher level of education report higher levels of burnout than less educated people (Maslach, Schaufeli and Leiter, 2001). However, this finding may be confounded by the fact that education level may be related to responsibility and other work-related stressors. Personality variables, such as level of hardiness, self-esteem and locus of control, have also been examined as predictors of burnout. Chang, Rand and Strunk (2000) found optimism to be a strong predictor of burnout and conclude that more optimistic workers are likely to feel less exhausted and cynical about their work and also experience a greater sense of accomplishment. Neuroticism, which includes anxiety, hostility, depression, self-consciousness and vulnerability has also been linked to burnout (Maslach, Schaufeli and Leiter, 2001).

16.3   THE CONSTRUCTION INDUSTRY CONTEXT

The construction industry is a demanding work environment and many job-related stressors are likely to be pertinent to the work of engineers engaged in this industry. For example, engineers share the responsibility for the delivery of projects that could cost lives or their company's reputation should the requisite quality, planned budget or specified completion date not be met. Companies also operate in a highly competitive market with relatively low profit levels to complete construction projects within tight deadlines. With the threat of significant penalties for time overruns, professionals and managers need to ensure their availability while work on site continues. Thus, work hours are often long and sometimes irregular. A recent survey has highlighted the time-related work pressures experienced by engineers in Australia (APESMA, 2000). The survey found that professional engineers work long hours, including significant amounts of regular unpaid overtime. Over the twelve-month period studied, engineers reported that the amount of work to be done had increased (63%), the pace of work had increased (62%), and that the amount of stress had increased (52%). APESMA also report that more than a quarter of respondents believed there had been an increase in health problems as a result of their working lives. The most common ailments they identified were those related to excessive workloads, such as continual tiredness (66%) and stress (70%). Bacharach, Bamberger and Conley (1991) also found that role conflict had a significant effect on burnout for engineers, but not for their comparison group of nurses. Role conflict occurs when two or more job demands cannot be met simultaneously. The management of construction projects typically involves a trade-off between cost, time and quality. Added to these are safety and environmental imperatives and the threat of personal liability in the event of unforeseen incidents. In this environment, role conflict is likely to be high. Bacharach, Bamberger and Conley (1991) suggest that, for engineers, role conflict (e.g. conflict between professional standards and budget constraints) may be more strongly associated with more severe, life and death' consequences, giving rise to burnout.

16.4　AIMS OF THE STUDY

Despite the fact that professionals working in the construction industry may be a high risk group for burnout, there has been little research into the burnout issue in the construction industry. The present study aimed to address this gap in our understanding. The study had several aims:

Firstly, the study sought to explore the relevance of the burnout phenomenon in the Australia construction industry. This included testing the reliability of the most widely used burnout scale, the Maslach Burnout Inventory (MBI) among a sample of Australian construction industry professionals;

Second, the study sought to identify job, demographic and personality characteristics associated with burnout. This included identifying the relative importance of these potential sources of burnout. It is important to ascertain the most important determinants of burnout since intervention strategies designed to prevent or treat burnout are more likely to succeed if they address the source of the problem; and Thirdly, the study aimed to explore respondents' experience at the work-family interface to determine whether family-related variables have a direct impact upon experienced burnout. This included considering the difference between demographic groups, such as respondents with or without children and respondents in dual compared to single-income families.

16.5　METHODOLOGY

16.5.1　Data collection and sampling

The data for this study were obtained from civil engineers engaged in professional practice in consulting and contracting organisations in New South Wales and Victoria. In the first instance, a register of companies employing civil engineers was obtained from the Institution of Engineers, Australia. General managers or human resources managers of listed companies were approached and asked whether the company would participate in the study. At this stage, the managers were thoroughly briefed on the objectives of the study. Questionnaires were randomly distributed through the internal mail systems of the companies that agreed to participate. Completed questionnaires were returned directly to the researchers in unmarked postage-paid envelopes provided for this purpose. Each questionnaire was distributed with a copy of a plain language statement describing the objectives of the study. The statement also explained the voluntary nature of respondent participation and assured anonymity of respondents and confidentiality of responses. Of 500 questionnaires distributed, 182 completed and usable ones were returned yielding a response rate of 36%.

16.5.2   Questionnaire design

Demographic information collected included the respondents' age, gender and marital/relationship status. Respondents were also asked to provide information about their partners' employment status, the number of children they have and the age of their youngest child. Respondents were asked how many years they have worked in the engineering profession and their current job title.

Job characteristics were measured using a 36-item instrument. The scale was designed to tap dimensions likely to be relevant to the work of construction industry professionals. These were: subjective overload (I never seem to have enough time to get everything done); responsibility (In this job, the safety of others depends on me); role clarity (Clear, planned goals and objectives exist for my job); role conflict (In my job, I can't satisfy everybody at the same time); satisfaction with promotion prospects (In this job, the chances for promotion are good); role clarity (In this job, I know exactly what is expected of me); social satisfaction (The people I work with are friendly); control over work pace (I determine the speed at which I work); and satisfaction with pay (I am very happy with the amount of money I make). Satisfaction with job security was assessed with a single item (In this job, the job security is good). The items for each dimension were taken from several previously deployed instruments, including the Michigan Organisational Assessment Questionnaire. Respondents were asked to rate items on a Likert-type scale ranging from 1 (strongly disagree) to 7 (strongly agree). Additional information about respondents' job/organisational situation was collected, including the size of the company for which they work, whether their job is site or office based and the average number of hours they work each week.

Satisfaction and conflict in respondents' relationships with their spouse or partner were measured using an eighteen-item instrument developed by Orden and Bradburn (1968). This instrument asks respondents to rate the frequency with which they engaged in nine satisfying activities, for example gone out together – to a cinema, to play sport or other entertainment, within the few weeks prior to their completion of the questionnaire. The scale also asks respondents to indicate the frequency with which they experienced conflict in nine key aspects of their relationship, for example being away from home, during recent weeks.

Personality traits, identified in the literature to be associated with experiences of occupational stress, were measured. These included neuroticism (I sometimes worry about awful things that might happen), extroversion (I enjoy mixing with people) and optimism (Generally, I feel that things will sort themselves out in the end). In addition to these traits, quick wittedness (I can put my thoughts into words quickly) and impulsiveness (I often do things on the spur of the moment) were also measured. These two traits were included because the requirement for construction industry professionals to respond quickly to unexpected circumstances is likely to be a stressor pertinent in their work. Relevant items from the Eysenck Personality Inventory (Eysenck and Eysenck, 1977) were incorporated into the measure. Respondents were asked to rate, on a seven point Likert scale, the extent to which they agreed or disagreed with each of the statements as a reflection of themselves.

Burnout, was measured using the Maslach Burnout Inventory - General Survey (Maslach, Jackson and Leiter, 1996). This 16-item inventory comprises three sub-scales assessing emotional exhaustion (I feel emotionally drained from my work), cynicism (I have become less interested in my work since I started this job) and lack of professional efficacy (At work, I feel confident that I am effective at getting things done). The items for the third dimension of burnout are framed in positive terms and thus a low score reflects a low sense of professional efficacy. The MBI was selected for use due to its brevity and proven reliability and validity of results (Corcoran, 1995; Shinn, 1982; Gaines and Jermier, 1983). Following Maslach, Jackson and Leiter (1996), since the response formats of intensity and frequency have been found to be highly correlated, only frequency ratings were used. Items are rated on a seven-point Likert scale ranging from Never (0) to Every day (6).

16.5.3 Analytical procedures

Principal components factor analysis was performed for all composite scales used in the study. Thus, the job characteristics, personality, burnout and turnover scales were each factor analysed. Factor analysis assumes that underlying dimensions or factors can be used to explain complex phenomena. On completion of factor analysis, factor scores for each factor can be computed for each case in a sample. These scores can then be used in further data analysis, such as correlation or regression analysis.

Having identified the factorial structure of the scales, Pearson correlation coefficients (denoted r) were calculated to determine the strength of association between variables. The correlation coefficients, presented in matrix format, indicate the strength of the linear association between two variables. Correlation analyses were undertaken as a precursor to the multiple regression analyses because, owing to the large number of job-related, personality and demographic variables, only those which were significantly associated with the dependent variable (e.g. the burnout dimensions) were included in the regression models.

Following the correlation analysis, multiple regression procedures were used to determine the extent to which the dependent variables, for example the burnout dimensions, could be predicted by the independent variables, for example the demographic, personality and job characteristics. Multiple regression analysis enables the relative importance of independent variables to be determined.

Reliability analysis allows an assessment of the measurement scales and items that make them up. For example, reliability analysis provides an answer to the question 'does my questionnaire measure burnout in a useful way?' Using reliability analysis, the extent to which items in the questionnaire scales are related to each other is considered and an overall index of the internal consistency of each scale is produced. Cronbach's alpha coefficients were calculated for each of the scales. Descriptive statistics for each scale, including the Cronbach's alpha coefficients are presented in Table 16.1.

16.6 RESULTS

16.6.1 The sample

The large majority of respondents were male (92.3%). Only 7.7% of the sample were female. Respondents were also predominantly employed in consulting firms (84.6%). Only 11% of respondents indicated they worked for a contracting organisation and 1.6% reported working for a supplier. Reflecting this, 81.3% of respondents indicated that they spent most of their time at work in an office environment. Only 7% reported that they were site-based and 10.4% indicated that their job involved both office and site work. The majority of respondents (65.4%) were married. A further 17.6% indicated that they were 'involved'. Only 15.9% of respondents were single and 1.1% indicated that they were divorced. Most respondents were participants in dual-income families. Of those respondents who were married or involved, 60.4% had a partner or spouse in paid employment. Only 22% of respondents in a relationship had a partner or spouse who was not in paid employment. The sample contained roughly equal numbers of parents (51.1%) and non-parents (48.4%).

16.6.2 Burnout in the engineering profession

Previous research has supported the convergent and discriminant validity of the Maslach Burnout Inventory which conceptualises burnout as comprising three factors: emotional exhaustion, cynicism and diminished personal accomplishment. However, several studies have found moderate to high correlations between emotional exhaustion and cynicism, suggesting the possibility of an alternative model (Koeske and Koeske, 1989; Wolpin, Burke and Greenglass, 1991). A two-factor model comprising emotional exhaustion/cynicism and personal accomplishment has been supported in some studies (Brookings, *et al.*, 1985, Dignam, Barrera and West, 1986). Golembiewski and Munzenrider (1981) have even suggested that burnout should be reduced to a unitary concept. One recent Australian study supports a five-factor model of burnout (Densten, 2001). In Densten's model, emotional exhaustion is broken into two factors, psychological strain and somatic strain. Personal accomplishment is also divided into a lack of accomplishment of personal goals (self) and a lack of accomplishment of organisational goals (others).

Table 16.1 Descriptive statistics for each of the measurement scales

	No. of items	Cronbach's alpha	Item mean	Scale mean	Scale SD	Item means variance
Job characteristics						
Quantitative overload	4	.87	4.73	18.91	5.02	.34
Responsibility	4	.66	5.62	22.48	3.34	.05
Role clarity	4	.68	4.53	18.10	4.33	.17
Pay satisfaction	2	.94	3.55	7.10	3.34	.01
Promotional satisfaction/worth	4	.59	4.30	17.19	4.00	.08
Role conflict	3	.65	4.56	13.67	3.33	.29
Social satisfaction	2	.49	5.37	10.74	1.89	.62
Personality factors						
Neuroticism	4	.77	4.38	17.52	4.89	.07
Extroversion/action	4	.64	4.78	19.12	4.05	.35
Optimism	3	.70	4.18	12.55	3.43	.61
Impulsiveness	3	.72	3.50	10.51	3.36	.53
Quick wittedness/confidence	3	.62	4.43	13.31	3.46	.02
Extroversion/social	2	.41	4.15	8.31	2.25	.17
Burnout						
Emotional exhaustion	5	.90	2.98	14.90	6.54	.38
Cynicism	5	.85	1.87	9.34	6.58	.22
Personal competence	4	.74	4.94	19.75	3.34	.08
Professional worth	2	.77	3.85	7.7	2.55	.00
Relationship quality						
Relationship satisfaction	7	.90	2.51	17.56	5.24	.05
Social relations	2	.80	2.27	4.55	1.87	.07
Relationship conflict	8	.85	1.30	10.43	5.83	.09
Turnover intention	3	.90	3.18	9.54	5.49	.08

In the light of the ongoing debate about the factorial structure of the burnout phenomenon, the first aim of our study was to explore the relevance of the three-factor burnout model, as conceptualised by Christina Maslach and her colleagues. Determining the factorial structure of the burnout phenomenon in the Australian construction industry context was also a pre-requisite to achieving the remaining aims of the study. Therefore, a confirmatory factor analysis was conducted to test the appropriateness of competing models of burnout, that is, the one, two, three or five-factor models, in our sample.

Table 16.2 shows the results of the principal components factor analysis for the burnout scales. As the item loadings indicate, the Emotional Exhaustion and

Cynicism scale items loaded as expected. However, the Personal Efficacy items broke down into separate factors, yielding a four-factor solution overall. This solution accounted for 69% of total variance. The four-factor structure differs from the more commonly produced three-factor model of burnout (Schutte, *et al.*, 2000). However, when the data were forced into a three-factor structure, the total variance explained by the solution fell to 61%. Furthermore, Table 16.2 shows a minimal problem with double loading of items, suggesting the discriminant validity of a four-factor burnout structure for our sample. The two items that loaded on the fourth factor were examined and both appear to relate to a sense of professional worth (I feel exhilarated when I accomplish something worthwhile in my job) as distinct from personal competence (At work, I feel confident that I am effective at getting things done). These dimensions were labelled 'Professional Worth' and 'Personal Competence' respectively, and considered separately in the remaining data analysis. The four-factor structure suggests that engineers experience burnout differently to other professional or occupational groups and may experience a sense of making a worthwhile contribution to society or feel a need to be professionally valued independently of believing in their individual competence or effectiveness at work.

16.6.3 Job and individual correlates of burnout

The second aim of the study was to determine whether burnout is best explained by individuals' demographic or personality traits or job-related variables. To investigate this question, correlation and regression analyses were conducted. Bivariate Pearson correlations between the demographic, personality and job characteristic variables and the four burnout factors are shown in Table 16.3. As Table 16.3 shows, the four dimensions of burnout were correlated with different demographic, job-related and individual characteristics.

Of the demographic variables, age of youngest child ($r=-.239$, $p=.025$) and tenure ($r=-.161$, $p=.034$) were negatively correlated with the emotional exhaustion. Age of respondent ($r=-193$, $p=0.011$) and number of children ($r=-240$, $p=0.001$) were negatively correlated with the cynicism dimension of burnout.

Of the job-related variables, three were positively correlated with emotional exhaustion. These were the number of hours worked per week ($r=.306$, $p=.000$), subjective overload ($r=.351$, $p=.000$) and role conflict ($r=.275$, $p=.000$). Four job-related variables were negatively correlated with cynicism. These included responsibility ($r=-.228$, $p=.003$), role clarity ($r=-.193$, $p=.011$), satisfaction with pay ($r=-.151$, $p=.048$) and satisfaction with promotion prospects ($r=-.381$, $p=.000$). Fewer job-related variables were found to be significantly correlated with the two personal efficacy dimensions. However, role clarity was significantly and positively correlated with personal competence ($r=.202$, $p=.008$) and satisfaction with promotion prospects was significantly and positively correlated with professional worth ($r=.197$, $p=.010$).

Table 16.2 Factor loadings for the four-factor model of burnout

Item	Factor 1 (emotional exhaustion)	Factor 2 (Cynicism)	Factor 3 (Personal competence)	Factor 4 (Professional worth)
I feel emotionally drained from my work	.859	.186	.021	-.028
I feel used up at the end of the work day	.839	.011	-.019	.122
I feel burned out from my work	.814	.305	-.110	-.065
I feel tired when I get up in the morning and have to face another day on the job	.742	.319	.007	-.050
Working all day is really a strain for me	.721	.384	-.077	-.015
I have become more cynical about whether my work contributes anything	.226	.767	-.335	.060
I have become less interested in my work since I started this job	.361	.760	.050	-.195
I have become less enthusiastic about my work	.391	.750	.081	-.182
I doubt the significance of my work	.189	.707	-.383	-.004
I just want to do my job and not be bothered	.126	.665	.054	-.090
At my work, I feel confident that I am effective at getting things done	-.029	-.081	.825	.155
I can effectively solve the problems that arise in my work	-.168	.041	.753	.017
In my opinion, I am good at my job	.077	-.009	.612	.422
I feel I am making an effective contribution to what this organisation does	.129	-.348	.570	.279
I feel exhilarated when I accomplish something worthwhile in my job	-.054	-.102	.139	.894
I have accomplished many worthwhile things in this job	.017	-.127	.287	.794

Of the personality traits, only neuroticism was positively correlated with emotional exhaustion ($r=.200$, $p=.008$). Both the action-oriented ($r=-.171$, $p=.025$) and social-oriented ($r=-.253$, $p=.001$) extroversion variables were negatively correlated with cynicism. Interestingly optimism was also positively correlated with cynicism ($r=.158$, $p=.038$), although this correlation was quite weak. Neuroticism was negatively correlated ($r=-.214$, $p=.005$) and quick-wittedness was positively correlated ($r=.238$, $p=.002$) with personal competence. No personality variables were correlated with the professional worth dimension of burnout.

One way analyses of variance (ANOVAs) were conducted to ascertain whether significant differences existed between the burnout scores of respondents who differed in terms of gender, marital status or work location (e.g. site or office). With an alpha level of .05, no significant differences in burnout were found between male and female respondents. However, this finding must be interpreted with caution owing to the very small number of women in the sample (7.8% of the total). Using the same alpha level, ANOVAs also revealed no significant differences between employees reporting different marital or relationship situations, nor was there a significant difference between site-based and office-based employees. On this basis, these variables were excluded from the following regression analyses.

16.6.4   Predictors of emotional exhaustion

Thirty-four percent of variation in emotional exhaustion can be "explained" by the specified regression model (adjusted R^2=.336). Significant predictors of emotional exhaustion included tenure (β=-.319, p=.001), overload (β=.302, p=.001) and role conflict at work (β=.255, p=.010). Consistent with this, long work hours and the feeling of having too much to do in the time available was a common theme emerging from the comments provided by some participants at the end of the questionnaire. For example, one participant wrote 'The consulting engineering industry has under-cut itself to the extent that projects rarely can be completed for the fee submitted - a profit can and still is made but at the expense of working long hours and not booking the time.' Another criticised unrealistic client programmes saying they resulted in 'insufficient time to do the work, hence long hours' The only personality characteristic that predicted emotional exhaustion was neuroticism although this variable had the least significant beta coefficient of all of the variables in the model (β=.204, p=.027).

16.6.5 Predictors of cynicism

Thirty-one per cent of variation in cynicism can be explained by the specified regression model (adjusted R^2=.314). Among the job-related variables, satisfaction with promotion prospects (β=-297, p=.000), responsibility (β=-.183, p=.009), role clarity (β=-.195, p=.003) and satisfaction with pay (β=-.150, p=.028) had significant predictive ability for cynicism. Consistent with these findings is the fact that dissatisfaction with the level of job-related rewards in comparison with the responsibility borne by engineers emerged as an important theme among comments written by participants at the end of the questionnaire. The responsibility for the health and safety of workers was of particular concern and there was a strong sense that engineers are undervalued by society as a whole. For example, one participant expressed this by saying 'Though OHS laws and heavy fines are now even more of a concern to engineers, than the already busy workload. All this responsibility is not worth the pay'. Of the personality variables, the social aspect of extroversion (β=-.233, p=.001) and action aspect of extroversion (β=-.211, p=.003) were significant predictors of cynicism. Age of respondent also had significant predictive ability for cynicism (β=-.157, p=.040).

Table 16.3 Pearson correlations between demographic, personality and job-related variables and the four burnout dimensions

	1	2	3	4	5	6	7	8	9	10	11	12
1. Age	-											
2. Number of children	.699***	-										
3. Age of youngest child	.808***	.242*	-									
4. Tenure	.968***	.703***	.824***	-								
5. Hours per week	.160*	.217**	-.178	.152*	-							
6. Subjective overload	.150*	.115	.024	.137	.366***	-						
7. Responsibility	.383***	.308***	.079	.391***	.266***	.000	-					
8. Role clarity	.022	.121	-.041	.008	.014	.000	.000	-				
9. Satisfaction (pay)	.293***	.219**	.312***	.324***	.059	.000	.000	.000	-			
10. Satisfaction (promotion prospects)	.021	.107	.138	.028	.011	.000	.000	.000	.000	-		
11. Role conflict	-.185*	-.055	-.237*	-.191*	.172*	.000	.000	.000	.000	.000	-	
12. Social support (work)	-.181*	-.112	.012	-.159*	-.122	.000	.000	.000	.000	.000	.000	-
13. Work pace control	.103	.136	.156	.134	.114	.000	.000	.000	.000	.000	.000	.000
14. Neuroticism	-.001	-.065	.024	-.005	-.005	.143	.104	-.101	-.063	-.030	.069	-.083
15. Extraversion (action)	-.304***	-.170*	-.329***	-.303***	.088	-.005	-.103	.037	-.206**	.166*	.000	.186*
16. Impulsiveness	-.160*	-.145	-.225*	-.147*	.027	.026	.049	-.046	-.062	.086	.013	-.046
17. Quick-wittedness/confidence	.156*	.079	.106	.160*	.202**	.137	.207**	.166*	.035	.119	.004	.045
18. Optimism	-.196**	-.130	-.105	-.176*	-.085	-.078	-.072	-.046	.043	-.078	.018	.189*
19. Extraversion (social)	-.089	-.038	.133	-.078	-.034	.037	.036	.166*	-.015	.207**	-.070	.173*
20. Emotional Exhaustion	-.128	-.051	-.239*	-.161*	.306***	.351***	.086	-.068	-.134	-.100	.275***	-.111
21. Cynicism	-.193*	-.240***	-.187	-.230***	-.090	-.041	-.228**	-.193*	-.151*	-.381***	.131	-.131
22. Personal competence	-.006	.043	.029	.002	.139	.034	.059	.202**	.049	-.038	.021	.115
23. Professional worth	-.011	.073	.091	.012	-.011	-.140	.092	.010	-.088	.197**	.105	.051

*** Correlation is significant at the 0.001 level (2-tailed)
** Correlation is significant at the 0.01 level (2-tailed)
* Correlation is significant at the 0.05 level (2-tailed)

Table 16.3 Pearson correlations between demographic, personality and job-related variables and the four burnout dimensions (cont'd)

	13	14	15	16	17	18	19	20	21	22	23
1. Age											
2. Number of children											
3. Age of youngest child											
4. Tenure											
5. Hours per week											
6. Subjective overload											
7. Responsibility											
8. Role clarity											
9. Satisfaction (pay)											
10. Satisfaction (promotion prospects)											
11. Role conflict											
12. Social support (work)											
13. Work pace control	-										
14. Neuroticism	-.052	-									
15. Extraversion (action)	.028	.000	-								
16. Impulsiveness	.116	.000	.000	-							
17. Quick-wittedness/confidence	.240***	.000	.000	.000	-						
18. Optimism	.022	.000	.000	.000	.000	-					
19. Extraversion (social)	-.053	.000	.000	.000	.000	.000	-				
20. Emotional Exhaustion	-.028	.200**	.013	.025	-.026	-.100	-.092	-			
21. Cynicism	-.041	.100	-.171*	.004	-.109	.158*	-.253***	.000	-		
22. Personal competence	.065	-.214**	.095	-.006	.238**	-.099	-.047	.000	.000	-	
23. Professional worth	.143	.080	.107	.043	.146	.072	.089	.000	.000	.000	

*** Correlation is significant at the 0.001 level (2-tailed).

** Correlation is significant at the 0.01 level (2-tailed)

* Correlation is significant at the 0.05 level (2-tailed)

Table 16.4 Demographic, personality and job-related predictors of burnout

Variables	Cumulative* R^2	Cumulative* Adjusted R^2	Standardised Beta (â)	p
Emotional exhaustion				
Tenure	.165	.155	-.319	.001
Overload	.262	.244	.302	.001
Role conflict	.327	.302	.255	.010
Neuroticism	.368	.336	.204	.027
Cynicism				
Promotional satisfaction	.151	.146	-.291	.000
Responsibility	.200	.190	-.166	.019
Extroversion/social	.236	.222	-.236	.000
Role clarity	.275	.257	-.201	.002
Tenure	.300	.279	-.196	.012
Extroversion/action	.337	.312	-.221	.002
Pay satisfaction	.353	.325	-.137	.047
Personal competence				
Neuroticism	.053	.048	-.216	.003
Quick-wittedness/ Confidence	.105	.094	.203	.006
Role clarity	.126	.110	.148	.046
Professional worth				
Promotional satisfaction/worth	.039	.033	.197	.010

* Note that R^2 values are cumulative and represent the change in R^2 as each variable is entered into the model

16.6.6 Predictors of professional efficacy

The regression model for personal competence explained only 13% of total variation (Adjusted R^2=.126). Personality traits that were significant predictors were neuroticism (β=-.216, p=.003) and quick-wittedness (β=.203, p=.006). Only one job-related variable, role clarity, was a significant predictor of personal competence (β=.148, p=.046). Only one job-related variable was a significant predictor of professional worth. This was satisfaction with promotion prospects (β=.197, p=.010). However, the very low adjusted R^2 (.033) indicates that this variable explains only 3% of overall variability in professional worth.

The results of these regression analyses suggest that the emotional exhaustion and cynicism dimensions of burnout are most readily explained by the demographic, personality and job-related variables in the study. Furthermore, the burnout dimensions are differentially predicted by these variables, suggesting that multiple intervention strategies may be needed to prevent or treat burnout. For the most part, job-related variables have greater predictive ability than personality variables, although some demographic variables also emerged as significant predictors of both emotional exhaustion and cynicism. The relative importance of job-related variables compared to personality variables suggests that jobs may be better designed to control the risk of burnout among engineers in the construction industry.

The direction of the relationship is indicated by the sign of the beta coefficient. Thus, tenure is a significant predictor of burnout and the negative beta coefficient indicates that as tenure increases, emotional exhaustion decreases. This may be indicative of the fact that senior managers have a greater authority to delegate tasks and may be able to overcome problems associated with overload or role conflict through this ability. In-keeping with other researchers' results, age was a significant predictor of cynicism and the negative beta coefficient indicates that younger employees are more cynical than older employees. This supports the argument that burnout occurs early, rather than late, in employees' working lives.

16.6.7   Burnout and the work-family interface

The third aim of the study was to explore respondents' experience at the work-family interface to determine whether family-related variables also have a direct impact upon experienced burnout. To achieve this, correlation coefficients were calculated between the burnout dimensions, job-characteristics and family variables. Family variables included the quality of relationship with spouse/partner. Following the factor analysis of the relationship satisfaction variables, an examination of the items loadings suggested that perceived relationship satisfaction should be subdivided into quality of relationship between partners and the extent to which a couple interact socially with others. These two dimensions of marital quality were labelled 'relationship satisfaction' and 'social relations' respectively and considered separately in the remaining analyses. As expected, perceived marital conflict was represented by a single factor. Parental demands were measured using the number of children and the age of respondents' youngest child. Finally, the employment status of respondents' spouse/partner was also included in the analysis to determine whether employees in dual income households differed from those in single-income households.

The correlation coefficients are presented in Table 16.5. The work and family variables were differentially correlated with the four dimensions of burnout. Emotional exhaustion was correlated with age of youngest child (r=-.193, p=.011), hours worked per week (r=.306, p=.000), subjective overload (r=.351, p=.000), job role conflict (r=.275, p=.000) and conflict in relationship with spouse or partner (r=.314, p=.000). Significant correlates of cynicism were number of children (r=-.240, p=.001), age of respondent (r=-.193, p=.011), satisfaction with job security (r=-.162, p=.033), perceived responsibility at work (r=-.228, p=.003), clarity of roles at work (r=-.193, p=.011), satisfaction with pay (r=-.151, p=.048), satisfaction with promotion prospects (r=-.381, p=.000) and satisfaction in relationship with spouse or partner (r=.181, p=.032). The personal competence dimension of burnout was correlated with satisfaction with job security (r=.182, p=.016) and clarity of roles at work (r=.202, p=.008). Professional worth was correlated with satisfaction with promotion prospects (r=.197, p=.010) and the social relations dimension of relationship satisfaction (r=.220, p=.009).

Next, regression analysis was conducted to determine the relative importance of the work and family variables in predicting burnout. Owing to the large number of independent variables, only those work and family variables that correlated significantly with the burnout dimensions were entered into the regression models for each dimension. Gender was not entered in any of the regression models owing to the fact that the one-way analyses of variance indicated that there were no significant differences between male and female respondents' experience of any of the burnout dimensions. Table 12.6 shows the results of stepwise multiple regression analyses for each of the burnout dimensions.

Of those variables entered into the model for emotional exhaustion, age of youngest child and hours worked each week were not found to be significant predictors. Conflict in relationship with spouse/partner emerged as the strongest predictor of emotional exhaustion (â=.351, p=.001), followed by subjective overload (â=.296, p=.002) and role conflict at work (â=.261, p=.008). Overall, the model predicted 32% of variation in emotional exhaustion (Adjusted R^2=.323).

Of the significant correlates of cynicism, only satisfaction with promotion prospects (â=-.423, p=.000), satisfaction in relationship with partner/spouse (â=.219, p=.005), clarity of role (â=-.179, p=.018), and responsibility (â=-.167, p=.030) at work emerged as significant predictors of cynicism. Overall, the regression model explained 25% of the variation in cynicism (Adjusted R^2=.246).

Only 5% of the variation in the personal competence dimension of burnout was explained by the work and family variables in regression model (Adjusted R^2=.052). Significant predictors were clarity of role at work (â=.177, p=.021) and satisfaction with job security (â=.150, p=.048). No significant predictors were found for the professional worth dimension of burnout. Owing to the low level of variation explained in both of the personal efficacy dimensions of burnout, these were excluded from further regression analyses.

16.6.8   Dual and single income couples

In order to test whether respondents who reported that their partners were in paid employment experienced different levels of burnout to those whose partners were not in paid employment, independent samples T-tests were conducted. No significant differences were observed between respondents in single and dual income households for any of the burnout dimensions.

Table 16.5 Pearson correlation coefficients for family, job demands and burnout

	1	2	3	4	5	6	7	8	9	10	11
1. Number of children	1.000										
2. Age of youngest child	.242*	1.000									
3. Age of respondent	.699***	.808***	1.000								
4. Hours worked each week	.217**	-.178	.160*	1.000							
5. Job security	-.021	.083	-.175*	-.177*	1.000						
6. Overload (work)	.115	.024	.150*	.366***	-.236**	1.000					
7. Responsibility (work)	.308***	.079	.383***	.266***	-.044	.000	1.000				
8. Role clarity (work)	.121	-.041	.022	.014	.143	.000	.000	1.000			
9. Satisfaction with pay	.219**	.312**	.293***	.059	.081	.000	.000	.000	1.000		
10. Promotion satisfaction	.107	.138	.021	.011	.242***	.000	.000	.000	.000	1.000	
11. Role conflict (work)	-.055	-.237*	-.185*	.172*	.062	.000	.000	.000	.000	.000	1.00
12. Social satisfaction (work)	-.112	.012	-.181*	-.122	.247***	.000	.000	.000	.000	.000	.000
13. Pace control	.136	.156	.103	.114	.120	.000	.000	.000	.000	.000	.000
14. Marital satisfaction	-.345***	.038	-.325***	-.223**	.040	-.175*	-.184*	-.009	-.049	.146	.089
15. Marital conflict	-.047	-.303**	-.170*	.244*	-.090	.117	-.051	-.008	-.219**	-.062	.236**
16. Social relations	.041	.063	.059	.027	-.007	-.061	.039	-.018	-.051	.214*	-.025
17. Emotional exhaustion	-.051	-.239*	-.128	.306***	-.140	.351***	.086	-.068	-.134	-.100	.275***
18. Cynicism	-.240***	-.187	-.193*	-.090	-.162*	-.041	-.228**	-.193*	-.151*	-.381***	.131
19. Personal competence	.043	.029	-.006	.139	.182	.034	.059	.202**	.049	-.038	.021
20. Professional worth	.073	.091	-.011	-.011-	-.028	-.140	.092	.010	-.088	.197**	.105

*** Correlation is significant at the 0.001 level (2-tailed)
** Correlation is significant at the 0.01 level (2-tailed)
* Correlation is significant at the 0.05 level (2-tailed)

Table 16.5 Pearson correlation coefficients for family, job demands and burnout (cont'd)

	12	13	14	15	16	17	18	19	20
1. Number of children									
2. Age of youngest child									
3. Age of respondent									
4. Hours worked each week									
5. Job security									
6. Overload (work)									
7. Responsibility (work)									
8. Role clarity (work)									
9. Satisfaction with pay									
10. Promotion satisfaction									
11. Role conflict (work)									
12. Social satisfaction (work)	1.000								
13. Pace control	.000	1.000							
14. Marital satisfaction	.029	-.001	1.000						
15. Marital conflict	.003	-.025	.000	1.000					
16. Social relations	-.013	-.004	.000	.000	1.000				
17. Emotional exhaustion	-.111	-.028	-.091	.314***	.016	1.000			
18. Cynicism	-.131	-.041	.181*	.057	-.115	.000	1.000		
19. Personal competence	.115	.065	.012	-.034	.002	.000	.000	1.000	
20. Professional worth	.051	.143	.117	.079	.220**	.000	.000	.000	1.000

*** Correlation is significant at the 0.001 level (2-tailed)
** Correlation is significant at the 0.01 level (2-tailed)
* Correlation is significant at the 0.05 level (2-tailed)

Table 16.6 Work and family predictors of burnout

Variables	R^2 (cumulative)*	Adjusted R^2 (cumulative)*	Beta (â)	p
Emotional exhaustion				
Relationship conflict	.211	.202	.351	.001
Overload	.288	.270	.296	.002
Role conflict (work)	.348	.323	.261	.008
Cynicism				
Satisfaction with promotion prospects	.150	.144	-.423	.000
Relationship satisfaction (spouse/partner)	.213	.201	.219	.005
Clarity of role (work)	.242	.225	-.179	.018
Responsibility (work)	.268	.246	-.167	.030
Personal competence				
Clarity of role (work)	.041	.035	.177	.021
Satisfaction with job security	.063	.052	.150	.048

*   Note that R^2 values are cumulative and represent the change in R^2 as each variable is entered into the model

In order to determine whether predictors of burnout differed between participants in single and dual income relationships, further regression analyses were conducted. First, the sample was split into two groups: those with partners in paid employment and those whose partners were not in paid employment. Secondly, the burnout dimensions were regressed on all of their significant correlates (as shown in Table 16.4). Separate regression models were generated for respondents in dual and single income relationships. The results of these analyses are presented in Tables 16.7 and 16.8.

Table 16.7   Predictors of emotional exhaustion: dual and single income relationships

Single income participants			Dual income participants		
Variable	Beta (â)	p	Variable	Beta (â)	p
Conflict in relationship with spouse/partner	.480	.003	Conflict in relationship with spouse/partner	.377	.005
			Hours worked per week	.376	.005
Adjusted R^2	.280		*Adjusted R^2*	.319	

The model for participants in dual income relationships explained a greater amount of variation in emotional exhaustion (32%) than that for participants in single income relationships (28%). The predictive ability of conflict in relationship with spouse and partner was common for both groups. However, the number of hours worked per week emerged as a significant predictor of emotional exhaustion among participants whose spouse or partner worked but did not predict exhaustion among participants in single income couples. This may be due to the fact that long hours interfere with the extent to which household and family chores can be shared between partners. The ability to share chores is more likely to be an issue when both partners are in paid employment.

Table 16.8 Predictors of cynicism: dual and single income relationships

Single income participants			Dual income participants		
Variable	Beta (â)	p	Variable	Beta (â)	p
Satisfaction with promotion prospects	-.369	.006	Satisfaction with promotion prospects	-.408	.000
Age	-.338	.011	Satisfaction in relationship with spouse/partner	.324	.000
Satisfaction with pay	-.333	.012	Clarity of role (work)	-.277	.002
Adjusted R²	.414		*Adjusted R²*		.280

Together, the work and family variables predicted 41% (Adjusted R^2=.414) of variation in cynicism among participants in single income couples, compared to only 28% (Adjusted R^2=.280) among those whose partners are also in paid employment. Satisfaction with promotion prospects was the strongest predictor for each group. However, other than satisfaction with promotion prospects, cynicism in the two groups was predicted by different independent variables. Age (â=-.338, p=.011) and satisfaction with pay (â=-.333, p=.012) were significant predictors of cynicism among participants whose partners did not work, while satisfaction in relationship with spouse (â=.324, p=.000) and clarity of role at work (â=-.277, p=.002) were significant predictors of cynicism among participants in dual income couples. The positive beta coefficient indicates that the more satisfied an employee is in his/her relationship, the more cynical he/she is likely to be about his/her work. This may reflect the fact that employees resent their excessive involvement in work more when they are in satisfying personal relationships. The negative impact of job demands on relationship quality emerged as an important theme in the comments respondents provided at the end of the questionnaire. For example, one participant wrote

'Due to the demanding and stressful atmosphere in which I work, I have found my relationship with my partner to be a release from the everyday grind. Despite relying on it as a release, I find the pressures and expectations of my work are always on my mind and often come up in conversation when with my partner. It is like having a dark grey cloud hovering over my head.'

Another commented

'In order to meet deadlines, work at home is often required. My wife believes I spend too much time at work. With both my partner and myself working, it is difficult to spend quality time together.'

It is possible that employees who place a high value on maintaining a satisfying relationship with their partner or spouse become cynical when their work interferes with the quality of this relationship.

16.6.9   Parents and childless respondents

In order to test whether respondents with children experienced different levels of burnout to those without children, independent samples T-tests were conducted. No significant differences were observed between respondents with or without children for the emotional exhaustion, personal competence or professional worth dimensions of burnout. However, respondents without children were more cynical than those with children and this difference was statistically significant ($t=2.537$, $p=.012$).

The same procedures were followed to identify differences in the burnout experience between respondents who are parents and those without children. Tables 16.9 and 16.10 show the results of these regression analyses for emotional exhaustion and cynicism respectively.

Table 16.9   Predictors of emotional exhaustion: parents and childless respondents

Childless			Parents		
Variable	Beta (â)	*p*	Variable	Beta (â)	*p*
Hours	.418	.002	Conflict in relationship with partner/spouse	.351	.001
			Subjective overload	.296	.002
			Role conflict (work)	.261	.008
Adjusted R²	.159		*Adjusted R²*		.323

The adjusted R^2 figures in Table 16.9 reveal that the work and family variables included in our study explain a much greater amount of variation in emotion exhaustion among participants with children (32%) than those without children (16%). Also, different variables predict emotional exhaustion among parents and non-parents. Emotional exhaustion in parents is predicted by conflict in relationship with spouse/partner ($â=-.351$, $p=.001$), subjective overload ($â=-.296$, $p=.002$) and role conflict at work ($â=-.261$, $p=.008$). Among non-parents, hours worked each week was the only significant predictor of emotional exhaustion. This suggests that the onset of emotional exhaustion is a much more complex issue among employees with children. In particular, the predictors of emotional exhaustion, such as relationship and role conflict and the feeling of having too many things to do in the time available (subjective overload) suggest that employees with children may struggle to balance the demands of their work and family lives.

Table 16.10   Predictors of cynicism: parents and childless respondents

Childless			Parents		
Variable	Beta (â)	_p_	Variable	Beta (â)	_p_
Satisfaction with promotion prospects	-.312	.016	Satisfaction with promotion prospects	-.490	.000
Clarity of role at work	-.307	.018	Satisfaction in relationship with spouse/partner	.265	.007
			Responsibility (work)	-.249	.010
Adjusted R²	.170		_Adjusted R²_	.274	

The adjusted R^2 figures in Table 16.10 reveal that the work and family variables included in our study explain a much greater amount of variation in cynicism among participants with children (27%) than those without children (17%). Also, different variables predict cynicism in each group. Predictors of cynicism among respondents who are parents were satisfaction with promotion prospects (â=-.490, p=.000), satisfaction in relationship with spouse/partner (â=.265, p=.007) and responsibility at work (â=-.249, p=.010). Cynicism in respondents without children was predicted by only two variables: satisfaction with promotion prospects (â=-.312, p=.016) and clarity of role at work (â=-.307, p=.018).

The combination of work and family predictors of burnout suggest that the burnout concept should not be viewed as occurring exclusively as a result of stressors in the work domain. Therefore, greater attention should be paid to issues at the work-family interface. Furthermore, differences between the predictors of burnout among dual and single income couples and parents and childless respondents suggest that the demographic characteristics of the workforce must be considered when devising strategies to prevent or treat burnout. The increasing diversity of the working population may require that multiple intervention strategies be developed to meet the needs of employees with different demographic characteristics.

16.7   DISCUSSION

16.7.1   Engineers' sense of professional worth

While previous burnout studies have yielded a three-factor structure to the burnout phenomenon, our study yielded a four-factor model. This four-factor model of burnout in the sample of engineers suggests that, as a professional group, engineers derive a feeling of accomplishment from their belief in the social worth or value of their work and that this exists independently of the engineers' beliefs in their own competence at work. This finding was also supported by the qualitative comments written on the back of some questionnaires. For example, one participant wrote *'engineering is a great profession. I, for one, am proud of the contribution my work makes to society'*. Another commented, *'engineering is a tough career choice...but we obviously chose what we did to try and provide benefits/service to the community'*. This sense of professional worth may have positive outcomes for individuals and organisations. For example one participant expressed a willingness to accept the demands of the job because he derived satisfaction from performing a useful role in society. However, it is also possible that this perceived inequity may contribute to other dimensions of burnout, particularly cynicism. In their written comments, the engineers expressed a strong belief that the rewards they enjoy are not commensurate with their contribution to society, leading to a sense of unfair treatment. Perceived equity has been linked with experienced burnout (Van Yperen, 1998; Van Dierendonck, Schaufeli and Buunk, 1998, Gurts, Schaufeli and De Jonge, 1998). According to equity theory, an individual perceives a situation as fair when their own ratio between outcomes and inputs is the same as that of a comparison other. Thus, when engineers compare their input-output ratio unfavourably with ratios enjoyed by other professions they may be more prone to burnout. In our sample, cynicism was associated with low satisfaction with pay and promotion prospects. The qualitative comments provided by a substantial number of respondents suggest that job-related rewards and professional worth are linked. Many respondents expressed the view that the rewards enjoyed by engineers are not commensurate with the role they play in society, relative to other professions. For example, one participant wrote *'Engineers are the most overworked and underpaid professionals on Australia'* another commented that *'The financial reward, compared to the required level of skill, the responsibility and hard work, for engineers, is ridiculously low'*.

It may be possible to manipulate perceptions of equity to help prevent employee burnout. For example, VanDierendonck, Schaufeli and Bunnk (1998) implemented and evaluated an intervention designed to improve the fit between individuals' goals and objectives and organisational rewards. The researchers conclude that cognitively oriented intervention programmes designed to reduce feelings of inequity can effectively reduce burnout. Interestingly, satisfaction with career prospects emerged among our sample as a stronger negative predictor of cynicism than satisfaction with pay. Therefore, it may be that the implementation of career development programmes and help with employees' career planning may be more valuable preventive measures than increased remuneration. However, the sense of inequity may be difficult for a single organization to change because the comparison group, i.e. the other profession, is external to the organisation.

Feelings of inequity were not directly measured in the study and neither does a cross-sectional study enable analysis of direction of any causal relationship between inequity and cynicism. It is therefore impossible to determine whether perceptions of inequity, deriving from a perceived imbalance between a sense of professional worth and reward, precede the development of a cynical attitude towards one's work. These relationships should be investigated further in future studies of burnout in the engineering profession.

16.7.2 Correlates of burnout

The results of the study lend support to the 'ecological' model of burnout, which views burnout as occurring as a result of a complex interaction between individual characteristics and issues in the work and non-work environments. Both personality and job characteristics were found to be significant predictors of burnout, although job characteristics appear to be more important, especially in predicting emotional exhaustion and cynicism. This is promising since it suggests that organisational interventions may successfully prevent burnout. For example, research has investigated the effect on burnout levels of removing job stressors by taking time off, going on vacation, or even taking small breaks from work. Westman and Eden (1997) measured the effect of respite on levels of burnout and found that levels of burnout decreased during a two-week vacation but had risen to almost the same levels as pre-vacation within three weeks of returning to work. The nature of contract-based work within the construction industry may mean that individuals perceive that they are unable to take their leave until the job is done, thus leading to ongoing emotional exhaustion. An APESMA study of Australian engineers found that around a third of engineers find it difficult to take their recreation leave entitlements (33%) and their accrued time-off leave entitlements (34%). Twenty-three percent stated that they were expected to work through their meal breaks. From an organisational perspective it would seem prudent for organizations to put into place measures to enable employees to take time off on a regular basis, rather than at long intervals. For example, efforts could be made within the company to encourage or even enforce the taking of regular holidays. Westman and Eden's (1997) finding that taking one or two days off has a beneficial effect, but that this effect is transitory, highlights the need to take frequent regular breaks. The use of rostered days off (RDOs) to enable professionals working in the construction industry to enjoy long weekends as a respite from job demands may be possible. Indeed, such an initiative has recently been introduced by the Melbourne City Council. The extension of RDOs to professional and white collar employees was suggested by several respondents in their comments at the end of the questionnaire.

Another finding that is consistent with the outcomes of previous burnout studies is that the burnout dimensions are differentially related to personality and job-characteristics. This suggests that multiple intervention strategies may be needed. The burnout dimensions were also predicted by both work and family variables, which suggests that theories of burnout that segregate experiences in the work and family domains are inadequate. Instead, studies should focus on employees' experience at the work-family interface and strategies designed to help employees to achieve a better balance between their work and family lives may be needed.

16.7.3   Workforce diversity

The work-related and family predictors of burnout differed for respondents in different demographic groups. For example, conflict in relationship with spouse or partner was a significant predictor of emotional exhaustion among respondents in both single and dual income couples, but in dual income couples, emotional exhaustion was also predicted by the number of hours the respondent worked each week. This may reflect the fact that it is longer work hours that may prevent respondents from helping with household chores or family care responsibilities, placing increased pressure on a spouse or partner who also works. No family variables predicted cynicism among respondents in single income relationships. Although, perhaps unsurprisingly, satisfaction with pay was a more important predictor of cynicism among respondents in single compared to dual income couples. Satisfaction in relationship with spouse/partner was a significant predictor of cynicism among respondents in dual income couples, although the association was a positive one. Thus, the more satisfied a respondent was with his/her relationship, the more cynical he/she tended to be. Reasons for this are unclear and this finding warrants further investigation. Work and family predictors of burnout also differed among childless respondents and those who are parents. As expected, family variables were more salient for respondents with children. For example, conflict in relationship with spouse/partner was a significant predictor of emotional exhaustion among respondents with children but not those without. Interestingly, hours predicted emotional exhaustion among respondents with no children while subjective overload, or a feeling of having too much to do in the time available, predicted emotional exhaustion in parents. Satisfaction in relationship with spouse/partner was a significant predictor of cynicism in respondents with children but not among those without children. Again, the association is a positive one. One possible reason for this is that parents derive greater satisfaction from their family life than non-parents and their work is less salient. An examination of the cynicism items in the burnout scale supports this interpretation as they suggest a lack of interest in work, which may be relative to an enhanced interest in family life. For example, items include 'I have become less interested in my work since I started this job' and I have become less enthusiastic about my work'. Again, this finding warrants further investigation and future studies may usefully consider work and family salience in the development of burnout.

The differences between employees with different family characteristics highlight the need to tailor intervention strategies to the needs of the workforce at any point in time. In the context of an increasingly diverse workforce, a staggering increase in the number of dual income couples and the increasing number of non-traditional households, such as lone parents and same sex couples, employees' needs will vary. Furthermore, these needs will change over time. For example, some writers suggest that elder care may eclipse the need for childcare as our population ages and birth rates diminish. It is likely that, as the profile of the workforce changes, so too will the family issues impacting on employees' experienced burnout. Preventive strategies might include implementing family friendly policies designed to help employees' balance work and family demands on their time. For example, in implementing more flexible work arrangements, providing for permanent part time work and providing assistance with childcare or elder care. Not only are these policies in-keeping with the need to manage diversity and ensure equal opportunities to minority groups but also may improve the retention rate among certain groups, such as women, who remain under-represented among construction industry professionals (Court and Moralee, 1995).

16.7.4 Organisational interventions

Individual-oriented approaches to the prevention of burnout include training employees to adopt effective coping methods, time management and negotiation skills (Gmelch and Gates, 1998). However, the fact that many situation and job factors play a significant role in the development of burnout means that strategies focusing on the individual might be of limited effectiveness. This limitation is probably compounded by employees' relatively low level of control over work-related sources of stress. Our findings confirm that the feeling of being valued has a positive effect on engineers' sense of professional worth, that is, having opportunities for personal fulfilment through promotion within the workplace. Maslach, Sahuafeli and Leiter (2001) suggest that interventions should focus on achieving a better job-person fit in six areas of organisational life. The six areas of workload, control, reward, community, fairness and values offer a broader range of options for organisational intervention than simply focusing on individual-oriented interventions. Thus, rather than train people to cope with overload and stress, this framework recognises that the negative effects of stress, such as burnout, might be contained where employees feel empowered, that they are doing something important or that their work is valued, that they are being well-rewarded for their efforts.

Despite the importance of family variables in predicting burnout, satisfaction in relationship with spouse or partner did not moderate the effect of burnout on turnover intention. The only exception to this outcome was the finding that the social relations dimension of relationship satisfaction had a significant moderating effect in the exhaustion-turnover association. These findings suggest that family-friendly work practices may be a useful preventive strategy for burnout but they are unlikely to influence the turnover intention of employees' suffering from burnout. If this is true, then early intervention aimed at prevention rather than treatment of burnout is recommended.

16.8   CONCLUSIONS

The results of the study suggest that, in the engineering profession, burnout arises as a result of a complex interaction of individual, work-related and situational factors. Burnout is not explained by any single 'cause'. This precludes the identification of a single, quick-fix solution to the problem. Despite this, job characteristics appear to be more important predictors of burnout than personality traits, suggesting that a preventive strategy may be to re-design engineers' work environment. In particular, long hours, unpaid overtime, responsibility, perceived inequity and job role conflict and clarity should be addressed. However, the job characteristics differentially predicted the burnout dimensions, suggesting that multiple intervention strategies are required. The results suggest that emotional exhaustion is associated with overload and role conflict and thus firms may consider ways in which the workload can be managed. Research evidence suggests that respite from work stressors also mitigates the emotional exhaustion and ensuring that employees take regular breaks, perhaps by offering 'rostered days off' to engineers who would otherwise not enjoy them, may be a viable preventive strategy. The cynicism dimension of burnout was related more to engineers' perceptions of inequity. Efforts may be made to ensure employees' contributions are recognised and rewarded. It is also possible that engineers' responsibilities can be made less onerous by the implementation of management systems, for example occupational health and safety or environmental management systems, establishing company standards and procedures. It is recommended that organizations identify their employees' priorities and be creative in formulating ways to respond to employees concerns.

The interaction between work-family variables in predicting burnout suggests that companies seeking to prevent burnout may need to address issues beyond the immediate work environment. This may include an assessment and management of the impact of work on employees' experiences of family or non-work life, including leisure time, sport and social activities. The implementation of work-life balance employment practices is not widespread in the construction industry. However, the provision of benefits, such as childcare assistance, parental leave, part-time work options, flexible work hours, or regular rostered days off for professional as well as non-professional employees, could yield benefits for both individuals and organizations. The finding that predictors of burnout varied by family characteristics (e.g. parent or non-parent, dual or single income household) also suggests that, if work-life balance initiatives are to effectively reduce employees' burnout, these must be tailored to the needs of the workforce. The increasing diversity of the workforce may require that flexible, 'cafeteria-style' benefit programmes be implemented.

Limitations and future research

The study was cross-sectional in nature. Cross sectional studies are unable to yield information about the causal relationships between variables. Furthermore, it is recognized that basing conclusions about the sources of burnout on the experience of different groups of individuals who have been in the industry for varying periods of time is problematic. Due to self-selection, it is likely that those who remain in the construction industry for a long period are those who are better able to cope with its demands, making survival bias a significant consideration. The response rate of only 36% could indicate that employees under a significant degree of time stress did not participate. Thus the study may underestimate the extent to which burnout is relevant to the work of engineers. The limitations of self-reporting and the healthy worker effect (i.e. the survival bias) should be noted. In addition, there was a strong reluctance amongst some firms' management for employees to take part. Managers expressed fears that the questionnaire itself may invoke burnout or cause industrial problems. Thus, firms that agreed to take part may have had environments that were less stressful or confrontational, or possibly had smaller workgroups with a degree of internal support. The under-representation of contractors' employees in the sample is of concern, as contractors are mainly site-based and thus exposed to significant issues of safety management and interpersonal conflict. This population could be seen to be at a higher risk for burnout and future studies should attempt to obtain data from this group.

The study made no attempt to identify the process by which burnout occurs. Research evidence suggests that burnout occurs as a process rather than the result of a discrete, acutely stressful event. There are two schools of thought concerning the intertemporal sequence of the burnout dimensions. For example, Maslach and her colleagues argue that employees' facing unrelenting, excessive job demands will experience emotional fatigue. As a mechanism for coping with this fatigue, the employees will distance themselves from their work. Cynicism is the manifestation of this distancing. As they devote less time and energy to their work, they experience a diminished sense of personal accomplishment and a lower sense of self-efficacy. A competing process model of burnout is purported by Golembiewski (1989) who suggests that depersonalisation or cynicism is the initial manifestation of burnout. This leads to diminished feelings of personal accomplishment, which, if experienced over a sustained period, culminates in emotional exhaustion. Studies addressing the debate as to the stages in the burnout process are inconclusive, though recent research lends support for Maslach's sequential model (Cordes, Dougherty and Blum, 1997). It is important to understand the process by which burnout occurs because the temporal order of burnout stages has important implications for the early detection and prevention of burnout. Further research, using a longitudinal design, should address this issue among construction industry professionals.

16.9   REFERENCES

APESMA, 2000, *APESMA working hours and employment security survey report*, PowerPoint presentation, (personal communication).

Appels, A. and Schouten, E., 1991, Burnout as a risk factor for coronary heart disease. *Behavioral Medicine*, 17, pp. 53-59.

Lee, R.T. and Ashforth, B.E., 1990, On the meaning of Maslach's three dimensions of burnout. *Journal of Applied Psychology*, 75, pp. 743-7.

Ashforth, B.E. and Lee, R.T., 1997, Burnout as a process commentary on Cordes, Dougherty and Blum. *Journal of Organizational Behavior*, 18, **6**, pp. 703-708.

Aryee, S., Dual-earner couples in Singapore: An examination of work and nonwork sources of their experienced burnout. *Human Relations*, 46, pp. 1441-1468.

Barnett, R., 1994, Home-to-work spillover revisited: A study of full time employed women in dual-earner couples. *Journal of Marriage and the Family*, 56, pp. 647-656.

Bacharach, S.B., Bamberger, P. and Conley, S., 1991, Work-home conflict aong nurses and engineers: mediating the impact of role stress on burnout and satisfaction at work. *Journal of Organizational Behavior*, 12, pp. 39-53.

Brookings, J.B., Bolton, B., Brown, C.E. and McEvoy, A., 1985, Self-reported job burnout among female human service professionals. *Journal of Occupational Behaviour*, 6, pp. 143-150.

Burke, R.J. and Greenglass, E., 1995, A longitudinal study of psychological burnout in teachers. *Human Relations*, 48, **2**, pp. 187-202.

Burke, R.J. and Greenglass, E.R., 2001, Hospital restructuring, work-family conflict and psychological burnout among nursing staff. *Psychological Health*, (in press).

Carayon, P., 1992, Longitudinal study of job design and worker strain: preliminary results. In *Stress and Well being at Work: Assessments and interventions for Occupational Mental Health*, edited by J.C. Quick, L.R. Murphy, L.R. and J.J. Hurrell (Washington: American Psychological Association, Washington).

Carroll, J.F.X. and White, W.L., 1982, Theory building: integration individual and environmental factors within an ecological framework. In W.S. Paine (ed.), *Job Stress and Burnout: Research, Theory and Intervention Perspectives* (Beverley Hills: Sage Publication), CA, pp. 41-58.

Chang, E.C., Rand K.L. and Strunk, D.R., 2000, *Personality and Individual Differences*, 29, pp. 255-263.

Cohen, S. and Wills, T.A., 1985, Stress, social support and the buffering hypothesis. *Psychological Bulletin*, 98, pp. 310-357.

Cooper, C. and Marshall, J., 1977, *Understanding Executive Stress*, (New York: PBI Books).

Corcoran, K., 1995, Measuring Burnout: An updated reliability and convergent validity study. In *Occupational Stress: A Handbook*, edited by R. Crandall, and P.L. Perrewe (Washington: Taylor and Francis).

Cordes, C.L. and Dougherty, T.W., 1993, A review and an integration of research on job burnout. *Academy of Management Review*, 18, pp. 621-656.

Cordes, C.L., Dougherty, T.W. and Blum, M., 1997, Patterns of burnout among managers and professionals. *Journal of Organizational Behavior*, 18, pp. 685-701.

Cotton, P., 1995, *Psychological Health in the Workplace: Understanding and Managing Occupational Stress* (Melbourne: Australian Psychological Society Ltd).

Court, G. and Moralee, J., 1995, *Balancing the Building Team: Gender Issues in the Building Professions*, The Institute for Employment Studies, Report 284.

Densten, I.L., 2001, Re-thinking burnout. *Journal of Organizational Behavior*, (in press).

Dolan, S.L., 1995, Individual, organizational and social determinants of managerial burnout: Theoretical and empirical update. *Occupational stress: A handbook*, edited by Rick Crandall, Pamela L. Perrewe, Philadelphia, PA, pp. 223-238 (US: Taylor & Francis).

Etzion, D., 1984, Moderating effect of social support on the stress-burnout relationship. *Journal of Applied Psychology*, 69, pp. 615-622.

Eckenrode, J. and Gore, S., 1990, *Stress between Work and Family*, (New York: Plenum Press).

Eysenck, H.J. and Eysenck, S.B.G., 1977, *Personality Structure and Measurement* (London: Routledge & Kegan Paul).

Freudenberger, H.J., 1974, Staff burnout. *Journal of Social Issues*, 30, pp. 159-165.

Gaines, J. and Jermier, J.M., 1983, Emotional exhaustion in a high stress organization. *Academy of Management Journal*, 26, pp. 567-586.

Geurts, S., Schaufeli, W. and De Jonge, J., 1998, Burnout and intention to leave among health-care professionals: A social psychological approach. *Journal of Social and Clinical Psychology*, pp. 341-362.

Gmelch, W.H. and Gates, G., 1998, The impact of personal, professional and organizational characteristics on administrator burnout. *Journal of Educational Administration*, 36, **2**, pp. 146-159.

Golembiewski, R.T. and Munzenrider, R., 1981, Efficacy of three versions of one burn-out measure: MBI as total score, sub-scale scores, or phases? *Journal of Health & Human Resources Administration*, 7, pp. 290-323.

Kamerman, S.B. and Kahn, A.J., 1981, *Child Care, Family Benefits and Working Parents*, (New York: Columbia University Press).

Koeske, G.F. and Koeske, R.D., 1989, Construct validity of the Maslach Burnout Inventory: a critical review and reconceptualization. *Journal of Applied Behavioral Science*, 25, pp. 131-132.

Lee, R.T. and Ashforth, B.E., 1993, A longitudinal study of burnout among supervisors and managers: comparisons between the Leiter and Maslach (1988) and Golombiewski *et al.*, 1986, models, *Organinizational Behavior and Human Decision Processes*, 54, pp. 369-398.

Lee, R.T. and Ashforth, B.E., 1996, A meta-analytic examination of the correlates of the three dimensions of job burnout. *Journal of Applied Psychology*, 81, **2**, pp. 123-133.

Leiter, M.P., 1990, The impact of family resources, control coping and skill utilization on the development of burnout: A longitudinal study. *Human Relations*, 43, pp. 1067-1083.

Maslach, C., Jackson, S.E. and Leiter, M.P., 1996, *Maslach Burnout Inventory Manual*, Palo Alto (CA: Consulting Psychologists Press), 3rd edition.

Maslach, C., Schaufeli, W.B. and Leiter, M.P., 2001, Job burnout. *Annual Review of Psychology*, 52, pp. 397-422.

Schaufeli, W. and Enzmann, D., 1998, *The Burnout Companion to Study and Practice: A Critical Analysis.*, London; Philadelphia, (PA: Taylor & Francis).

Schutte, N., Toppinen, S., Kalimo, R. and Schaufeli, W., 2000, The factorial validity of the Maslach Burnout Inventory-General Survey (MBI-GS) across occupational groups and nations. *Journal of Occupational & Organizational Psychology*, 73, (Part 1), pp. 53-66.

Shinn, M., 1982, Methodological issues: Evaluating and using information. In *Job stress and burnout*, edited by W.S. Paine (Newbury Park: Sage).

Tennant, C., 1996, Experimental stress and cardiac function. *Journal of Psychosomatic Research*, 40, pp. 569-583.

Van Dierendonck, D., Schaufeli, W.B. and Buunk, B.P., 1998, The evaluation of an individual burnout intervention program: the role of inequity and social support. *Journal of Applied Psychology*, 83, pp. 392-407.

Van Yperen, N.W., 1998, Informational support, equity and burnout: the moderating effect of self-efficacy. *Journal of Occupational and Organizational Psychology*, 71, pp. 29-33.

Westman, M. and Eden, D., 1997, Effects of respite from work on burnout: vacation relief and fade-out. *Journal of Applied Psychology*, 82, pp. 516-527.

Westman, M., Etzion, D. and Danon, E., Job insecurity and crossover of burnout in married couples. *Journal of Organizational Behavior* (in press).

Wolpin, J., Burke, R.J. and Greenglass, E.R., 1991, Is job satisfaction an antecedent or a consequence of psychological burnout? *Human Relations*, 44, pp. 193-209.

Wright, T.A. and Bonett, D.G., 1997, The contribution of burnout to work performance. *Journal of Organizational Behavior,* 18, **5**, pp. 491-499.

CHAPTER 17

Health - Safety's Poor Cousin - The Challenge for the Future

Alistair Gibb

17.1 CONSTRUCTION'S ILL-HEALTH PROBLEM

WHO, the World Health Organisation, defines health as a state of complete physical, mental and social well-being, and not merely the absence of disease or infirmity (WHO, 1946). The APaCHe[1] team has developed this argument to apply to organisations as well as individuals. Healthy organisations operate to meet their objectives efficiently, providing healthy, fulfilling employment for their workers. In promoting healthy individuals and organisations, prevention is considerably more effective than cure. To be effective, prevention requires a holistic, whole-system, complete life-cycle approach recognising and addressing the complex influences arising from individual and collective attitudes and actions.

To date significant effort has been directed towards the more immediate, high profile (and perhaps more easily solvable) problem of safety rather than the more elusive health matters. Health and safety is often used by experts and non-experts alike to represent only safety. Most health and safety managers, supervisors or inspectors throughout the world have little more that a very rudimentary knowledge of occupational health issues. However, ill-health continues to kill and disable many construction workers worldwide. In fact, the delay in the effects becoming evident is one of the key reasons why the subject should be taken more seriously.

Every year many thousands of construction workers suffer from work-related ill health. In 1995, the UK's Self-reported Work-related Illness Survey found an estimated 134,000 construction-related workers reported a health problem caused by their work, resulting in an estimated 1.2 million days lost in a work-force of 1.5 million. The construction sector prevalence rate for self-reported work-related illness was 6.8% compared with an all-industry average of 4.7%. In particular for construction, asbestos-related disease resulted in about 700 deaths annually; there were 96,000 musculoskeletal disorder cases; 15,000 respiratory disease cases; 6,000 cases of skin disease and 5,000 noise induced hearing loss cases. Recent research also indicates that construction workers are particularly at

---

[1] APaCHe is a partnership of industrialists, academics and government with the aim of achieving a step change in the physical, psychological and social health of all those involved in the industry.

risk from hand-arm vibration although this was not evident in the 1995 figures. At a recent conference in Illinois, Chuck Bailey of the US CPWR (Center to Protect Worker's Rights) stated that US construction employed 6% of the country's workforce but accounted for 26% of the hearing loss claims. He also explained that 8.2% of the incidences of carpal tunnel problems in the US occur in construction compared to only 3% in keyboard operators even though key-board use tends to be the main area of attention by the media. Many of these conditions have a long latency period between exposure and onset of symptoms, leading to difficulties attributing the root causes of the ill-health condition.

The UK Government estimates that asbestos-related diseases will kill up to 10,000 people each year by the year 2020 (Croner, 1998) and 36,000 people will suffer from 'vibration white finger' with over a million considered to be at risk (HSE, 1998). UK and US construction workers have the third highest rate of skin disorders (Burkhart, et al., 1993) and cement is believed to cause up to 40% of the UK's occupational dermatitis (Beck, 1990). The problem is just as bad elsewhere in Europe. For example, in 1993, the 2000 largest Portuguese enterprises lost more than 7.7 million working days as a result of illness, representing 4.5% of all working days for these companies (Gründemann and van Vuuren, 1998). 20% of sickness absence among 15 to 66 year-olds in Denmark is caused by the working environment (Gründemann and van Vuuren, 1998). Dermatitis is becoming the main reason for absence from work in Germany (Berger, 1998).

The European construction sector is aware of these ill-health issues, but they have difficulty managing them because of the highly mobile workforce, lack of continuity of employment, predominance of self-employed workers, and short-term, rapidly changing workplaces. There are also other, more general, life-style-health issues for construction workers, often linked with the peripatetic aspects of their employment. Furthermore, it has been estimated that 80% of construction workers in the UK have not even registered with their local doctor.

There are variations in the way that health and safety legislation and control are handled in the various European States (European Commission, 1995 and 1997), however, in general, health and safety are considered together, with health very much the poor relation.

Recent work in Ireland has confirmed occupational health as a major problem for construction (Brenner and Ahern, 2000). In this study of Irish construction workers between 1981 and 1996 found that the top three reasons for absence were:

1.	Injury	4734
2.	Infectious disease	3981
3.	Musculoskeletal disorder	2057

The top three causes of permanent disability were:

1.	Cardiovascular disease	954
2.	Musculoskeletal disorder	928
3.	Bronchitis/emphysema/asthma	415

The pace is increasing to seriously consider health issues and there are moves in the UK to set up national occupational health support scheme. Two reports have been produced (Amey Vectra, 2001 and UCATT, 2001) supporting this proposal. However, the plans are currently stalled pending resolution of issues such as who should pay for this service and what will be done with the sensitive information gathered. The government funded Amey Vectra report established the following focus areas requiring action (Amey Vectra, 2001):

* Musculoskeletal   (Manual handling, upper limb disorders,
                             repetitive or forceful movements)
* Hazardous substances   (Asbestos, solvents, silica, MDF, wood dust)
* Noise exposure   (Power tools, machinery)
* Vibration   (Hand held tools)
* Biological exposure   (Bacteria, fungi)
* Radiation exposure   (Ultra violet, lasers, infra-red)
* Stressors   (Work pressure/duration, life style)

Ill health costs are increasing, both to the industry and to the broader society. The industry must meet the costs from lost production of ill workers, insurance claims and compensation following forced absences or early retirement, recruitment of replacement personnel and retraining of affected workers who need to be assigned new tasks. Society must cover the increased costs from social security payments, health care and early retirement. The precise nature of how these costs will present themselves will differ from country to country but the overall problems remain: annual costs for sickness benefits are becoming more significant nationally:

* 2.4 Bn Euro/US$ (Belgium in 1998)
* 3.7 Bn Euro/US$ (Denmark in 1998)
* 20 Bn Euro/US$ (UK in 1997)
* 30.5 Bn Euro/US$ (Germany in 1998).

17.2   MANAGING CONSTRUCTION HEALTH RISKS

Clients (owners) and contractors have a moral and legal (in Europe) duty to develop effective health risk management systems, based on full and careful appraisal of the health risks to which all their employees (including subcontracted workers) will be exposed. But occupational health management is not easy, the main difficulties include:

* Health is a complex issue.

- Health traditionally has a low profile compared to safety.
- Health management is associated with potentially large set-up costs.
- Benefits are not immediate and are consequently difficult to demonstrate.
- The latent effect of illness is difficult to quantify.
- Exposure to health risks can be multiple with changes in the nature and level.
- Long term strategies are required.
- The workforce is sizeable, temporary and mobile.
- Many workers are not directly employed.
- The workers are often not interested, perhaps due to possible loss of livelihood or the endemic 'macho culture' of construction workers.
- There is a lack of health expertise within the industry (Gibb, Haslam and Gyi, 2001).
- There is even more under-reporting than for safety breaches.

These are real problems, however, construction must avoid using them as excuses and grasp the opportunity to do something about this unacceptable situation.

17.3   STRATEGIC HEALTH MANAGEMENT SHOULD LEAD TO BENEFITS

The main aim of health management is prevention rather than cure after the event. Effective, strategic management action by senior managers will reduce accidents and incidents leading to injury and ill-health. However, despite this knowledge, even though construction managers understand their role in safety and the prevention of injury they are less clear regarding the prevention of ill-health (Gyi, Haslam and Gibb, 1998). This 'reality-gap' must be acknowledged and eradicated. There is little available guidance on health management in construction, at least in the UK, in fact, one of the only health management manuals is the *ECI Guide to Managing Health in Construction* (Gibb, Gyi and Thompson, 1999). The UK HSE have been planning to release a manual for some years but this has still not materialised. However, there are indications that the topic is starting to turn the corner with various key parties beginning to show considerable interest.

Whilst the first action for senior management is to address these issues strategically, they must also be worked out at the project-level as explained in the ECI health manual (Gibb, Gyi and Thompsonl, 1999). Health protection remains a line management responsibility with doctors, medical professionals and other experts assisting in the development and implementation of health management programmes. It is also essential to determine the need for and type of medical support on site before an appropriate health risk management system can be developed.   The management of both client and contractor must ensure that appropriate control measures are provided, and their effectiveness evaluated.

The Exploration & Production Forum (1993), benefiting from the strong health and safety culture in the offshore sector, stated that a successful health management system should cover the following areas:

- Formulating policy and developing the organisation. This includes identifying health objectives and reviewing progress towards their achievement.
- Planning, implementing and auditing of health activities and standards.
- Performance measurement and reporting.

They also advise that at least the following actions should have been taken:

- Roles defined.
- Resources allocated.
- Means of achieving objectives specified.
- Performance monitored.
- Specific task targets and procedures defined.
- Compliance checked.

Health risk management tends to concentrate on the identification of health hazards that may arise. The client (owner) or designer can do this assessment, often called 'Hazcon 1' at this stage of a project, particularly in the Engineering Construction sector, however the client must take the ultimate responsibility. The Hazcon approach is described further in the ECI SHE manual (ECI, 1995). Hazcon should include the following:

- Overall project method statement.
- Experience gained from previous projects including close out reports.
- Similar activities in relevant locations.
- Health audits carried out on similar project.
- General background information on planned areas of operation.
- Environmental issues.
- Major risk factors.
- Substances already at site and those that will be taken to site.
- Emergency response procedures.
- Incident reporting procedures.
- Published guidance by enforcing authorities (ECI, 1995).

In the UK, a number of leading organisations are working hard to 'turn the tide' on construction health, for example the Channel Tunnel Rail Link team have a number of occupational health initiatives running. They are also looking to the long term by supporting strategic research efforts such as the Hand-Arm Vibration study by Loughborough University (expecting an early 2004 start). This work will look at use of tools and equipment to establish the actual usage profile to improve predictors and inform action plans for HAVS.

Table 17.1 shows the sort of items that could effect the health of an employee.

Physical	Chemical	Biological	Mechanical	Psycho-social
Noise	liquids	insects	posture	stress
Vibration	dusts	fungi	movement	work pressure
ionising radiation	fumes	bacteria	repetitive tasks	monotony
non-ionising radiation	fibres	viruses		unsociable hours
heat and cold	mists			ergonomic
Electricity	gases			
Pressure	vapours			

Figure 17.1   Factors affecting workers health (Gibb, Gyi and Thompson, 1999)

17.4   APACHE – A PARTNERSHIP FOR CONSTRUCTION HEALTH

The success of the construction industry in any country depends on the commitment, innovation and productivity of its people at all levels, managerial, professional, skilled and unskilled. Construction needs workers who are healthy and motivated. APaCHe is a UK-based partnership between industrialists, academics and government with the aim of achieving a step change in the physical, psychological and social health of all those involved in the industry. The vision is a construction industry with:

- healthy workers
- healthy work systems and practices
- healthy work organizations.

The key industrial partner in APaCHe is the Construction Confederation, providing cross-sector representation of construction organisations of all sizes, including small to medium sized enterprises (SMEs). Academic hub partners are Loughborough University and the Institute of Work, Health and Organisations ((I-WHO, University of Nottingham). Together, the academic partners offer construction, health, ergonomics, psychology, work organisation, and design expertise, along with a range of medical specialities. The final hub partner is the UK Health and Safety Executive, the key government body responsible for policy and enforcement issues.

17.5   HEALTHY CONSTRUCTION PRODUCTS AND PROCESSES

The UK's Health and Safety Executive (HSE) has started to raise the profile of construction health following the launch of their 'Backs for the Future' guidance and a conference in October 2000 entitled 'Tackling Health Risks in Construction'. Health and safety is beginning to be seen by employers and employees as an important issue to be addressed collaboratively, as exemplified by the trades union/ employers (TUC/CBI) Partnerships in Prevention programme. There are potential synergies from the recent 'people first economy' strategy developed by the UK Department for Trade and Industry (DTI, 2001), with Respect for People, Investors in People, the Learning and Skills Councils and the commitment to lifelong learning becoming more prominent. With health issues closely linked to construction processes, techniques and technology, an industry/academia/ government collaboration is much needed, providing the essential support required for real improvement to be achieved.

The construction process leads to health consequences in the way in which construction is undertaken and the risks that arise. Project teams make choices about how to build and these directly affect the health of workers, for example decisions taken whether to use pre-assembly or to fabricate on-site. Pre-assembly should lead to better safety and health. However, at present, influences on health are not understood sufficiently to support endorsement of the pre-assembly on the basis of health. Furthermore, the industry invariably uses the changing, one-off nature of sites as an excuse for not planning and organising the workplace efficiently and this is to the detriment of the health of operatives. Construction projects are becoming more complex, with time and cost constraints more severe and increased reports of stress amongst construction personnel are being associated with this (Madine, 2000). Recent initiatives such as Haliburton's Stress Toolkit training pack are beginning to address this issue.

Construction technology and techniques, tools and equipment are rarely designed with the health of users in mind. Several leading organisations working with Loughborough University are addressing some of these challenges. These include:

- piloting an innovative electric breaker tool to reduce hand/arm vibration;
- improving glove design to provide enhanced grip and dexterity (making it more likely that the PPE will be used); and
- developing seat design for construction plant as many drivers spend their whole working day, including breaks, in their seats leading to a significant risk of increased back pain.

Current research at Loughborough has identified that poorly designed and inappropriately used tools and equipment may be a significant contributory factor to accident causality (Gibb, et al., 2002 and Torrens et al., 2000).

17.6   HEALTHY CONSTRUCTION ORGANISATIONS AND INDUSTRY

The ECI health management manual claims that larger organisations are just starting to take health seriously but, to date, many small to medium sized enterprises (SMEs) have not had the infrastructure, resources or awareness to address these matters. The development of organisations with a health orientation has been severely hampered by construction's sub-contract and sub-sub-contract culture and recent 'right-sizing' exercises in many organisations. There is a significant opportunity for expertise and technology transfer, as well as a need for development of appropriate strategies and practices.

The UK's Construction Design & Management (CDM) regulations (derived from the EU Directive on Temporary and Mobile Construction Sites) seek to eliminate hazards through the design process. However, there have been concerns about the effectiveness of legislation implementation questioning the extent to which it has resulted in a real reduction of health and safety hazards following consideration by design teams. Also, most recent evaluation has concentrated on safety rather than health. Furthermore, taking a human factors perspective, design extends to include the design of the work place, tools and techniques as well as the finished building or structure which goes beyond the traditional use of the term design implicit in CDM.

A new research project funded by the UK Department of Trade and Industry and the Health & Safety Executive is developing risk assessment tools for designers to effectively consider the occupational health of construction workers (Designing for Healthy Construction, 2002). This work is looking first at the Engineering Construction sector (Petro-chemical/power generation) as they have a significantly better attitude towards occupational health issues and generally a more developed health and safety culture.

A major challenge for UK construction is the current shortfall between the supply of qualified new recruits and increased industrial demand. The growth rate of UK construction output is expected to vary between 2.2% and 3.5% by 2005, with 64,000 new workers needed each year. At the same time, a 'major cause for concern is the sharp decline in the proportion of younger workers in the industry and an increase in those aged 45 and over' (DTI, 2000). To make matters worse, low levels of unemployment in most sectors mean that construction will have to compete harder to fulfil its labour requirements. There are already skill shortages as replacement has fallen below the yearly forecasted requirements especially in the largest occupations such as carpenters and joiners, bricklayers, plumbers and electricians. The steady loss of workers due to ill health is something the industry must now address urgently.

The recent 'Revitalising' initiative from the UK Health and Safety Commission (HSC) has set construction targets to:

- reduce the number of working days lost per 100,000 workers from work-related injury and ill health by 30% by 2010;
- reduce the incidence rate of fatal and major injuries by 10% by 2010;
- reduce the incidence rate of cases of work-related ill health by 20% by 2010;
- achieve half of the improvement under each target by 2004.

These targets will not be achieved unless there is a considerably greater emphasis on occupational health by the construction industry than at present.

17.7   CONCLUSION - HEALTHY CONSTRUCTION WORKERS

Whilst healthy products, processes, organisations and industry are key objectives, the ultimate aim of any health strategy is that the health of individuals will improve. To be successful, any industry depends on the commitment, innovation and productivity of its people at all levels, managerial, professional, skilled and unskilled. Therefore, construction needs healthy and well motivated workers. The challenge for the future is a construction industry with:

- healthy workers
- healthy work systems and practices
- healthy work organizations.

Construction's desperate heritage of worker ill-health is not going to change overnight, but at least some organisations seem to be starting to take it seriously – the rest of industry must mobilise to ensure that health is no longer safety's poor cousin.

17.8   REFERENCES

Amey Vectra, 2001, National occupational health support scheme for the construction industry: An initial feasibility study. *HSE Report*, 300-2009 R1, September, 88pp.

Beck, M.H., 1990, Letter to editor. *British Medical Journal*, 301, pp. 932.

Berger, J., 1998, Cement dermatitis - An international problem. In *Proceedings of CIB W99 Conference - Environment, Quality and Safety in Construction*, Lisbon, June, 1-8.

Brenner, H. and Ahern, W., 2000, Patterns of ill-health in Irish construction workers. *Construction Employees Health Trust*, Eire.

Burkhart, G. et al, 1993, Job tasks, potential exposures and health risks of labourers employed in the construction industry. *American Journal of Industrial Medicine*, 24, pp. 413-425.

Croner, 1998, TUC urges asbestos switch. Construction safety, *Briefing*, (Croner Publications Ltd), Iss 35, August, p.3, http://www.healthandsafety-centre.net

Designing for Healthy Construction, 2002, *DTI Partners in Innovation Research Project*, European Construction Institute and Loughborough University (Project manager: Alistair Gibb).

DTI, 2000, *Excellence and Opportunity, A Science and Innovation Policy for the 21st century*, Department of Trade and Industry, para. 33, p. 10.

DTI, 2001, *Opportunity for All in a World of Change – A White Paper on Enterprise, Skills and Innovation*, Department of Trade and Industry, section 2.

E&P Forum, 1993, *Health Management Guidelines for Remote Land-based Geophysical Operations*, Report No. 6.30/190 (Oxford: Words and Publications).

European Commission, 1995, *Labour Inspection (Health & Safety) in the European Union*, Senior Labour Inspectors Committee, Copies available from: Unit 5 Accident/Injuries, The European Commission, The Plateau de Kirchberg, L-2920 Luxembourg.

European Commission, 1997, *Labour Inspection (Health & Safety) in the European Union - A Short Guide (Austria, Finland, Sweden)*, Senior Labour Inspectors Committee, Copies available from: Unit 5 Accident/Injuries, The European Commission, The Plateau de Kirchberg, L-2920 Luxembourg.

ECI, 1995, *Total Project Management of Construction Safety, Health and Environment*, European Construction Institute, edited by A.G.F. Gibb, D. Tubb and Thompson, T., 2nd edition (London: Thomas Telford), ISBN 0-7277-2082-1. pp. 206.

Gibb, A.G.F., 2002, Health*, Safety's Poor Cousin – Keynote presentation, In CIB W99 Triennial International Conference - One Country Two Systems*, edited by S. Rowlinson, Hong Kong, May, Spon (in press).

Gibb, A.G.F., Haslam, R.A., Gyi, D.E. and Torrens, G., 2002, APaCHe – A partnership for construction health. *In Proceedings of the W99 Triennial International Conference - One Country Two Systems*, edited by S. Rowlinson, Hong Kong, May, CIB Publication 274, pp. 55-60, ISBN 9627757047.

Gibb, A.G.F., Haslam, R.A., Gyi, D.E., Duff, R., Hide, S. and Hastings, S, 2002, ConCA – Preliminary results from a study of accident causality. *In Proceedings of the W99 Triennial International Conference - One Country Two Systems*, edited by S. Rowlinson, Hong Kong, May, Hong Kong, CIB Publication 274, pp. 61-68, ISBN 9627757047.

Gibb, A.G.F., Haslam, R.A. and Gyi, D.E., 2001, Cost and benefit implications and strategy for addressing occupational health in construction. In *Conseil Internationale de Batiment, Working Commission W99*, Barcelona, October, pp. 29-38, ISBN 84-7653-790-5.

Gibb, A.G.F. Gyi, D.E. and Thompson, T., 1999, *The ECI Guide to Managing Health in Construction* (London: Thomas Telford), 170 pp. ISBN 0-7277-2762-1.

Gründemann, R.W.M. and van Vuuren, C.V., 1998, Preventing absenteeism at the workplace: A European portfolio of case studies. *European Foundation for the Improvement of Living & Working Conditions*, Dublin, Ireland.

Gyi, D.E., Haslam, R.A. and Gibb, A.G.F., 1998, Case studies of occupational health management in the engineering construction industry. *Occupational Medicine*, 48, **4**, pp. 263-271, ISSN 0962 7480/98

HSE, 1998, Ignore workers health and profits will suffer warns chairman at launch of new HSE health campaign. *Health and Safety Executive Press Release*, 22 May, E114, p. 98.

Madine, V., 2000, The stress time bomb, Building. *The Builder Group*, London, 3 November, pp. 20-22.

Torrens, G.E., Mansfield, N.J., Newman, A., Gyi, D.E. and Gibb, A.G.F., 2000, Task and product analysis of an electric breaker tool, Managing workplace injuries: and ergonomic approach. In *Proceedings of International Conference MSD 2000*, Dublin, June, Health & Safety Authority & Irish Ergonomics Society, pp. 1-10.

UCATT, 2001, *The Case for a National Occupational Health Scheme for the Construction Industry*, Michelle Tampin, Sypol, Union for Construction Workers, London, November, 23pp.

WHO (World Health Organization), 1946, *Official Records*, **2**, p. 100, Preamble to the Constitution.

CHAPTER 18

Indoor Air Quality Problems in Buildings in the United States

Leslie O'Neal-Coble

INTRODUCTION

In recent years indoor air quality (IAQ) problems leading to the so-called sick building syndrome have fed considerable litigation. Personal injury claims, which utilise remedies under insurance disability and workers' compensation laws, or the Occupational Safety and Hazard Act, as well as property damage, constructive eviction, and design and construction defect claims may relate to IAQ problems.

The World Health Organisation estimates that nearly 30% of all new and remodelled buildings worldwide may be afflicted with indoor air quality problems (US Environmental Protection Agency, 1991). This and studies performed by the US Environmental Protection Agency rank IAQ concerns among the top environmental risks to public health (US Environmental Protection Agency, 1999). Not surprisingly, commentators expect IAQ litigation to proliferate (Cross, Frank B., 1996). Perhaps 8% of commercial buildings in the U.S. fall short of compliance with engineering standards for acceptable indoor air according to the National Energy Management Institute (Kelly, Thomas J., 1999). This chapter outlines the various types of IAQ problems related to buildings, and addresses some of the practical and legal problems stemming from IAQ issues.

18.1   SOURCES OF IAQ PROBLEMS

The two primary sources of IAQ-related problems are biological contamination and chemical contamination. Biological contamination can result from microbial bacterial or fungal growth or both, on building materials, carpets, furniture and the like or contaminated central air handling systems, for example, and is usually caused by persistent excessive moisture. In the context of building construction, excess moisture may result from one or more of the following: improper drying out during construction; improper design, construction or maintenance of the building envelope or improper design, construction or maintenance of the heating, ventilation and air conditioning (HVAC) system, poor material selection, or construction mistakes (Icard, Thomas and Wright, W. Cary, 1999).

Chemical contamination may result from releases of organic compounds fumes or gases emitted from plastics, fibres, coatings, or chemicals used in building components or furnishings, office operations or building cleaning. Volatile organic compounds (VOCs) have been identified as possible sources of indoor air quality problems. New buildings typically have higher concentrations of VOCs because of releases from construction materials, building elements, and new furnishings; however, VOCs can also be emitted by bacteria or fungi (Heady, Gene, 1995).

Buildings may have both biological and chemical contamination. IAQ problems often have multiple causes, which make their diagnosis and solutions more difficult.

18.2   CONSEQUENCES OF IAQ PROBLEMS

Either type of contamination may trigger occupant complaints of health effects. Symptoms of IAQ-related complaints include breathlessness, dry cough, bronchial asthma, chest tightness, rashes, itching, eye irritation, drowsiness, dizziness, to more serious effects such as Legionnaire's disease or cancer (Icard and Wright).

One study indicated that as many as 20% of all office buildings in the United States had IAQ problems or were 'sick' (Katz, Robert W., 1991). Indoor air quality problems have become one of the principal environmental problems in the United States in this decade.

The increased public awareness and concern with IAQ issues has resulted in building occupants demanding compensation for adverse health effects and 'building teams' pointing fingers over economic damages. Most IAQ lawsuits contain accusations of a sick building injury by occupants. Building owners, faced with costly solutions to these problems, also make claims against all who had anything to do with the design, construction, or maintenance of the building, using a variety of legal theories. Owners, landlords, tenants, designers, and contractor, should understand the legal aspects of these claims.

18.3   MEDICAL ASPECTS OF INDOOR AIR QUALITY

Although full coverage of the medical aspects of indoor air quality is beyond the scope of this presentation, certain medical issues are important for the legal practitioner.

Over the last ten years there has been an increase in lawsuits over indoor air quality. Several of these suits have resulted in multimillion dollar verdicts for remediation, reconstruction, and relocation costs (Centex-Rooney Construction Co., Inc., 1997). The issues causing these verdicts are the risks to health caused by the presence of contaminants in indoor air. Those individuals affected by these contaminants may be workers who are present in the building during their workday or they may be temporary occupants. One aspect of indoor air quality problems is that the symptoms tend to occur when the person is in the building yet moderate or disappear after the person leaves the building. Another problem is the difficulty with establishing the cause of the symptoms. Some of the contaminants affecting indoor air quality are: mould, environmental tobacco smoke, volatile organic compounds (VOCs), dust, dust mites, bacteria, and just about any matter than can affect human comfort or health (Godish, 1995).

Two common terms used in discussing indoor air quality include 'sick building syndrome' which generally means a broad range of symptoms which have no identifiable cause, and 'building related illness' which means that a unique set of symptoms have been linked to an identifiable cause (Godish, 1995). Some of the health effects that may be linked to exposure to indoor air chemical, biological, and physical agents are listed in the following table:

Table 18.1 Health Effects Caused by IAQ Failures

Allergic bronchopulmonary mycosis (ABPM)	Infection
Allergic bronchopulmonary aspergillosis (ABPA)	Infectious disease
Allergy	Influenza
Asthma	Inhalation fever
Bronchitis	Legionnaires' disease
Building-related illness (BRI)	Opportunistic fungal infection
Building-related symptom (BRS)	Organic dust toxic syndrome (ODTS)
Common cold	Pneumonia
Conjunctivitis	Pontiac fever
Cryptococcosis	Primary amebic meningoencephalitis (PAM)
Dermatitis	Rhinitis (coryza)
Granulomatous amebic encephalitis (GAF)	Sick building syndrome
Hantavirus pulmonary syndrome (HPS)	Sinusitis
Histoplasmosis	Sore throat
Humidifier fever	Toxic effect
Hypersensitivity disease (allergy)	Tuberculosis (TB)
Hypersensitivity pneumonitis (HP) (extrinsic allergic alveolitis)	

Source: ACGIH, *Bioaerosols, Assessment and Control (1999)*

Although the link between a particular illness and the indoor air contaminant may be clear in some cases (i.e. Legionnaires disease is clearly caused by a bacteria, Legionella pneumophila); in many cases neither science or the medical profession has established a definitive link between the contaminant and the disease (ACGIH). Despite this lack of a direct causal link to disease, the risk of such a link is so great that it cannot be ignored and may not be necessary to establish liability.

One additional risk factor with indoor air quality is that of psychogenic effects of indoor air contaminants. Once an individual complains about health concerns due to indoor air, other occupants may begin to experience similar symptom or believe they are suffering similar symptoms. This reaction has several names, the most common is 'Mass Psychogenic Illness'. The symptoms may exist but have no physical sign nor laboratory findings of disease (Godish, 1995).

Regardless of the source of the complaints, building owners/managers, contractors, subcontractors and designers should not ignore them. Complaints related to hot/cold temperatures, excess humidity, unusual odours (chemical or musty) or health complaints of headaches, sinus problems, lethargy, shortness of breath, and similar types of health issues must be taken seriously. There may be a pattern to these complaints that indicate a problem with indoor air quality. If a pattern is shown or the complaints exceed a minimum number, the building owner/operator should hire qualified experts to investigate. The options for investigation will depend on the nature and severity of the symptoms. They may range from isolating the specific area to a large scale epidemiological survey and evacuation of the building. Indoor air quality problems can be extremely difficult to pinpoint due to the multiple factors involved. For example, if the evidence indicates that mould is the potential source of the symptoms there are a number of ways to test for and remediate the mould with conflicting theories on the best method (ACGIH, 1999). The remedy will depend on the type of mould present and the extent of contamination (ACGIH, 1999). This remediation may range from cleaning a limited area with bleach to fullscale containment with decontamination suits, similar to that involved in asbestos abatement (ACGIH, 1999 and Godish, 1995). There are only limited standards available for the removal of mould contaminated building materials, with the City of New York Health Department and ACGIH being the most recent to provide such standards.

One critical aspect of investigating indoor air quality problems is to keep the building occupants informed of the status and results of the investigation. In fact, if the contaminants are toxigenic fungi, occupants may have to be informed under the Hazardous Communication Standard (Godish, 1995).

18.4   THEORIES OF LIABILITY IN IAQ CASES

18.4.1   Design and construction claims

When confronted with a 'sick building' most owners look to the original designer and constructor to recover the cost of repair on the basis that the owner did not contract for a 'sick building', the owner did not cause the problems, and the owner should not be required to finance correcting the problems.

There are numerous legal theories available to building owners in pursuing such claims: breach of contract; breach of express warranty; breach of implied (both common law and statutory) warranty; breach of warranty of fitness for a particular purpose; common law indemnity; negligence; and strict liability. Each of these theories has certain benefits and limitations, which are generally discussed below.

18.4.2   Contract claims

Breach of contract claims require proof of a written or an oral contract, failure to perform some aspect of the contract, and damages resulting from the failure to perform, subject to any contractual damage limitations. Since owners usually have contracts with the architect and with the general contractor or construction manager they typically make breach of contract claims against those parties for IAQ-related problems.

Contract theories, obviously, are limited to defendants with contractual relationships. This may reduce the number of possible defendants and the possible sources of funds to pay settlements or judgments. The lack of contractual privity may be overcome by direct beneficiary or third party beneficiary arguments (In re Masonite Corp., 1998). Additional limitations in some contracts are notice requirements for claims and some contracts expressly limit the recoverable damages.

If the general contractor provided a performance bond, the owner may be able to make a bond claim in addition to the construction contract claim. However, if the IAQ problems do not arise until after the building is completed, there is a split of authority as to whether such latent defects are covered under the surety bond (Florida Bd. of Regents v. Fidelity & Deposit Co. of Maryland). As with warranty claims, an owner considering a surety bond claim should carefully review the bond for notice requirements and should comply with such requirements as soon as possible.

18.4.3   Warranty claims

Many construction contracts contain express warranties for the overall building. Subcontractors and manufacturers often warrant specific building components as well. These warranties may provide additional bases for owners' claims.   However, warranties may be so limited both in scope and in time as to have little value. Owners considering claims on written warranties should carefully review them for notice provisions and time limits. Notice letters should be sent as soon as possible.

Many states have created implied warranties of fitness for residential construction.   Although such implied warranties have generally not been applied to commercial construction, some courts have questioned why there should be such a distinction in the legal remedies available to purchasers of different types of property (Florida Eastern Properties, Inc. v. Southeast Commercial Developers, Inc., 1985). Perhaps commercial owners will be able to make such claims in the future; at present, their viability is questionable.

18.4.4   Negligence claims

Negligence claims require proof of four elements: (1) defendant owed plaintiff a duty to act in some way; (2) defendant did not perform its duty; (3) defendant's failure to perform its duty caused plaintiff to suffer some injury; (4) the injury resulted in the plaintiff suffering a loss.

Claimants may be able to recover greater damages under a negligence theory than under a contract theory.   Until recently, it was common to find both theories pursued in the same case. In the past few years, the 'economic loss rule' has been used to defeat negligence claims for purely economic damages (that is, damages other than for personal injuries and property damage).   The economic loss rule may prevent recovery on a negligence theory where the damage is to the "product" itself (In re Masonite Corp., 1998).   Some courts have held that a building is a single product, so the owner could not recover for damages to one building component caused by another building component (Casa Clara Condominium Ass'n., Inc. v. Charley Toppinso & Sons, Inc., 1993). In this analysis, one court has stated that, 'one must look to the product purchased by the plaintiff, not the product sold by the defendant. (Casa Clara Condominium Ass'n., Inc. v. Charley Toppinso & Sons, Inc.)

Application of the economic loss rule may prevent an owner from pursuing claims from a responsible third party manufacturer, supplier, or subcontractor to recover the cost to remediate and reconstruct the building.

Some jurisdictions have allowed plaintiffs to bring actions based on negligence for indoor pollution claims, finding asbestos contamination sufficient to invoke the property damage exception to the economic loss rule (Northridge Co. v. W.R. Grace & Co.). However, this rule remains a significant bar to negligence claims in these cases.

18.4.5   Strict liability

Strict liability theory holds a defendant strictly liable for a defective product without proof of negligence, without an intent to guarantee, without privity of contract, and without consideration of contractual liability disclaimers (Reisman, David, 1995). This theory is widely used in products liability cases. The policy considerations underlying such cases are that a seller who places unreasonably dangerous products in the stream of commerce should be liable for physical harm its products cause. Applying these policy considerations to buildings is difficult because buildings are not usually thought of as products. The courts which have held that a building may be a 'product' for strict liability purposes have considered mobile homes or mass produced homes, not occasional sales of homes (Blagg v. Fred Hunt Co., Inc.). Some courts have found that portions of structures, such as defective precast panels or facing tiles, may be considered products (Chicago Bd. of Educ. v. A,C, & S, Inc.). A Georgia court declined to apply the doctrine of strict liability to an owner's claim against a homebuilder because the builder was not involved in the manufacture of personal property (Seely v. Loyd H. Johnson Construction Co., Inc.).

Courts have reached differing conclusions regarding whether strict liability can apply to economic losses alone, without physical injury (School Dist. of City of Independence, 1988). Some courts have held where there is a risk or death or personal injury, tort remedies are available (United States Gypsum Co. v. Mayor of Baltimore).

Strict liability theory may not apply to architects and engineers unless it can be shown that the design or the system was standardised or mass marketed (N.D., 1996).

18.4.6   Other theories

Some plaintiffs have argued that health effects caused by a sick building caused or contributed to their 'disability' under the Americans With Disabilities Act (Keck v. New York State Office of Alcoholism and Substance Abuse Service, 1998). A federal court in Colorado reversed an insurance company's denial of long-term disability benefits. In this case of a sick building injury, the plaintiff relied on insurance and disability law for a remedy. The plaintiff used a federal statute, ERISA, to enforce her rights under the terms of a long-term disability insurance plan (Clausen v. Standard Insurance Co., 1997). Additionally, as workers' compensation laws are broadened, IAQ litigation could include more worker compensation claims.

18.4.7   Damages

As noted elsewhere in this chapter, anyone involved with the ownership, management, construction, design, or maintenance of a building may become involved in a lawsuit over indoor air quality. Depending on the degree and source of the indoor air quality problem, the damages can range from minor cleanup costs to total evacuation and reconstruction of the buildings. The damages may not be directly linked to a disease. For example, in establishing damages for remediation of a building due to the presence of toxigenic mould, it may not be necessary to prove that the mould caused the health effects, rather it may only be necessary to prove that it was reasonable to incur the remediation, relocation, and reconstruction costs (Centex-Rooney Construction Co., Inc.,1997).

Reasonable remediation damages can range widely from the mere costs of the bleach for a simple mould problem, or the cost of a ventilation fan if the problems are due to volatile organic compounds, or million dollar containment costs. The issue will be whether the cost of the remediation is reasonably necessary. If the building has to be evacuated (either partially or totally), the cost of temporary facilities, moving costs, and related damages may also be recovered. Plus, the costs to reconstruct the building may be recoverable. The actual cost incurred in reconstructing the building may be reduced to the present value at the time of the original construction, if the damages are assessed several years after original construction. In addition, if upgrades are required due changes in the law, i.e. installation of handicapped facilities that were not required in the original construction, these costs may also be recoverable.

18.5   COLLECTION PROBLEMS IN INDOOR AIR QUALITY CASES

Having a viable legal theory is only one step in recovering costs related to repairing or remediating an IAQ problem. Finding sources of funds to pay judgments or settlements can be a challenge for all parties. As IAQ claims increase, the insurance industry seeks cost containment through provisions such as pollution exclusion clauses.

18.5.1   Design professional insurance

Assuming that the design professional has insurance, the question remains the amount and extent of the coverage. Surprisingly, design professionals on large projects frequently have relatively low coverage limits, sometimes $1 million or less. In addition, many owners are unhappy to find that the insurance policy is written so that the cost of litigating the dispute reduces the amount of coverage. This is called a declining balance policy. With this type of policy, it is possible that there will be little or no insurance left after a trial or an appeal.

In addition, to limit their exposure some insurers include pollution exclusion clauses in their policies, such as the following:

It is hereby understood and agreed that such insurance as is afforded by this policy does not apply to any claim based upon, arising out of or in any way involving the discharge, dispersal, release or escape of smoke, vapours, soot, fumes, acids, alkalis, toxic chemicals, liquids or gases, waste materials or other irritants, contaminants or pollutants.

Pollution exclusion clauses exclude from coverage emissions that may be the cause of IAQ problems. Since the mid-1980s, insurers have attempted to avoid the costs of environmental pollution by the use of pollution exclusion clauses, and have placed IAQ claims under the same exclusionary blanket as 'traditional' environmental pollution sources in order to avoid payment on these claims. Court decisions are mixed as to whether insurance companies can use these clauses to refuse coverage of IAQ-related claims (Donaldson v. Urban Land Interests, Inc.).

Another concern with design professional's liability insurance is whether the policy is a 'claims made' or an 'occurrence' policy. 'Claims made' policies cover only those claims made during the policy period, regardless of when they arose. 'Occurrence' policies cover claims which arose during the policy period, regardless of when they are made. It is not uncommon for owners to require design professionals to have errors and omissions coverage during a project, only to find that the coverage has been dropped by the time the claim arises, leaving the owner with no 'deep pocket'.

18.5.2 Contractor's commercial general liability insurance

Commercial general liability insurance generally does not cover the cost to repair or to replace defective work or the material itself, but covers consequential damages arising from defective work. Consequential damages may include loss of use of the building, damages to furniture and fixtures and even the cost to repair other portions of the work. General contractor's CGL policies may cover defective work performed by subcontractors, but not work the general contractor performed itself.

The owner, the general contractor and the subcontractors should carefully examine the CGL policy and its exclusions. From the owner's standpoint, the CGL insurance may be a good source for paying a settlement or a judgment. From the general contractor's and subcontractor's standpoint, the CGL carrier may help pay costs to defend a lawsuit.

18.6   PERSONAL INJURY CLAIMS

18.6.1   Duty to abate

The owner, the contractor and the designers may share in the duty to abate known hazards. If the indoor air quality problem is of such magnitude that there is a health threat or risk of physical harm to the occupants, the owner or employer may have no other choice than to abate the hazard. There are a number of cases finding a duty to abate in situations involving toxic chemical spills (E.D. Mich. 1992) formaldehyde foam insulation (Alaska, 1983), and asbestos, so it is not unlikely a court could reach a similar finding in an indoor air quality case.

18.6.2   Theories of liability in personal injury claims

Personal injury claimants generally use negligence or strict liability theories for their claims. The economic loss rule does not apply to personal injury claims. The most difficult hurdle in personal injury suits is proving causation. The plaintiff must show that contaminants in the building caused his or her symptoms. 'Scientific cause and effect relationships are generally hard to prove and precise diagnosis of certain diseases is possible only with an autopsy.' It is difficult to discover which of many possible agents caused illness and to identify the precise cause of that agent. "Proving causation becomes particularly difficult because a sick building may contain a multiplicity of suspect contaminant. Accordingly, individual contaminants might not be conclusively or exclusively linked to the alleged harm'. (Heady). Also, occupational diseases may take a long time to arise, making it difficult to determine at what point the worker contracted the disease.

Despite these difficulties, some plaintiffs have been successful in obtaining large verdicts for IAQ-related injuries. In the 'Waterside Mall' case, five plaintiffs were awarded just under $1 million for injuries allegedly caused by exposure to various airborne toxins.

18.7   CONCLUSION

Indoor air quality problems present substantial risk to building owners, design professionals, contractors, subcontractors, and their insurers. When faced with such claims, the parties are better served by focusing on the solution rather than on affixing blame. Because litigating these cases is extremely expensive, parties should look for creative alternative dispute mechanisms to try to resolve the case if possible.

18.8 REFERENCES

ACGIH, 1999, *Bioaerosols, Assessment and Control.*

Blagg v. Fred Hunt Co., Inc., 612 S.W.2d 321 (Ark. 1981); Kaneko v. Hilo Coast Processing, 656 P.2d 343 (Haw. 1982) (holding that buildings may be products); Oliver v. Superior Ct., 259 Cal. Rptr. 160 (Cal.Ct. App. 1989) (holding strict liability not applicable to occasional sales).

Casa Clara Condominium Ass'n., Inc. v. Charley Toppinso & Sons, Inc., 620 So. 2d 1244 (Fla. 1993).

Centex-Rooney Construction Co., Inc., St. Paul Ins. V. Martin County, Florida, 706 So. 2d 20 (Fla. 4th DCA 1997).

Chicago Bd. of Educ. v. A,C, & S, Inc., 525 N.E. 2d 950 (Ill. App. Ct. 1988); Trustees of Columbia v. Mitchell/Giurgola Assoc., 109 A.D.2d 449 (N. Y. App. Div 1985); but see Casa Clara Condominium Ass'n., Inc. v. Charley Toppino & Sons, Inc.,620 So. 2d 1244 (Fla. 1993).

Clausen v. Standard Insurance Co., 961 F. Supp. 1446 (D. Col. 1997).

Cross, Frank B., *Legal Responses to Indoor Air Pollution*, 169, 1990, Geisler, Robert, The Fungusamoungus: Sick Building Survival Guide, 8 St. Thomas L. Rev. 511, 513 (1996)

Donaldson v. Urban Land Interests, Inc., 564 N.W. 2d 728 (Wis. 1997); Terramatrix, Inc. v. United States Fire Insurance Co., 939 P. 2d 483 (Colo. Ct. App. 197); Evanston Ins. Co. v. Treister, 794 F. Supp. 560 (D.V.I. 1992).

Florida Bd. of Regents v. Fidelity & Deposit Co. of Maryland, 416 So. 2d 30 (Fla. 5th DCA 1992) (surety not liable for.latent defects); School Board of Pinellas County v. St. Paul Fire & Marine Ins. Co., 449 So. 2d 872 (Fla. 2d DCA), review denied, 458 So. 2d 274 (Fla. 1984) (surety liable for latent defects).

Florida Eastern Properties, Inc. v. Southeast Commercial Developers, Inc., 479 So. 2d 793 (Fla. 5th DCA 1985).

Gastel, Ruth, 1994, *Occupational Disease: Insurance Issues*, Ins. Info. Inst. Rep., June.

Godish, Thad, 1995, *Sick Buildings, Definition, Diagnosis and Mitigation*, pp. 1, 31-34, 37-42, 93-203, 333.

Heady, Gene, 1995, *Stuck Inside These Four Walls: Recognition of Sick Building Syndrome Has Laid the Foundation to Raise Toxic Tort Litigation to New Heights*, 26 Texas Tech. L. Rev. 1041, 1048.

Heady, supra, note 4.

Icard and Wright, supra, note v.

Icard, Thomas and Wright, W. Cary, *Sick building Syndrome and Building-related Illness Claims: Defining the Practical and Legal Issues*, 14, *The Construction*

Lawyer, 1 (1994); Post, Nadine M., Containing Noxious Mold, *Engineering News-Record* (May 1999) at 32.

In re Masonite Corp. Hardboard Siding Products Liability Litigation, 21 F. Supp. 2d 593 (E.D. La. 1998).

Katz, Robert W., How to Prove a Sick Building Case, *Trial*, (Sept. 1991) at 58.

Keck v. New York State Office of Alcoholism and Substance Abuse Services, 10 F. Supp. 2d 194 (N.D.N.Y. 1998).

Patrick v. Southern Co. Services, 910 F. Supp. 566 (N.D. Ala. 1996).

Kelly, Thomas J., Measuring the ROI of IAQ, *Buildings*, (March 1999) at 2.

Northridge Co. v. W.R. Grace & Co., 471 N.W. 2d 179 (Wis. 1991); 80 South Eighth Street Ltd. Partnership v. Carey-Canada, Inc., 486 N.W.2d 393 (Minn. 1992).

Reisman, David, Strict Liability and Sick Building Syndrome: Defining a Building as a Product Under Restatement (Second) of Torts, Section 402A, 10, J. Nat. Resources & Envtl. L. 35 (1995).

Roseville Plaza Ltd. v. United States Gypsum Co., 811 F. Supp. 1200 (E.D. Mich. 1992).

School Dist. of City of Independence, Missouri v. United States Gypsum Company, 750 S.W. 2d 442 (Mo. App. 1988).

Seely v. Loyd H. Johnson Construction Co., Inc., 470 S.E.2d 283 (1996); See Golden, Brian M., 'Strict Liability Applied to the Homebuilder: A Defect in the Law of Defective Products', 14 The Construction Lawyer 11, (October 1994).

Shooshanian v. Wagner, 672 P.2d 455 (Alaska 1983).

Sime v. Tvenge Assoc. Architects & Planners, 488 N.W. 2d 606 (N.D. 1992).

United States Environmental Protection Agency, 1991, *Building Air Quality*, ORL1 #709997 v1.

United States Environmental Protection Agency, Air & Radiation Pub. No. ANR-445-W, Indoor Air Facts No. 4 (revised): Sick Building Syndrome 1 (April 1991).

United States Environmental Protection Agency, Office of Radiation and Indoor Air, Why should you be concerned about the quality of the air that you breathe? (last modified June 8, 1999) <http.//www.epa.gov/iaq/>.

United States Gypsum Co. v. Mayor of Baltimore, 647 A.2d 405 (Md. 1994); Council of Co-Owners Atlantis Condominium, Inc. v. Whiting-Turner Contracting Co., 517 A.2d 336 (Md. 1986).

Section 4

Training and Education

CHAPTER 19

Health & Safety Training and Education: A Case of Neglect?

Steve Rowlinson

Is training and education adequately planned, implemented and monitored in the construction industry? What are the drivers for them and the impediments to them? How can an integrated approach be developed. These are questions that are addressed in this section.

19.1 SAFETY TRAINING AND EDUCATION

Taking Hong Kong as an example, the normal construction site worker has gone through an informal apprenticeship under a master (*sih fuh*) and this training has been based on common trade practice rather than safe, modern methods of working. Thus, with many workers there is a lot of bad practice to unlearn. This problem needs to be approached in a structured manner. Three basic types of training can be identified:

- induction training;
- refresher training;
- ongoing training.

19.2 INDUCTION TRAINING

Induction training is aimed at the worker just about to join the construction site and its aim is to improve basic safety awareness. This type of training can be very general in nature and would cover items such as personal protective equipments, working at heights, fall protection, etc. The Hong Kong Construction Association (HKCA), Works Branch (WB) and Hong Kong Housing Authority (HKHA) have all introduced such training along with a green card which allows a worker to enter the site; without the card the worker is deemed not to be prepared for work. Two basic problems need to be addressed with this training:

1. the content and retention by the worker of the course materials;

2. the timing and provision of courses so that a temporary labour shortage does not develop as workers await an opportunity to be inducted.

19.3   REFRESHER TRAINING

Refresher training is aimed at those employees who have been doing the same job for many years. People such as these have many years of experience in traditional methods of construction. However, they may well not be up-to-date with modern or innovative methods. Also, it is quite possible that they may have slipped into bad practice or simply forgotten some of the things that they have learned in the past. For example, a steel fixer may have been working on bridge deck construction for many years and may have grown unfamiliar with some of the hazards and technology used in high-rise construction (such as cast in place formwork). Hence, before being transferred to a new building project it would be sensible to send such a tradesman on a refresher course. These courses can cover a whole range of topics but they should be targeted at particular tradesmen and at new projects coming up.

19.4   ONGOING TRAINING

Ongoing training addresses the needs of salaried workers for re-training in new techniques and the needs of new workers to be trained from scratch. At this point one might make a distinction, what is really required is not just training but education, i.e. an understanding of the need for safety procedures and a perception of one's responsibility to work safely for the sake of oneself and co-workers. Typical training topics might be;

*   craft and technical training for specific trades which teaches safe practice and procedures. This might include signalling training for banksmen and trench shoring for excavation workers. General safety precautions for working in confined spaces might also come into this category;
*   supervisory training for gangers and foremen which would include elements of craft training indicated above but also more general oversight such as hazard hunts and toolbox talks;
*   managerial training aimed at site and project managers and head office staff which deals with the details and philosophy of safety management and includes the compliance with and need for statutory requirements such as safety meetings and inspections.

These issues, specific to Hong Kong, are reflected to a large extent in many industries around the world. Specific examples are given in this section of safety training and education provided to young and aspiring professionals in the United Kingdom and South Africa.

Cameron and Fairlie note that the CIOB (Chartered Institute of Building) sets the standard by which the 'chartered builder' is qualified. Further, the CIOB details the experience required to achieve corporate membership. In January 1995, the CIOB ran a higher education project, under contract from the Department of Employment, which provided a basis for establishing a Higher Education Discipline Network for Building Management (CIOB, 1995). The project had three objectives:

- to identify excellence in the delivery of building management education, with a particular focus on innovative practices and vocational learning;
- to improve the standard of education by bringing teaching staff together to share information and learn from the experiences of others; and
- to disseminate examples of good practice through publications and seminars.

They describe how this project culminated in the generation of the Education Framework (CIOB, 1996). The Framework is designed to provide best practice for the delivery of construction management degrees in the UK.

Smallwood describes construction health and safety education for the Tertiary built environment and notes that although health and safety education is both essential and a pre-requisite for management commitment, South African construction management programmes address health and safety to varying degrees. Those that do so to a lesser degree invariably focus on legislation and do not address procurement and other dynamic issues of health and safety. However, no particular health and safety curriculum should be viewed as optimum, as new management strategies, systems, processes and research findings amplify the need for amendment and improvement of a curriculum.

Given the aforementioned, two surveys were conducted. The first survey was conducted to determine the extent to which construction management programmes addressed health and safety. The second survey was conducted among a group of 'health and safety best practice' general contractors (GCs) to determine the important issues relative to a construction health and safety curriculum. Findings of the surveys include recommendations on the following: health and safety/productivity/quality, role of management, legislation, worker participation, role of unions, and programmes predominate among construction management health and safety curricula subject areas, and legislation, management of subcontractors (SCs).

Smallwood also notes that construction is a multi-stakeholder process and consequently all stakeholders, architectural designers included, influence the construction process. Design influences and impacts on construction health and safety directly and indirectly. Directly through the: design concept; selection of structural frame; detailed design; selection of cladding, and specification of materials. Indirectly through: the selection of procurement system; related interventions such as prequalification; decision regarding project duration, and selection of contractor.

Given the aforementioned, Smallwood argues that architectural designers should be empowered to contribute to construction health and safety. This chapter reports on a descriptive survey conducted among architectural departments at Technikons and Universities in South Africa, which determined that: architectural programmes address construction health and safety to a limited extent; construction H&S is perceived to be fairly important to the discipline of architecture, and design related activities have a moderate impact on construction health and safety.

Lingard presents a different perspective, whilst arguing for an integrated training approach to first aid. Drawing on theories of social psychology which have postulated a causal relationship between attitudes and behaviour (Fishbein and Ajzen, 1975; Ajzen, 1991) she uses the work of Harvey et al. (2001) to argue that this theoretical connection is likely to apply to occupational health and safety (OHS) behaviours, as it does to other behaviours. There is some empirical research evidence to support this link. For example, Donald and Canter (1994) identified a significant correlation between safety attitudes and accident rates. Attitude theories acknowledge that attitudes can be changed and therefore, it is possible that beliefs about OHS risk and the salience of OHS in one's job can be influenced by organisational interventions, including training. If one accepts that attitudes and behaviour are causally linked, then this attitudinal change could bring about enhanced OHS performance. This chapter investigates the effectiveness of first aid training in changing the OHS attitudes of operatives working in small construction firms in Australia.

19.5   REFERENCES

Ajzen, I., 1988, *Attitudes, Personality and Behaviour*, Open University Press, Milton Keynes.

Chartered Institute of Building, 1995, *Higher Education Discipline Network in Building Management*, CIOB (Berkshire: Ascot).

Chartered Institute of Building, 1996, *Education Framework*, CIOB (Berkshire; Ascot).

Fishbein, M. and Ajzen, I., 1975, *Belief, Attitude, Intention and Behavior: An Introduction to Theory and Research* (Reading, MA: Addison-Wesley).

CHAPTER 20

The Exposure of British Construction Management Degree Students to Health and Safety Learning and Assessment: Are we Fulfilling our Duty of Care?

Iain Cameron and Norman Fairlie

INTRODUCTION

The CIOB sets the standard by which the 'chartered builder' is qualified. Further, the CIOB details the experience required to achieve corporate membership. In January 1995, the CIOB ran a higher education project, under contract from the Department of Employment, which provided a basis for establishing a Higher Education Discipline Network for Building Management (CIOB, 1995). The project had three objectives:

(i)   to identify excellence in the delivery of building management education, with a particular focus on innovative practices and vocational learning;
(ii) to improve the standard of education by bringing teaching staff together to share information and learn from the experiences of others; and
(iii) to disseminate examples of good practice through publications and seminars.

This project culminated in the generation of the Education Framework (CIOB, 1996a). The Framework is designed to provide best practice for the delivery of construction management degrees in the UK.

20.1   RESEARCH OBJECTIVE

The objective of this chapter therefore is to evaluate a single important question: To what extent does health and safety feature in this model of best practice?'.

20.1.1   CIOB Regulations

Those who aspire to become a chartered builder in the UK generally require a CIOB-accredited honours degree in construction management in addition to at least three years' experience at an appropriate level.

This results in the achievement of corporate member grade (MCIOB). Section 6 (regulation 40) of the CIOB's 'Membership regulations' (CIOB, 1994a):

'An applicant must have had a minimum of three years' experience at professional level in a firm, consultant practice, department or other organisation engaged in the construction, alteration, maintenance, repair, provision, design, inspection or management of buildings, other construction work or in building education'.

It lists the following acceptable activities: 'Building control; building asset management; building surveying; estimating; construction management; cost and production control; design; inspection and maintenance management; project management; purchasing; quantity surveying; site engineering; teaching; planning; quality; facilities management; and any other function acceptable to the Institute.' However, safety is not explicitly noted. At the very least, this seems to contradict the CIOB's own 'Regulations of professional competence and conduct' (CIOB, 1994b), which detail the standard of competence that corporate members must achieve.

These Regulations, unlike the Education Framework, place a strong emphasis on health and safety. Rule 15 specifically targets health and safety:

'15   Members shall at all times have due regard for the safety, health and welfare of themselves, colleagues and any others likely to be affected:

    15.1   a knowledge of health and safety risks in the industry and the main principles and strategies for control;
    15.2   an understanding of the responsibilities for the safety, health and welfare placed on all parties involved in the building process;
    15.3   a working knowledge of the current legislation and advisory information;
    15.4   a recognition of the importance of keeping themselves up to date'.

20.1.2   The Safety, Health and Welfare Committee

The requirements of Rule 15 are supported by the work of the CIOB's Safety and Health Committee (CIOB, 1998a). The Committee consists of ten members whose objectives are to act as a focus for health and safety within the CIOB. In particular, the Committee:

i)   provides advice on health and safety training, including revising and reviewing the Institute's syllabus for corporate membership examinations;
ii) liaises on health and safety matters with other professional institutions and similar bodies, government departments and other committees and groups within the CIOB;
iii) examines, consults on and responds to consultative documents received by the Institute from government departments, the Health and Safety Executive (HSE), the Health and Safety Commission (HSC) and other organisations;
iv) produces articles for inclusion in the Institute's monthly journal and other national construction media;
v)   reports regularly to the Institute's Professional Services Board and receives its guidance on matters to be investigated and reported on.

In a Safety, Health and Welfare Committee report (CIOB, 1998a), the chairman commented on a pilot study the Committee had undertaken in relation to the health & safety content of university construction degrees:

> *For several years the Committee has been looking at the Education Framework with regard to the health, safety and welfare content. ... It is apparent that there is a lack of direction and standards, with some universities and colleges taking the subject of health and safety more seriously than others.*

The Committee concluded that there was a lack of guidelines in relation to health and safety from the CIOB towards universities. In particular, the Committee recommended that the CIOB should issue, as a matter of urgency, firm directives to the universities in order to make health and safety a significant component of a student's education, preferably via a specific health and safety module, hence ensuring that basic health and safety practices are taught to every student undertaking a CIOB-accredited construction degree.

20.1.3   Overview of CIOB's objectives

A CIOB-accredited course is designed to equip those undertaking it with a formal education that provides this initial preparation. This core knowledge can then be developed by industry (Riggs, 1988). The CIOB's Professional Development Programme for graduate members (CIOB, 1998b) attempts to structure this training in order that graduates will possess the level of competence required by the Institute's regulations to become a corporate member. If the CIOB expects prospective members to comply with health and safety legislation (as laid down in its 'Regulations of professional competence and conduct') then it is reasonable to suppose that its Education Framework should provide this initial preparation by ensuring that accredited degree programmes adequately address health and safety issues.

Rule 15 gives an indication of the importance that the CIOB places on health and safety for corporate membership and yet the Education Framework's (CIOB 1996a) module descriptors make no specific demands on universities to include health and safety (Figure 20.1).

There are currently 30 higher education institutions in the UK offering CIOB-accredited construction management degrees. The left-hand side of Figure. 20.1 shows a lack of specific reference to health and safety within the structure of undergraduate programmes, although the subject receives treatment within some modules. The detailed syllabuses of taught modules reveal that health and safety receives little attention within levels 1 and 2. For example, 'Design and technology 1' (at level 1) has a passing reference to 'safety precautions' in relation to excavations, and 'Management' (at level 2) has a reference to 'health, safety and welfare policies' and 'audit procedures'. In total, fewer than two lines of a two-page module descriptor are devoted to safety for any module within the CIOB's Education Framework. The only module that makes any notable reference to safety is one within the 'Construction management' option at level 3. Indeed, none of the modules within the Education Framework makes detailed reference to risk assessment, arguably an essential process.

Figure 20.1 therefore demonstrates an anomaly between CIOB education and CIOB training.

20.2   EDUCATION

There is a comprehensive body of literature that attempts to define education. The Department of Employment (1971) describes education as activities which aim at developing the knowledge, moral values and understanding required in all walks of life rather than knowledge and skill relating to only a limited field of activity. However, Peters (1972) claimed that education was too complicated to define. He regarded being educated as a state that individuals achieve, with education being a set of processes that lead to this state. This indicates a broad learning base during the initial stages of education with the field narrowing and becoming more specialised during the final stages of learning. With regard to construction health and safety, the importance of formal education was recognised by Davies and Tomasin (1996):

> It is recommended that time is devoted on the relevant courses at universities or to the subject of health and safety in the construction industry. This should cover an outline of safety legislation and the duties of employers and employees. It should describe the hazards that professional staff face in their work and outline the problems facing designers in ensuring that their work is safe to build, operate and maintain.

Garavan (1997) suggested that educational achievements are viewed as a prerequisite for a job. Indeed, health and safety is now beginning to be recognised as an important part of business management education (see, for example, Confederation of British Industry 1990).

20.3   TRAINING

Training of recent graduates in construction management is required in order that they might effectively put theory into practice. Training then becomes an integral part of continuing professional development. The educational foundation on which such training builds should be able to support both the needs of the organisations employing the graduates and the professional needs of the individuals concerned in terms of their career development. The CIOB has, through its Professional Development Programme (CIOB, 1998b), emphasised the importance of continuing professional development for graduate construction managers. Gilley and Eggland (1989) elaborate on career development by describing it as 'an organised, planned effort comprising structured activities that result in a mutual career plotting effort between employees and organisations'.

It must be borne in mind that company directors are answerable to their shareholders, who are looking for a return on their investment. A positive corporate image will help to achieve this. In addition, an organisation with a poor health and safety record is unlikely to be a desirable option for potential clients. Therefore, it is in a company's interests to view health and safety training as a long-term investment rather than an immediate cost. However, employers and graduates should not be

expected to rely on training to fill gaps in basic knowledge that have been left by
inadequate provision on degree programmes.

Education Framework	Professional Development Training
Level 1 Modules (formation studies):	Twelve Competencies (attained over 3 years):
Design and Technology 1	Decision Making
Science for Building	Communication
Materials	Managing Information
Site Surveying	Planning
Business Environment 1	Quality
Information and Decision Making	Managing Resources
Structures	Costs and Valuations
Other CIOB Certificates & Diplomas	Personal Management
	Health and Safety (see below)
Assessed Experiential Learning	
Level 2 Modules (core studies):	Health and Safety: This unit is designed to
Design and Technology 2	assess overall knowledge and understanding
Business Environment 2	of safe working practices and principal
Legal Studies	hazards in construction activities and to
Management	assess competence in risk assessment in your
Management of Building Production	own working environment following
Building Services	satisfactory assessment of the required
Pre-contract Services	knowledge and understanding.
Other CIOB Certificates & Diplomas	
	Element 1: Identify, describe, explain
Assessed Experiential Learning	principal hazards and appropriate health and
	safety working practices
	Element 2: Identify hazards, and assess risk
	from work
Level 3 Modules (professional studies):	
Four Option Modules	Element 3: Identify and describe
Project Evaluation & Development	implementation of risk control measures.
Assessed Experiential Learning/Thesis	
'UNIVERSITY'	'INDUSTRY'

Figure 20.1 The elements of the CIOB's Education Framework (University Education) and its
Professional Development Programme (Industry Training)

20.4   RESEARCH METHODOLOGY

The second authors' study is similar in nature to one undertaken previously by the
CIOB's Safety and Health Committee in 1996 (CIOB, 1996c). This earlier study
generated a low response. The authors' follow-up study therefore attempted to enhance
the validity of its findings. The CIOB's earlier study was based on a questionnaire
('Health and safety - academic courses'), which was sent to all higher education

institutions running CIOB-accredited construction management degree courses. A response rate of only 30% was achieved. The authors considered that a semi-structured telephone interview format would be a better method of data collection.

A list of British CIOB-accredited university honours degree programmes in construction management was obtained from the CIOB's 'Directory of Construction Courses 1996/97' (CIOB, 1996b). It was considered that to obtain satisfactory answers, the programme manager should be interviewed. Thereafter, the programme managers were contacted and interviewed during March 1999. The length of each interview varied between 10 and 30 minutes, depending on each respondent's interest in the subject, their level of involvement with the CIOB and the time available. Out of a total of 30 CIOB-accredited universities, the second author secured 25 telephone interviews representing an 83% response.

20.5   RESULTS - RESPONSES TO QUESTIONS

Q1: How many students are currently in their honours year of the course?

Table 20.1 shows that the size of courses varies considerably. Indeed, there are certain CIOB-accredited universities that are major suppliers of managers to the construction industry. In Scotland, with 23 students in their final year, Glasgow Caledonian University represents the major provider. In England and Wales, major providers are Leeds Metropolitan University, the University of Glamorgan, Oxford Brookes University and Northumbria University. The module content and teaching methods of these programmes are therefore particularly important in judging the standard of graduates.

Table 20.1   Number of students in the final year of each CIOB-accredited BSc(Hons) construction management degree (March 1999)

University	No. Students	University	No. Students
Paisley	12	Coventry	23
Luton	12	Loughborough	23
Heriot-Watt	13	Sheffield Hallam	25
Reading	15	Southampton Institute	30
De Montfort	16	Brighton	30
Napier	20	UMIST	32
Wolverhampton	20	Ulster	33
Bolton Institute of HE	21	West of England	34
Liverpool John Moores	22	Leeds Metropolitan	50
Salford	22	Glamorgan	70
Central Lancashire	22	Oxford Brookes	80

Glasgow Caledonian	22	Northumbria	87
Anglia Polytechnic	23		

Q2: How many honours dissertations per year address health and safety issues?

Table 20.2 indicates the wide variations that exist between universities in the number of health and safety dissertations, ranging from Sheffield Hallam University with 6 from 26 (23%), to Leeds Metropolitan University with only 1 from 50 (2%). The latter has a large number of students (50 in the final year), and it was recognised by the interviewee that the two per cent figure should be higher. The syllabus is now being reviewed to increase safety.

Table 20.2 Number of students within each university selecting a dissertation topic relating to health and safety (March 1999)

University	No. Students	University	No. Students
Paisley	1 from 12	Coventry	2 from 25
Luton	2 from 12	Loughborough	3 from 25
Heriot-Watt	1 from 13	Sheffield Hallam	6 from 26
Reading	1 from 15	Southampton Institute	1 from 30
De Montfort	1 from 16	Brighton	1 from 30
Napier	1 from 20	UMIST	2 from 32
Wolverhampton	1 from 20	Ulster	1 from 33
Bolton Institute of HE	2 from 20	West of England	3 from 35
Liverpool John Moores	1 from 21	Leeds Metropolitan	1 from 50
Salford	1 from 23	Glamorgan	5 from 70
Central Lancashire	4 from 23	Oxford Brookes	2 from 80
Glasgow Caledonian	5 from 23	Northumbria	7 from 87
Anglia Polytechnic	1 from 25		

Q3: What is the safety input as a percentage of the total hourly input per year?

Table 20.3 shows the programme managers' estimate of current safety input. Universities teaching safety solely on an ad hoc basis within modules found it difficult to quantify a response. Programme managers seemed unable to identify prescribed hours per module dedicated solely safety. Accurate figures could be given only where safety was taught as a dedicated series of lectures within a general module, or as a stand-alone module.

Q4: Which of the following forms does the health and safety input take?

(i)   a completely separate health and safety module;
(ii) separate and identifiable lectures within another module; or
(iii) discussed ad hoc as safety measures and procedures.

Table 20.3   Estimates of safety input as a percentage of total course input (March 99)

University	Percentage	University	Percentage
Paisley	10%	Coventry	10%
Luton	12%	Loughborough	12%
Heriot-Watt	10%	Sheffield Hallam	12%
Reading	3%	Southampton Institute	10%
De Montfort	15%	Brighton	5%
Napier	10%	UMIST	2%
Wolverhampton	5%	Ulster	5%
Bolton Institute of HE	5%	West of England	5%
Liverpool John Moores	10%	Leeds Metropolitan	1%
Salford	10%	Glamorgan	5%
Central Lancashire	10%	Oxford Brookes	2%
Glasgow Caledonian	10%	Northumbria	8%
Anglia Polytechnic	12%		

Responses showed that the majority (48%) provide health and safety input by separate lectures within modules. Forty-four percent of universities, on the other hand, teach health and safety if and when it is deemed necessary, whereas only 8% (two universities) deliver a specific health and safety module. A sizeable number therefore cover health and safety topics, presumably to varying extents, within lectures on different subjects. Southampton Institute of Higher Education confirmed this by stating that health and safety is integrated at all times where appropriate within lectures. The University of Brighton considered that separate health and safety input was not necessary as continuous ad hoc input achieved the same results. This is, of course, a good argument, providing universities are sure that staff actually do this.

Q5: Is there identifiable safety coursework that the students must undertake?

Seventeen of the 25 universities contacted (68%) do not set separate, identifiable health and safety coursework. Southampton Institute of Higher Education pointed out that, while its coursework is not specifically health and safety-oriented, students must address the relevant regulations, where required, in all coursework. This is an opinion that was echoed by most of the programme managers, regardless of whether specific health and safety coursework was set. However, most were unable to exemplify the point and the few who did mentioned site-based technical issues (for example, scaffolding, earthwork supports and use of plant). No interviewee offered responses that included, for example, developing a safety management system, or promoting a positive safety culture.

Q6: Are there specific health and safety examination questions?

Of the 25 universities questioned, 13 (52%) acknowledged that they had set a separate, identifiable health and safety question in at least one examination, with the other 12 relying on health and safety being addressed where appropriate in other questions.

Q7: Is it possible to sit final exams without answering a health and safety question?

Eleven of the 13 universities responding 'yes' to question 7 also answered 'yes' to this question. Only two (Loughborough University and Glasgow Caledonian University) structure their examinations so that health and safety questions must be answered.

This means that students attending the other 23 universities do not have to answer a question specifically designed to test their knowledge of health and safety legislation or practice. Comments from the other universities resembled those in replies to question 6 (i.e. that health and safety should be addressed at all times where appropriate). Again, this point was not illustrated when prompted by the interviewer. The fact that programme managers could not, in the main, outline safety legislation covered (aside from the CDM Regs) or elaborate on taught safe systems of work, suggests a less than desirable safety content.

Q8: Should health and safety be compulsory within the CIOB Education Framework?

Only one programme manager believed that health and safety should not be a compulsory discrete subject within the CIOB's Education Framework. Views were mixed on the impact that making health and safety compulsory would have on students' learning and the level of guidance that would have to be issued to programme organisers. However, the consensus was that inclusion of safety within the Framework would give standards in the form of prescription that would provide a consistent level of health and safety education.

Q9: If the CIOB included safety in its Education Framework, would this:

   (i)   raise students' overall awareness of health and safety issues; or
  (ii) make no difference as the current input is sufficient?

Although 96% of programme organisers stated in response to question 8 that they thought that the CIOB should include health and safety in its Education Framework, only 32 % considered that it would raise their students' awareness of health and safety. The remainder suggested that their current input is sufficient, regardless of the absence of formal obligations. The conclusion from responses to questions 8 and 9 is that nearly all programme managers want safety course requirements to be prescribed, in order to ensure consistency and to raise standards in other institutions, but two-thirds of those interviewed believed that such prescription would make little difference to their own students.

Q10: Are you aware of the specific health and safety module within the CIOB's Professional Development Programme for graduate training?

The obligations placed on the graduate construction manager with regard to health and safety are extensive (as outlined in the Professional Development Programme - see Figure 20.1). However, four programme managers (16%) were unaware of the Programme's existence. Despite this, once the Programme had been explained, respondents believed that students were suitably prepared to tackle the training framework, even though the survey tends to suggest that undergraduates do not receive sufficient safety education.

Q11: Are you aware of Rule 15 [safety] of the CIOB's 'Regulations of professional competence and conduct'?

Twelve programme managers (48%) were not aware of Rule 15.

20.6   DISCUSSION

There is mounting pressure on the CIOB to include health and safety in its Education Framework. Two key issues extracted from the survey are outlined below:

i) Legislation: The CDM Regulations have forced the construction industry to face health and safety issues. Health and safety must now be considered at all stages of the building process. It is argued that recent legislation has raised the profile and importance of health and safety to such an extent that universities must address these changing needs now.

(ii) Safety culture: Egan (1998) has criticised the culture of the industry and, in particular, its poor safety culture. It is necessary for the industry to embrace a culture whereby safe working practices are the norm.

20.7   CONCLUSIONS

The following conclusions have been drawn from the survey results:

1. CIOB-accredited construction degree programmes in the UK play an important role in the initial preparation of construction managers. However, in general, a lack of direction exists within CIOB-accredited construction degree programmes in the UK regarding the depth of health and safety education that undergraduates receive. The survey suggests that around half of the British universities recognise its importance and devote appropriate time to it, while others view it as an area of study that should be broached only when necessary.

2. Opinions vary considerably on the role that universities should play in undergraduate health and safety education. Respondents implied, with only a couple of exceptions, that industry has a much greater role to play than universities in the development of an individual's safety expertise. As a result, the health and safety requirements that the CDM Regulations have placed on the industry, for instance, are not being taught uniformly across British CIOB-accredited construction degree programmes.

3. The level of knowledge of health and safety legislation and practices among students is not assessed adequately. In all but two of the accredited universities, candidates can sit examinations and achieve a pass mark without having to answer a single question on health and safety. Eight of the 25 British universities set health and safety coursework. Some respondents suggested that health and safety was part of production coursework but, when asked to exemplify, no persuasive responses were offered.

4. Forty-eight percent of CIOB programme managers were unaware of the CIOB's Rule 15 (outlining competence in health and safety). This is not acceptable.

5. The CIOB's interpretation of adequate health and safety input within its Education Framework differs from that of its Safety and Health Committee. This is compounded by the unreconciled relationship between the Education Framework and the Professional Development Programme with regard to safety.
6. The lack of primary reference to health and safety in the CIOB's Education Framework has led to a situation where health and safety education is at the discretion of programme managers or even individual lecturers.
7. This chapter had one objective: 'To what extent does the CIOB's Education Framework include safety?'.

Evidence from the survey of programme managers of British CIOB-accredited construction degrees suggests that the Education Framework does not totally exclude safety, as passing references can be found to the term 'safety considerations'. However, the survey does indicate that sufficient attention is not currently being paid to health and safety within degree programmes. Although production lectures occasionally include health and safety issues, the form that this takes is less than transparent. The CIOB must place health and safety education in a more central role if the health and safety improvements envisaged by the Construction Task Force (Egan, 1998) are to be achieved.

20.8   RECOMMENDATIONS

Similar types of study should be conducted within other countries represented at this conference. For instance, studies within Hong Kong educational institutions who offer qualifications in construction, and in particular those Hong Kong Universities who offer CIOB accredited Construction Management degrees, would offer an indicator of the educational importance of safety. This type of replication research will assess students'' Health and Safety education experience (extent and depth of safety learning). This will give a good indicator of the status of safety within each respective countries Higher Education system.

20.9   POSTSCRIPT

This research was completed over one year ago and since the completion of this study, several high profile safety culture initiatives have been launched. Ambitious targets have been set to reduce accidents across all industries ('Revitalising Health and Safety') and in construction in particular ('Turning Concern into Action'). These initiatives have specifically targeted '*increased education levels within safety-critical professions*' and this would include Construction Management. It is understood that the CIOB intend to responded to this demand by putting together an action-plan to address the shortfall in safety content within their Education Framework. At August 2001, the details of this are not publicly known. However, it is encouraging to see that CIOB and therefore accredited degree are at last beginning to 'see the light' after years of indifference.

The UK Health and Safety Executive have also recently published the findings of HSE Contract Research Report 4117/R72.060 'Identification and Management of Risk in Undergraduate Construction Courses' by John Carpenter (Symonds Group) and Peter Williams (Liverpool John Moores University). This report has sent shock-waves through most of the UK's Professional Institutions and has really spurred the Royal Institute of Chartered Surveyors (RICS) and the Chartered Institute of Building (CIOB) into action by forcing them to pressurise their partnership degree providers within the Higher Education sector to embed safety studies within the curriculum. The main findingss of the HSE subcontractors' report are:

Recent statistics for accidents and fatalities in construction indicate a serious and worrying decline in the health and safety performance of the industry.

In this respect, the Safety Summit (Turning Concern into Action) called by the Deputy Prime Minister in February 2001, laid down a clear marker to the construction industry that it needs to drastically improve its health and safety performance and image. It was also made clear, although this should have been apparent, that all those involved in the construction industry in its widest sense have a role to play in this respect.

One element of the industry's response to this challenge concerns the education of aspiring architects, engineers, surveyors and builders who will eventually manage and lead this necessary change. Consequently, when they enter the world of work as the industry's future professionals, construction undergraduates need to be equipped with not only a basic appreciation of health and safety management but also a fundamental understanding of its essential ethos and role in the overall management of risk.

The need for this educational underpinning is specifically stipulated in the joint Government/Health and Safety Commission Strategy Statement 'Revitalising Health and Safety' which requires 'safety critical professionals' to have an 'adequate education in risk management'. This initiative is further emphasised by the Corporate Governance Group of the Institute of Chartered Accountants ('Implementing Turnbull'), which identifies the important role of risk management, and the health and safety element in particular, in the strategic management of business.

The academic phase in professional education is crucial in all of this and the contribution and support of Academia is vitally important. It is at the formative

Educational Base that attitudes and perceptions are developed and the necessary analytical and critical abilities are inculcated enabling currently accepted industry practices to be challenged and reformed.

This report therefore describes a study into the current provision of health and safety teaching within undergraduate construction courses, and also touches upon the part played by Government, accreditation bodies, institutions, and industry organisations in the educational framework. The study covers all the major construction disciplines: architecture, building, engineering and surveying.

During the study, a total of thirty-one Higher Education Centres were visited in order to meet both teaching staff and students, and these visits revealed that:

- There are some excellent examples of good practice and endeavour, and individuals who are making significant contributions. However, these individuals tend to be few in number.
- There is a significant number of Centres where the inclusion of health and safety into the curriculum is not actively supported by the head of department.
- Generally there is a willingness by staff to encompass health and safety risk management within courses but assistance is needed in respect of teaching aids and industrial input.
- Health and safety risk management is not yet widely recognised as an intellectual subject with a central role in construction risk management.
- Health and safety risk issues are generally not well integrated into the curriculum and undergraduates are not adequately assessed in this particular area of study.
- The management and provision of professional development opportunities for teaching staff in health and safety risk management topics is at best poor and generally non-existent.

As a consequence of these findings the report recommends that Academia:

- Embraces health and safety risk management as an integral and intellectual component part of the curriculum equivalent in all respects to the study of other risk management aspects of the construction process.
- Actively promotes the concept of a 'health and safety champion' within their staff complement who will initiate and lead the integration of health and safety risk management within all construction courses.

- Provides appropriate professional development for all construction professional staff to enable them to deliver the input required and provide active support for the 'champion'.
- Works to maximise the links with Industry in order to develop intellectual exchange and learning opportunities for staff, students and practitioners alike, and in order to instil the business case for health and safety risk management in construction courses.
- Considers the concept of 'health and safety awareness' days to supplement the above.

The study also found, however, that the difficulties faced by Academia are exacerbated by the fact that accreditation bodies have widely varying, and mostly inadequate, requirements in respect of health and safety curriculum content. It also concludes that other influential bodies such as Government, professional institutions and industry organisations are not giving this subject the attention deserved.

The report therefore also recommends that:

- Government and Funding Agencies use their position to raise the profile of health and safety in the Higher Education sector.
- Those accreditation bodies that currently have minimal reference to health and safety review their requirements, and that all accreditation bodies actively work towards a pan industry approach.
- Health and safety be presented and taught as integral to the topic of construction risk management.
- Those institutions involved in the built environment work to significantly raise the profile of health and safety within their area of influence, particularly in respect of its relationship to the Education Base.
- An industry umbrella body such as the Construction Industry Council leads in developing a pan industry, standard curriculum template for health and safety risk management delivery in Undergraduate construction courses.
- Industry and Academia work together to maximise each other's skills and experience.
- That consideration is given to the introduction of a 'health and safety passport' for Undergraduates such that all those entering the industry have a common and adequate level of understanding that has been appropriately assessed.

In summary, therefore, all parties need to make a concerted and unified effort to actively develop and promote health and safety risk management as integral to the education of future construction professionals. This report strongly concludes that this can only be achieved by the joint endeavours of Academia and the wider construction industry working together with unity of purpose and a genuine desire to bring about the improvements demanded at the Safety Summit.

It is hoped that this HSE report will act as a catalyst to this end as well as a useful reference document for all those involved in this important area. This is already happening.

20.10 REFERENCES

Chartered Institute of Building, 1994a, *Membership Regulations and Graduate Notes on the Professional Interview*, CIOB (Berkshire: Ascot).

Chartered Institute of Building, 1994b, *Regulations of Professional Competence and Conduct*, CIOB (Berkshire: Ascot).

Chartered Institute of Building, 1995, *Higher Education Discipline Network in Building Management*, CIOB (Berkshire: Ascot).

Chartered Institute of Building, 1996a, *Education Framework*, CIOB, (Berkshire: Ascot).

Chartered Institute of Building, 1996b, *Directory of Construction Courses 1996/97*, CIOB (Berkshire: Ascot).

Chartered Institute of Building, 1996c, *Health & Safety Academic Course Questionnaire and Analysis of Results*, CIOB (Berkshire: Ascot).

Chartered Institute of Building, 1998a, *Chairman's Report from the Members of the Safety, Health and Welfare Committee to the Members of the Professional Services Board*, CIOB (Berkshire: Ascot).

Chartered Institute of Building, 1998b, *Professional Development Programme*, CIOB, (Berkshire: Ascot).

Confederation of British Industry, 1990, *Developing a Safety Culture* (London: CBI).

Davies, V.J. and Tomasin, K., 1996, *Construction Safety Handbook* (London: Telford).

Department of Employment, 1971, *Glossary of Training Terms* (London: HMSO).

Egan, J., 1998, *Rethinking Construction*, Report of the construction task force to the Deputy Prime Minister, John Prescott, on the scope for improving the quality of UK construction, (London: DETR).

Garavan, T.N., 1997, Training, development, education and learning: different or the same? *Journal of European Industrial Training*, 21, **2**, pp. 39-50.

Gilley, J.W. and Eggland, S.A., 1989, *Principles of Human Resource Development* (Reading: Addison-Wesley Publishing).

Peters, R.S., 1972, *Education and the Educated Man* (London: Routledge and Kogan Page).

Riggs, L.S., 1988, Educating the construction manager. *Journal of Construction Engineering and Management*, 114, **2**, pp. 279-285.

CHAPTER 21

Tertiary Built Environment Construction Health and Safety Education

John Smallwood

INTRODUCTION

Construction occupational fatalities, injuries and disease result in considerable human suffering and affect, not only the workers directly involved, but also their families and communities, and contribute to the national cost of medical care, and rehabilitation. However, occupational fatalities, injuries and disease also contribute to variability of resource, which increases project risk. Such risk can manifest itself in damage to the environment, reduced productivity, non-conformance to quality standards and time overruns, and ultimately in an increase in the cost of construction. Other possible manifestations include damage to client property and, or impaired production processes, and a poor client and contractor image as a result of accidents (Smallwood, 1996).

The focus on cost, quality and schedule by project management is probably attributable to client measurement of project performance on the basis of cost, quality, schedule, and utility - utility includes constructability (Rwelamila and Savile, 1994). A further aspect is that traditionally, cost, quality and time have constituted the parameters within which projects have been procured and managed (Smallwood, 1998). Although this traditional approach has been perpetuated by tertiary construction education, clients, designers, project leaders and the construction industry, it has not been successful, with the greater percentage of projects not being completed within budget, and to quality and time requirements (Allen, 1999). The need for a paradigm shift and focus on health and safety is amplified by the complementary role of health and safety in overall project performance cited by various authors – health and safety enhances productivity, quality, time and ultimately, cost (Hinze, 1997; Levitt and Samelson, 1993).

Anderson (1999) attributes the non-improvement in the UK construction industry accident rate to seven factors, *inter alia*, lack of education and training. Anderson maintains that management education and training, particularly that provided by tertiary institutions, 'fails to give the necessary emphasis to the subject, and those new to the industry have to fall back on 'learning on the job' as opposed to gaining experience on the job.' According to Anderson (2002a), although the UK construction industry's poor accident record has led to calls for more stringent legislation, one essential aspect that has received little attention is the effective teaching of health and safety to industry professionals when they are undergoing their tertiary education.

During research conducted in South Africa, 95.8% of PMs maintained that inadequate, or the lack of health and safety increased project risk (Smallwood, 1996). Inadequate, or the lack of health and safety results in both variability of resource output, and consequently an increase in risk, and also in the probability of an accident. Given that, risk is a function of probability and impact, and that the outcome of accidents is largely fortuitous, the potential risks as a result of inadequate, or the lack of health and safety, are substantial.

During 2001, the Health & Safety Executive (HSE) (2001), United Kingdom (UK), commissioned research into the provision of health and safety teaching in construction related undergraduate courses, namely architecture, building, engineering and surveying. Recommendations include, *inter alia*, that academia should recognise that health and safety risk is part of construction risk management and an essential intellectual element of all construction related courses, and all courses/programmes should be audited with a view to including health and safety risk management in all built environment programmes as an integral and cross curricula element.

Kevin Myers, Chief Inspector, HSE, UK, wishes to draw architects and designers into the health and safety debate. He maintains that many have not properly understood the health and safety implications while working under the Construction (Design and Management) Regulations. He continues and says: 'Educational institutions, of course, can play a key role in hammering home this safety message' and maintains that health and safety modules should become an essential part of training. 'It is an essential requirement and should not be reduced to a two-hour lecture in the course of a seven-year accumulation of professional competence. It should be marbled into the way people go about their business'. (Crates, 2002).

Anderson (2002b) in his comment on the *Revitalising Health and Safety in Construction* document maintains that there should have been a whole chapter on the role of education and educators in health and safety as education is one of the few genuine levers for long-term improvement in health and safety.

21.1   COST OF ACCIDENTS

The cost of accidents (CoA) is frequently cited as a major motivation for addressing health and safety (Hinze, 1997). The CoA can be categorised as being either direct or indirect. Direct costs tend to be those associated with the treatment of the injury and any unique compensation offered to workers as a consequence of being injured and are covered by workmen's compensation insurance premiums. Indirect costs, which are borne by contractors, include *inter alia*, reduced productivity of the workforce, clean-up costs, and wages paid while the injured is idle (Hinze, 1997). Research indicates the total CoA to constitute 6.5% of the value of completed construction (The Business Roundtable, 1995) and approximately 8.5% of tender price (Anderson, 1997). A further motivation is the synergy between health and safety and other project parameters of: cost; environment; productivity; quality, and schedule.

21.1.1 Synergy

Research conducted among PMs in South Africa investigated the impact of inadequate health and safety on various project parameters. Productivity (87.2%) and quality (80.8%) predominated, followed by cost (72.3%), client perception (68.1%), environment (66%), and schedule (57.4%). Health and safety is a prerequisite for productivity and quality as, housekeeping, *inter alia*, complements access and ergonomics. Accidents result in increased cost, damage to the environment and can substantially retard project progress as a result of either, decreased productivity, or a cessation of the works. Client perception may be adversely affected by accidents, as accidents are not project requirements, and/or clients may schedule specific health and safety related contractual requirements, particularly in the case of projects in or adjacent to an existing facility (Smallwood, 1996).

21.1.2 Legislation/International agencies

The Occupational Health and Safety Act (OH&S Act) (Republic of South Africa, 1993) schedules comprehensive requirements for employers such as contractors. However, Section 10 allocates responsibility to designers to ensure that any 'article' is safe and without risks to health.

The draft Construction Regulations (Republic of South Africa, 2002) schedule important requirements with respect to clients and designers. Clients shall, *inter alia*: allow sufficient time for the completion of projects; pre-qualify contractors, and ensure that where design changes are made, sufficient health and safety information is provided to the contractors. Designers shall, *inter alia*: make available all relevant information about the design such as the soil investigation report, design loadings of the structure, and methods and sequence of construction, and inform the principal contractors of any known or anticipated dangers or hazards or special measures required for the safe execution of the works.

The International Labour Office (ILO) (1992) specifically states that designers should: receive training in health and safety; integrate the health and safety of construction workers into the design and planning process; not include anything in a design which would necessitate the use of dangerous structural or other procedures or hazardous materials which could be avoided by design modifications or by substitute materials, and take into account the health and safety of workers during subsequent maintenance.

21.1.3 The status quo

Research undertaken to investigate the extent to which construction health and safety is addressed in tertiary built environment programmes in South Africa determined that construction management programmes placed the most emphasis on construction health and safety, followed by quantity surveying. Approximately half of civil engineering programmes included construction health and safety, and the minority of architectural programmes did (Smallwood, 2002a; Smallwood, 2002b; Smallwood, 2002c).

21.1.4 The built environment disciplines

Architectural graduates generally provide design services, and depending upon procurement practices in their respective countries, may fulfil the function of principal agent.

Civil Engineering graduates invariably fulfil a range of functions in terms of the built environment. These include, *inter alia*, construction management, design, project management for clients, and materials manufacturing, which individually and collectively influence, and have a role to play in health and safety. Depending upon procurement practices, they too may fulfil the function of principal agent (Smallwood, 2002b).

Construction management programmes are invariably structured to prepare graduates to fulfil a range of functions in terms of the built environment. These include construction management, project management for clients, property development and administration, management consulting and materials manufacturing (Smallwood, 2002a).

Given that all project stakeholders clients, designers, project managers (PMs) and contractors influence and contribute to construction health and safety, PMs, in their capacity as project leaders and coordinators, are uniquely positioned to integrate health and safety into all aspects of the design and construction processes (Smallwood, 1996; Hinze, 1997).

Quantity surveying graduates also fulfil a range of functions in terms of the built environment. These include quantity surveying consulting services for clients, project management, quantity surveying within contracting organisations, and management within a range of industry stakeholder organisations.

21.1.5 Optimum Health and Safety requires a multi-stakeholder effort

The Health and Safety Executive (HSE, 2002) maintains that more can be achieved by working together than separately. Cooperation and coordination between all parties is crucial to improving health and safety and integrated teams are an effective way of achieving this.

Research conducted by Smallwood (2002a) among South African general contractors (GCs) investigated, *inter alia*, the degree of importance of the inclusion of construction health and safety in the tertiary education programmes of various construction related disciplines. Given that the importance index (II) values were all above the midpoint value of 2.0 the inclusion of health and safety in built environment tertiary education programmes can be deemed important. II values > 3.2   4.0 indicate that the inclusion of health and safety education in the related disciplines' programmes can be deemed to be between more than important to very important/very important (Table 21.1).

Table 21.1: Perceived importance of the inclusion of construction health and safety in the tertiary education programmes of various construction related disciplines according to GCs (Smallwood, 2002a)

Discipline	Response (%)					II	Rank
	Very important.... Unimportant						
	1	2	3	4	5		
Construction managers	92.8	7.2	0.0	0.0	0.0	3.93	1
Project managers	71.4	21.4	7.2	0.0	0.0	3.64	2
Civil engineers	57.1	42.9	0.0	0.0	0.0	3.57	3
Electrical engineers	50.0	50.0	0.0	0.0	0.0	3.50	4=
Structural engineers	50.0	50.0	0.0	0.0	0.0	3.50	4=
Mechanical engineers	50.0	42.8	7.2	0.0	0.0	3.43	6
Architects	35.7	50.0	14.3	0.0	0.0	3.21	7
Quantity surveyors	14.3	35.7	50.0	0.0	0.0	2.64	8

21.1.6   Issues relative to Health and Safety

Laakanen (1999) relates the main developments and key issues arising from studies conducted in construction health and safety relative to programme content: new regulations; the level of musculoskeletal injuries; new approaches to health and safety; occupational health; accidents; work experience; rehabilitation; promotion of employment; need for development in the work environment, and new health and safety measures.

The promulgation of the EU Construction Directive 92/57/EEC on the implementation of minimum health and safety requirements at temporary or mobile construction sites, required development of knowledge and skills with respect to the participation of clients and designers in health and safety, and the integration of design and construction in terms of health and safety.

Ergonomic interventions have major potential to mitigate the high number of musculoskeletal injuries. Physical training also contributes to the mitigation of such injuries.

A holistic approach requires the integrated development of work organisation, physical environment and rehabilitation.   Work organisation and physical environment requires an appreciation and understanding of the role of planning and pre-planning of health and safety to realise optimum ergonomics.

Rehabilitation forms an integral part of health and safety and needs to be integrated with other interventions to ensure feedback. Relative to health and safety, the role of programmes, awareness in the form of information, motivation and goal setting, training, campaigns, audits and enhanced vocational education in health and safety performance, amplify the need for the inclusion of such subject areas in a tertiary programme.

The risk of occupational diseases amplifies the need for related education. The need for expertise related to induction and other forms of training is reinforced by the incidence of accidents involving new workers. The disproportionate number of accidents involving falls indicates a need for expertise relative to accident prevention. Similarly, inadequate housekeeping indicates a need for expertise relative to planning, pre-planning, systems, and audits.

21.1.7   The influence of design

According to Jeffrey and Douglas (1994) it has to be accepted that in terms of causation there is a link between design decisions and safe construction. This is based on research carried out by the European Foundation for the Improvement of Living and Working conditions, which concluded, that of site fatalities, 35% were caused by falls, which could have been reduced through design decisions.

Designers influence health and safety, directly and indirectly. Design, supervisory, and administrative interventions constitute direct forms of influence. Design interventions include: concept design; general design; selection of type of structural frame; site location; site coverage; details; method of fixing, and specification of materials and finishes. Supervisory and administrative interventions include: reference to health and safety upon site handover, and during site visits and inspections; inclusion of health and safety as an agenda item during site meetings, and the requiring of health and safety reporting by contractors. Indirectly, as a result of: type of procurement system used; pre-qualification; project time; partnering, and the facilitating of pre-planning (Smallwood, 2000a).

Research conducted in South Africa among architectural practices investigated, *inter alia*, the frequency at which health and safety is considered/referred to relative to various aspects. Given the range of responses, an importance index (II) with a minimum value of 0, and a maximum value of 4.0, was computed to enable a comparison and ranking of the aspects (Smallwood, 2000b). Table 21.1 indicates that eleven of the II values of the sixteen aspects are above the midpoint value of 2.0, which indicates that consideration/reference to the aspects, can be deemed to be prevalent. The top three ranked aspects, position of components, method of fixing, and specification, predominate in terms of II values.

A further role identified for designers is that of optimal interaction with clients, particularly at the design brief stage. This is the most crucial phase for the successful, and healthy and safe completion of any project. Deviations from it at a later stage resulting in variation orders (VOs) can be the catalyst that triggers a series of events from designer through to workers that culminate in an accident on site. Consequently, clients must know exactly what they require and develop a comprehensive brief for the design team (Jeffrey and Douglas, 1994).

Designers, and consequently PMs, also influence the pre-planning of health and safety. Pre-planning health and safety realises a structured approach to health and safety related issues by both designers and contractors. Liska (1994) maintains that there are two parts to pre-planning: pre-project and pre-task, and that pre-planning provides the foundation for project health and safety programmes. Pre-planning identifies all the ingredients of and resources required for the health and safety programme to be effective and efficient. However, the design of a project is a great influence on determining the method of construction and the requisite health and safety interventions. Consequently, sufficient design related information, needs to be available at pre-project stage to facilitate budgeting for adequate resources. According to a study conducted by Oluwoye and MacLennan (1994) among PMs and site managers of multi-storey projects in Sydney, drawings, legislation and site inspections are the sources of information most frequently consulted for health and safety planning.

Table 21.2: Frequency at which architectural practices consider/refer to health and safety relative to various aspects (Smallwood, 2000b)

Aspect	Response (%)					II	Rank
	A	O	S	R	N		
Position of components	17.3	31.7	20.2	14.4	7.7	2.52	1
Method of fixing	22.1	32.7	18.3	15.4	5.8	2.51	2
Specification	26.9	24.0	19.2	16.3	5.8	2.49	3
Content of material	21.2	24.0	21.2	19.2	7.7	2.29	4
Edge of materials	26.0	21.2	14.4	19.2	10.6	2.26	5
Details	21.2	23.1	21.2	14.4	11.5	2.24	6
Finishes	20.2	22.1	19.2	20.2	8.7	2.18	7
Type of structural frame	19.2	24.0	18.3	21.2	8.7	2.17	8
Plan layout	21.2	18.3	20.0	18.3	11.5	2.10	9
Texture of materials	18.3	19.2	25.0	17.3	12.5	2.08	10
Design (general)	16.3	17.3	26.9	22.1	10.6	2.05	11
Schedule	14.4	20.2	20.2	19.2	13.5	1.91	12=
Surface area of materials	16.3	20.0	18.3	17.3	16.3	1.91	12=
Elevations	17.3	14.4	21.2	22.1	14.4	1.88	14
Site location	18.3	13.5	19.2	25.0	17.3	1.84	15
Mass of Materials	11.5	15.4	25.0	25.0	12.5	1.78	16

(A=Always; O=Often; S=Sometimes; R=Rarely; N=Never)

Constructability is a further design related issue. 'Design for safe construction' is one of 16 constructability design principles listed by Adams and Ferguson (McGeorge and Palmer, 2002). However, most of the other 15 principles are indirectly related to, and consequently influence health and safety. Method of fixing, size, mass and area of materials, position of components, *inter alia*, amplify the relevance of constructability to health and safety. Consequently, PMs and designers should assess constructability throughout the design stage of a project. Research conducted in South Africa determined identified 'evaluating constructability' as the occasion throughout all project stages when PMs deliberated health and safety most frequently (Smallwood, 1996).

21.1.8   Procurement related issues

Procurement systems and related issues are important as they affect, among other, contractual relationships, the development of mutual goals, the allocation of risk, and ultimately, provide the framework within which projects are executed (Dreger, 1996). Evidence gathered suggests incorrect choice and use of procurement systems has contributed to neglecting of health and safety by project stakeholders (Rwelamila and Smallwood, 1999). The traditional construction procurements system (TCPS), which entails, *inter alia*, the evolution of a design by designers, the preparation of bills of quantities and related documentation by quantity surveyors, the engagement of a contractor through competitive bidding, invariably on the basis of price, does not complement health and safety. Primarily due to: the separation of the design and construction processes, the incompleteness of design upon both preparation of documentation and the commencement of construction, and the engagement of contractors on the basis of price (Rwelamila and Smallwood, 1999). However, Design-Build complements health and safety as a result of the integration of the design and construction processes (Meere, 1990).

Competitive tendering marginalises health and safety. Market conditions in South Africa are such that contractors frequently find themselves in the iniquitous position that should they make the requisite allowances for health and safety, they run the risk of losing a tender or negotiations to a less committed competitor. During research conducted in South Africa, approximately 50% of PMs advocated the inclusion of a provisional sum for health and safety, which would ensure that all tenderers allocate an equitable amount of resources to health and safety (Smallwood, 1996).

South African contract documentation does not engender health and safety. Although references are made to health and safety in standard contract documentation, they are generally indirect, hardly coercive, and depending upon the level of commitment, contractors continue to address health and safety to varying degrees (Smallwood, 1996).

Project duration also impacts on health and safety, as a shortened duration may be incompatible with the nature and scope of the work to be executed (Hinze, 1997). Hinze (1997) cites pressure to meet unrealistic deadlines as a common source of mental diversion, which diversion increases the susceptibility of injury.

Various authors advocate the pre-qualification of GCs and SCs on health and safety by clients and GCs respectively. The purpose of pre-qualification in the health and safety sense is to provide a standardised method for the selection of contractors on the basis of demonstrated safe work records, health and safety commitment and knowledge, and the ability to work in a healthy and safe manner. This will ensure that only health and safety conscious contractors are selected (Hinze, 1997; Levitt and Samelson, 1993).

21.1.9 Role of quantity surveyors (QSs)

QSs, when providing a quantity surveying consulting service, are uniquely positioned to contribute to health and safety as they may advise regarding the type of procurement system, advise regarding project duration, prepare contract documentation, compile bills of quantities or schedules of rates, and assess pre-qualification and tender submissions.

21.1.10 Role of project managers (PMs)

PMs in their capacity as project leaders and coordinators influence and are uniquely positioned to integrate heath and safety into all aspects of the design and construction processes (Smallwood, 1996; Hinze, 1997). Their influence is similar to that of designers, and is both direct and indirect. Direct influence includes design and related interventions that consider health and safety, and supervisory and administrative interventions.

21.1.11 Construction

According to Krause (1993) incidents, accidents included, occur at the end of the heath and safety upstream/downstream sequence: culture → management system → exposure → incidents. Culture includes values, vision, mission, purpose, goals and assumptions. Management system includes education and training, practices such as the prequalification of contractors and the implementation of a management system, programme that ensures the use of safe work procedures, site layout, behaviour consequences, accountability, priorities, attitude, measurement system, improvement model such as total quality management (TQM), and resource allocation. Exposure encapsulates behaviour, conditions, plant and equipment, and facilities.

21.1.12 Course content

Research conducted by Smith and Arnold (1999) among GCs in the USA to investigate the optimum health and safety course content for construction students at Pennsylvania State University determined the following to be the significant skills required of employees with between one and five years' experience: pre-project hazard analysis; preparation of accident reports; conducting tool-box talks; participating in project health and safety meetings; performing hazard analysis; recognising common hazards; conducting health and safety audits; maintaining material safety data sheet (MSDS) files, and the managing of permits. Experience modification rating (EMR), incident ratings, and the cost of accidents (CoA), in particular the indirect CoA, was also considered to be important.

The South African GCs surveyed by Smallwood (2002a) were requested to indicate the extent to which they agreed/disagreed with respect to the inclusion of various subject areas in an honours level (final year) construction management health and safety curriculum (Table 21.3). It is notable that the values of all the importance indexes (IIs) are above the midpoint value of 2.0, which indicates that GCs agree that all subject areas should be included. However, II values > 3.2 ≤ 4.0 indicate that GCs agree to strongly agree/strongly agree regarding such inclusion, and II values > 2.4 ≤ 3.2 indicate that their concurrence is between neutral to agree/agree. It is significant that the 'OH&S Act and regulations' achieved a ranking of first, as this legislation provides the framework within which health and safety occurs, or does not occur. The second ranked management of subcontractors (SCs) is possibly attributable to the frequent citing of SCs as a problem relative to health and safety by GCs. The third ranked 'Health and Safety/Productivity/Quality' indicates the importance of addressing synergy, which is the optimum basis for the promotion of health and safety. The fourth ranked 'role of management', and joint fifth ranked 'worker participation', constitute the 'two pillars' of any health and safety programme/process. The other joint fifth ranked subject area, 'culture', is at the upstream end of the 'upstream/downstream' sequence: culture → management system → exposure → incidents, i.e. optimum health and safety culture is a pre-requisite for realising a healthy and safe project and, or an improvement in health and safety performance. 'Programmes', 'education and training' and 'pre-planning' achieved a joint ranking of seventh. Given that invariably health and safety courses and curricula primarily address 'programmes', the ranking of seventh is notable. International research indicates there to be an inverse relationship between education and training, and the occurrence of incidents. 'Best practice' health and safety requires that both designers and contractors need to engender and focus on the pre-planning of health and safety. The Compensation for Occupational Injuries and Diseases (COID) Act, which achieved a joint ranking of tenth with environment and role of project managers, provides the legislative framework for workers' compensation. The joint tenth ranking of 'environment' is possibly attributable to the current level of focus thereon. Although the 'role of project managers' only achieved a joint ranking of tenth with an II of 3.29, the ranking is nevertheless higher than that achieved by 'role of clients' and 'role of designers', namely joint sixteenth.

It is notable that only four of the ten subject areas, which achieved a ranking between thirteenth and twenty-second, had an II with a value lower than 3.00. Although the importance of these ten subject areas to construction management can be debated, 'project plans', 'economics of health and safety', 'health and hygiene', and 'ergonomics', are of particular importance. 'Project plans' entail a risk assessment and the development of strategies to realise healthy and safe projects. 'Health and hygiene' is concerned with the welfare of workers and 'ergonomics' is concerned with the relationship between workers and their workplace. Both these aspects indirectly affect productivity, quality and schedule.

Table 21.3: Extent to which GCs support the inclusion of various subject areas in an honours level (final year) construction management health and safety curriculum (Smallwood, 2002a)

Subject area	Response (%)*					II	Rank
	SA	A	N	D	SD		
OH&S Act and Regulations	78.6	21.4	0.0	0.0	0.0	3.79	1
Management of subcontractors	71.4	21.4	7.2	0.0	0.0	3.64	2
Health and safety/productivity/quality	57.1	35.7	0.0	0.0	0.0	3.62	3
Role of management	61.5	38.5	0.0	0.0	0.0	3.57	4
Culture (values, vision, purpose, mission, goals, policy)	57.1	35.7	7.2	0.0	0.0	3.50	5=
Worker participation	50.0	50.0	0.0	0.0	0.0	3.50	5=
Programmes	50.0	42.8	0.0	7.2	0.0	3.36	7=
Education and training	50.0	35.7	14.3	0.0	0.0	3.36	7=
Pre-planning	50.0	35.7	14.3	0.0	0.0	3.36	7=
COID Act (Workers' compensation)	35.7	57.1	7.2	0.0	0.0	3.29	10=
Environment	50.0	28.6	21.4	0.0	0.0	3.29	10=
Role of project managers	42.8	42.8	14.4	0.0	0.0	3.29	10=
Project plans	35.7	50.0	14.3	0.0	0.0	3.21	13
Economics of health and safety	23.1	69.2	7.7	0.0	0.0	3.15	14
Health and hygiene	28.6	50.0	21.4	0.0	0.0	3.07	15
Measurement and statistics	14.3	78.5	0.0	7.2	0.0	3.00	16=
Role of clients	28.6	42.8	28.6	0.0	0.0	3.00	16=
Role of designers	35.7	28.6	35.7	0.0	0.0	3.00	16=
Influence of procurement systems	14.3	57.1	28.6	0.0	0.0	2.86	19
Role of the media and awareness	21.4	42.8	28.6	7.2	0.0	2.79	20
Ergonomics	7.2	64.2	21.4	7.2	0.0	2.71	21
Role of unions	7.2	50.0	35.7	0.0	7.2	2.50	22

* (SA = Strongly agree; A = Agree; N = Neutral; D = Disagree; SD = Strongly disagree)

Table 21.4 indicates the extent to which South African construction management, civil engineering and quantity surveying programmes address various construction health and safety subject areas. The extent to which the programmes address the subject areas is presented relative to a ranking based upon the perceived need for their inclusion in a final year construction management programme. In general, construction management programmes address subject areas more comprehensively than quantity surveying and civil engineering programmes, the latter addressing the subject areas to the least extent.

21.1.13 Form of presentation

Forty-five percent of College and University respondents to a study conducted in the USA, stated that their curriculum includes a subject wholly devoted to construction health and safety. Of the 55% of respondents that responded in the negative, some stated that health and safety is either addressed in a generalised manner in other subjects, or that a certain group of subjects address health and safety relative to the subject material (Coble, *et al.*, 1999). However, Coble *et al.* (1999) recommend that all construction management programmes seriously consider specifically addressing health and safety in their curricula.

Research conducted in South Africa determined that the majority (66.7%) of construction management programmes at tertiary institutions include construction health and safety as a component of the subject construction management. Although 50% of GCs advocate the presentation of construction health and safety as a separate subject there is greater convergence between the TIs' and GCs' response relative to 'component of the subject construction management'.

The majority (66.7%) of Departments of Construction Management surveyed by Smallwood (2002a) include construction health and safety as a component of the subject construction management. However, during the survey of South African GCs, 50% of GCs advocate the presentation of construction health and safety as a separate subject.

Table 21.4: Extent to which various programmes address construction health and safety subject areas (Smallwood, 2002a; Smallwood, 2002b)

Subject area	GC Rank	Construction Management	Civil Engineering	Quantity Surveying
OH&S Act and Regulations	1	88.9	77.8	85.7
Management of subcontractors	2	75.0	33.3	57.1
Health and safety/productivity/quality	3	100.0	55.6	100.0
Role of management	4	100.0	88.9	100.0
Culture (values, vision, purpose, mission, goals, policy)	5=	57.1	33.3	28.6
Worker participation	5=	87.5	77.8	71.4
Programmes	7=	87.5	44.4	85.7
Education and training	7=	62.5	66.7	42.9
Pre-planning	7=	75.0	11.1	57.1
COID Act (Workers' compensation)	10=	75.0	44.4	71.4
Environment	10=	62.5	22.2	42.9
Role of project managers	10=	75.0	55.6	57.1
Project plans	13	75.0	11.1	57.1
Economics of health and safety	14	42.9	22.2	28.6
Health and hygiene	15	50.0	44.4	42.9
Measurement and statistics	16=	25.0	11.1	28.6
Role of clients	16=	50.0	11.1	28.6
Role of designers	16=	28.6	33.3	28.6
Influence of procurement systems	19	57.1	0.0	57.1
Role of the media and awareness	20	28.6	11.1	28.6
Ergonomics	21	14.3	22.2	14.3
Role of unions	22	87.5	22.2	71.4

21.2   SUMMARY

Fatalities, injuries and disease result in waste, contribute to the cost of construction and impact on society as a whole.   Conversely, optimum health and safety engenders overall project performance as a result of the synergistic impact on other project performance measures.

A range of international recommendations and national legislation advocates and requires multi-stakeholder contributions to construction health and safety. In practice, all built environment disciplines influence and can contribute to construction health and safety. Consequently, the thrust of tertiary construction health and safety education should be the role of the various project stakeholders, the benefits of optimum construction health and safety and the role of pre-planning.

21.3   RECOMMENDATIONS

Table 21.5 presents the recommended construction health and safety subject areas for tertiary built environment programmes in terms of an importance rating out of maximum of 3. A rating of 0 indicates that it is not recommended that the subject area be included in the programme, 1 that it is recommended, 2 that it is important, and 3 that it is very important.

Ideally, construction health and safety should be afforded separate subject status. However, it is unlikely to be afforded such status, in which case it should be included as an identifiable component in a subject construction management in programmes such as construction management, civil engineering, project management, and quantity surveying. With respect to architectural programmes, it should be included as an identifiable component of design and professional practice. This would ensure that the design and procurement related issues are addressed within their respective contexts. In general, the inclusion of construction health and safety as an identifiable component will prevent the potential dilution that could result from it being integrated into various subjects.

Table 21.5: Recommended construction health and safety subject areas for tertiary built environment programmes

Subject area	Architecture	Construction Management	Engineering	Project Management	Quantity Surveying
			Discipline		
OH&S Act and Regulations	2	3	3	3	2
Management of subcontractors	2	3	3	3	2
Health and safety/productivity/quality	3	3	3	3	3
Role of management	1	3	3	3	1
Culture (values, vision, purpose, mission, goals, policy)	2	3	3	3	2
Worker participation	0	3	3	2	1
Programmes	2	3	3	2	2
Education and training	1	3	3	2	1
Pre-planning	3	3	3	3	3
COID Act (Workers' compensation)	1	3	3	2	1
Environment	3	3	3	3	3
Role of project managers	3	3	3	3	3
Project plans	3	3	3	3	2
Economics of health and safety	3	3	3	3	3
Health and hygiene	3	3	3	3	3
Measurement and statistics	3	3	3	3	2
Role of clients	3	3	3	3	3
Role of designers	3	3	3	3	3
Influence of procurement systems	3	3	3	3	3
Role of the media and awareness	1	3	3	3	1
Ergonomics	3	3	3	3	2
Role of unions	0	3	3	2	1

21.4   REFERENCES

Allen, J.D., 1999, Measuring performance. *Construction Manager*, May, p. 18.

Anderson, J., 1997, The problems with construction. *The Safety & Health Practitioner*, May, pp. 29-30.

Anderson, J., 1999, Construction Safety: Seven factors which hold us back. *The Safety & Health Practitioner*, August, pp. 16-18.

Anderson, J., 2002a, Back to School. *The Safety & Health Practitioner*, February, pp. 23-24.

Anderson, J., 2002b, Health and safety needs new ideas. *Construction News*, September 19, p. 13.

Coble, R.J., Hinze, J.H., McDermott, M.J. and Elliott, B.R., 1999, College's emphasis on safety. In *Proceedings of the Second International Conference of CIB Working Commission W99*, Honolulu, Hawaii, edited by A. Singh, J.H. Hinze and R.J. Coble (Rotterdam: Balkema), pp. 257-264.

Crates, E., 2002, The ball's in your court. *Construction News*, September 19, p. 13.

Dreger, G.T., 1996, Sustainable development in construction: Management strategies for success.   In *Proceedings of the 1996 CIB W89 Beijing International Conference Construction Modernization & Education*, Beijing, CD - file://D1/papers/160-169/163/p.163.htm.

Health & Safety Executive (HSE), 2001, *Identification and Management of Risk in Undergraduate Construction Courses* (London: Sage).

Health and Safety Executive (HSE), 2002, *Revitalising Health and Safety*, (London: HSE).

Hinze, J.W., 1997, *Construction Safety* (New Jersey: Prentice-Hall, Inc.).

International Labour Office (ILO), 1992, *Safety and Health in Construction* (Geneva: ILO).

Jeffrey, J. and Douglas, I., 1994, Safety Performance of the UK Construction Industry. In *Proceedings of the 5th Annual Rinker International Conference focusing on Construction Safety and Loss Control*, Gainesville, Florida, edited by Issa, R., Coble, R.J. and Elliott, B.R (Gainesville: University of Florida), pp. 233-253.

Krause, T.R., 1993, Safety and quality: Two sides of the same coin. *Occupational Hazards,* April, 47, **50**.

Laakanen, T., 1999, Construction work and education: occupational health and safety reviewed. *Construction Management and Economics*, 17, **1**, pp. 53-62.

Levitt, R.E. and Samelson, N.M., 1993, *Construction Safety Management*, 2nd edition (New York: John Wiley & Sons Inc.).

Liska, P., 1994, Zero injury techniques. In *Proceedings of the 5th Annual Rinker International Conference focusing on Construction Safety and Loss Control*, Gainesville, Florida, edited by R. Issa, R.J. Coble and B.R. Elliott (Gainesville: University of Florida), pp. 213-232.

McGeorge, D. and Palmer, A., 2002, *Construction Management New Directions*. 2nd edition (Oxford: Blackwell Science).

Meere, R., 1990, Building can seriously damage your health. *Chartered Builder*, December, pp. 8-9.

Oluwoye, J. and H. MacLennan, 1994, Designing for safety and the environment. *Proceedings of the 5th Annual Rinker International Conference focusing on Construction Safety and Loss Control*, Gainesville, Florida, edited by R. Issa, R.J. Coble and B.R. Elliott (Gainesville: University of Florida), pp. 175-185.

Republic of South Africa., 1993, *Government Gazette No. 14918. Occupational Health & Safety Act: No. 85 of 1993* (Pretoria).

Republic of South Africa, 2002, *Government Gazette No. 23310. Draft Construction Regulations* (Pretoria).

Rwelamila, P.D. and Savile, D.W., 1994, Hybrid value engineering: The challenge of construction project management in the 1990s. *International Journal of Project Managemen*, 12, **3**, pp. 157-164.

Rwelamila, P.D. and Smallwood, J.J., 1999, Appropriate project procurement systems for hybrid TQM.   In *Proceedings of the Second International*

Conference of CIB Working Commission W99 Implementation of Safety and Health on Construction Sites. Honolulu, Hawaii., edited by A. Singh, J.H. Hinze and R.J. Coble (Rotterdam: Balkema), pp. 87-94.

Smallwood, J.J., 1996, The role of project managers in occupational health and safety. In *Proceedings of the First International Conference of CIB Working Commission W99 Implementation of Safety & Health on Construction Sites,* Lisbon, Portugal, edited by Dias, L.A. and Coble, R.J., (Rotterdam: Balkema), pp. 227-236.

Smallwood, J.J., 1998, Health and safety and the environment as project parameters. In *Proceedings of the CIB World Building Congress 1998 Symposium C: Legal and Procurement Practices- Rights for the Environment,* edited by Fahlstedt, K., (Gavle: Sweden: Sverige AB), pp.1587-1594.

Smallwood, J.J., 2000a, The holistic influence of design on construction health and safety (H&S): General contractor (GC) perceptions. In *Proceedings of the Designing for Safety and Health Conference,* London, Loughborough, edited by A.G.F. Gibb (Loughborough: European Construction Institute), pp. 27-35.

Smallwood, J.J., 2000b, *A study of the Relationship between Occupational Health and Safety, Labour Productivity and Quality in South African Construction,* [Unpublished Ph.D. (Construction Management) thesis, University of Port Elizabeth].

Smallwood, J.J., 2002a, Construction management health and safety (H&S) course content: Towards the optimum. In *Proceedings of the 3rd International Conference of CIB Working Commission W99 Implementation of Safety and Health on Construction Sites,* Hong Kong, edited by S. Rowlinson (Hong Kong: Department of Real Estate and Construction, The University of Hong Kong), pp. 193-200.

Smallwood, J.J., 2002b, Civil engineering education: Towards the optimum construction health and safety (H&S) course content. In *Proceedings of the 3rd South African Conference on Engineering Education,* Durban, edited by Case, J., (Durban: Durban Institute of Technology), pp. 175-180.

Smallwood, J.J., 2002c, The need for the inclusion of construction health and safety (H&S) in architectural education. In *Proceedings of the 3rd International Conference of CIB Working Commission W99 Implementation of Safety and Health on Construction Sites,* Hong Kong, edited by S. Rowlinson (Hong Kong: Department of Real Estate and Construction: The University of Hong Kong), pp. 207-212.

Smith, G.R. and Arnold, T.M., 1999, Safety education expectations for construction engineering and management students. In *Proceedings of the Second International Conference of CIB Working Commission W99,* Honolulu, Hawaii, edited by A. Singh, J.H. Hinze and R.J. Coble (Rotterdam: Balkema), pp. 265-272.

The Business Roundtable, 1995, *Improving Construction Safety Performance. Report A-3* (New York: The Business Roundtable).

CHAPTER 22

First Aid and Preventive Safety Training: The Case for an Integrated Approach

Helen Lingard

INTRODUCTION

An increasing awareness of the poor performance of the construction industry with regards to occupational health and safety (OHS) and welfare has focused attention on the importance of health and safety training for construction operatives. OHS training seeks to achieve two objectives. The first is to improve individuals' awareness, knowledge, attitudes and skills in relation to health and safety and the second is to effect positive behaviour change (Cox and Tait, 1998). Simply put, the aim of training is to alter permanently, the behaviour of employees in a way which will further the achievement of organisational goals, such as improved OHS performance. However, the paths leading from training to improved performance in OHS are neither direct nor automatic (DeJoy, 1994). Vojtecky and Schmitz (1986) highlight the need for the systematic evaluation of training a programme's effectiveness in bringing about a desired outcome. This evaluation should cover two interrelated outcomes of the training programme:

- how effective was the training approach in delivering the necessary learning? and
- how has the training impacted on the performance of the organisation in terms of achieving its strategic objectives? (Loosemore, *et al.*, 2003).

The focus of traditional OHS training has been on providing workers with the knowledge, skills and abilities to do their job safely. However, Goldstein (1993) has observed a low correlation between learning an ability to do something and actual job behaviour. With regard to OHS, this low correlation has been explained by the moderating effect of motivational factors (Lindell, 1994). An important element in the effectiveness of a training programme is the extent to which learning is transferred from training interventions into the workplace. This directly relates to the ability of training to impact upon organisational performance and thereby fulfill the second objective cited above. Holton (1996) developed a theoretical model of factors affecting the transfer of training to the workplace. An adaptation of Holton's model is depicted in Figure 22.1. According to Holton, there are three distinct outcomes of a training programme. These are: learning;

individual performance; and organisational performance. Assuming that learning occurs, the transfer of learning is determined by three factors: motivation to transfer; the transfer climate; and transfer design.

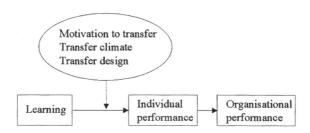

Figure 22.1 A theoretical model of factors affecting the transfer of training to the workplace

Source: Adapted from Holton's model

Motivation to transfer is influenced by employees' expectations of the benefits to be gained from the transfer of learning, for example a reduction in the likelihood of an injury. Holton identifies transfer design as a cause of employees' failure to transfer learning to the workplace. This refers to the possibility that intellectual learning may occur but trainees are not provided with the opportunity to practice the training in the work context or not be taught how to transfer this learning. Finally Holton's model suggests that offering a suitable transfer climate is an important determinant of transfer. The importance of organisational environment was highlighted in a recent study by Tracey, *et al.* (2001) who report that supervisor, job and organisational supportiveness was significantly related to employees' training motivation. Thus, on completion of a training course, trainees will return to the workplace and respond to cues in the environment. Cues that remind trainees of their training can facilitate transfer of that learning. Also, when the learning is put into practice it should be reinforced with praise or positive feedback to ensure that the desired behaviour is maintained. It is important to note that cues and feedback can emanate from supervisors or managers and that conflicting cues could prevent the successful transfer of training outcomes. For example, the motivational effects of incentive schemes that reward fast production may counterbalance the benefits of an OHS training programme. Holton's model used to predict training outcomes seems to offer valid explanations as to why some training programmes do not achieve the desired results (Donovan, Hannigan and Crowe 2001).

22.1   PSYCHOLOGICAL MODELS OF ACCIDENT CAUSATION

Workers' ability to perceive and recognise OHS risks is understood to be essential in the prevention of workplace accidents. However, psychological models of accident occurrence also postulate that, assuming that injury avoidance requires that humans are able to recognise warning signs and avoidance mechanisms, that they make a decision to avoid a danger and have the physiological ability to avoid the danger. For example, Surry's (1974) decision model of danger build-up and release, depicted in Figure 22.2, is based on a series of questions which detail the human response process to danger. Thus, risk-taking behaviour will still occur, even in situations where humans recognise an imminent danger or the immediate risk of an injury when they fail to make the decision action to avoid the danger in the build-up phase or attempt to avoid the injury in the danger release stage. Human responses at the 'decision to avoid' stage are to a large degree determined by their level of safety motivation.

22.1.1   OHS motivation

Motivation has been defined as the arousal, direction and persistence of behaviour (Steers and Porter, 1991). Safety motivation can therefore be defined as the arousal, direction and persistence of behaviour that reduces the likelihood of injury or illness According to contemporary motivation theories, intended behaviour, effort or choice are the primary mechanisms for explaining behaviour (Vroom, 1964; Ajzen, 1988). These models generally agree that an individual's behaviour is influenced by beliefs about behaviour-outcome contingencies. The problem with safety motivation is that accidents are infrequent and many occupational illnesses have long latency periods. Thus, direct personal experience of negative OHS consequences is rare. In the absence of immediate and certain punishments (pain and suffering arising from injury or illness) workers' beliefs about behaviour-outcome contingencies tend not to have a strong motivational effect upon OHS behaviour. Furthermore, information provided in OHS training can be undercut by personal experience. Workers may hold the view that 'I've done it this way for twenty years and never had an accident yet' or observe other workers who do not follow safety procedures with no apparent adverse consequences. Research also shows that repeated experience of unsafe behaviour without an injury or illness leads to systematic desensitisation and diminished fear (Job, 1990).

22.1.2   First-aid training

Previous research suggests that first-aid training has a positive motivational effect on workers' OHS performance. For example, first-aid training has been linked to lower incidences of work-related injury (Miller and Agnew, 1973; McKenna and Hale, 1981) and people trained in first aid have expressed a greater willingness to take personal responsibility for safety and a willingness to adopt safe behaviour (McKenna and Hale, 1982). These findings suggest that first-aid training may have a positive preventive effect, over and above meeting the traditional objective of providing laypersons with the skills to manage the consequences of incidents once they have happened. If this is the case, there could be value in providing first-aid training to all employees in a workplace, rather than to a limited number of designated 'first aiders'. The research reported in this chapter sought to explore the motivational effect of first aid training on workers' OHS behaviour in the Australian construction industry.

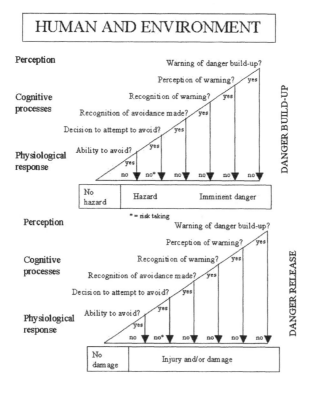

Figure 22.2 Decision model of danger build-up and occurrence here>

22.2   THE STUDY

An experiment, lasting twenty-four weeks, was conducted on Australian construction sites. Participants' subjective understandings of OHS risks, options for risk control and motivation to control OHS risks were explored during in-depth interviews before and after receipt of a first-aid training intervention. In addition to the interview data collection, objective measurement of participants' safety performance was conducted before and after participants received first-aid training.

22.2.1   Small business

The focus of the study was on construction industry small businesses. The majority of Australian construction firms are small businesses with 97% of general construction businesses employing less than 20 employees and 85% employing less than five employees (ABS, 1998). Research suggests that Australian small businesses do not manage OHS risk as effectively as larger businesses in the industry (Holmes, 1995) and may be unaware of their responsibilities under OHS law (Mayhew, 1995). Recent research shows that small business construction industry participants understand risk in terms of factors internal to workers, such as a lack of knowledge or carelessness (Holmes, *et al.*, 1999), and conceptualise risk control as a matter for the individual rather than the firm (Lingard and Holmes, 2000). These factors present difficulties for the prevention of occupational injuries and disease and may explain the higher incidence of occupational injury in small construction firms (McVittie, Banikin and Brocklebank, 1997).

22.2.2   The intervention

All participants attended a generic emergency first-aid training course. This course provides the knowledge and skills to enable participants to provide initial assistance or treatment to a casualty in the event of an injury or sudden illness before the arrival of specialist medical assistance. The course contained both theoretical and practical components, both of which were examined at course completion. Skills taught included the performance of cardio-pulmonary resuscitation and appropriate treatment for fractures and wounds, including bandaging. The course did not contain any information specific to the construction industry. Nor did it contain any information about occupational health and safety risks or their prevention. Training was held off-site at training venues and participants attended the training outside normal work hours. Training course content was standard but participants were able to select which course they attended from among a list of pre-scheduled courses. Thus participants could opt to attend a course close to their worksite on the evenings of their choice. Training courses were all 21 hours long and were delivered over a period of three and a half weeks or seven weeks, depending on whether participants opted to attend once or twice each week. Training courses were delivered by professional employees of

St John Ambulance Australia, one of the largest providers of first aid training in Australia. Trainers were not involved in data collection, which was undertaken by researchers.

22.2.3 The sample

A purposeful, typical case sampling strategy was used to recruit the study participants. The sampling strategy involved the use of key informants to identify what was typical of a small business construction firm, followed by the selection of a small homogeneous sample to describe this sub-group. A set of criteria for typical construction industry small businesses was developed through consultation with an employer organisation (the Master Builders Association of Victoria) and an employees' organisation (the Construction, Forestry, Mining and Energy Union).
These criteria included:

- Long-term involvement in the industry.
- Undertaking of a range of work.
- Self-owned/managed or family-owned and run.
- Consistent operators in the industry; and
- Employment of between three and ten people.

Businesses that matched these criteria were then identified and 25 participants from 14 different construction industry small businesses were recruited. The maximum number of participants from any one firm was four. Two participants were builders whose three subordinates also participated in the study. Three participants were subsequently withdrawn from the study, leaving a total of 22. Two participants were withdrawn because they failed to complete the first-aid training courses into which they enrolled and a third was withdrawn because his occupation rendered him unsuitable. This participant was a farm manager, employed by a small business to oversee the construction of a large stable complex. The age of participants ranged from 19 to 52 years with an average age of 31.2 years. The number of years that the participants had spent in the construction industry ranged between three months and 34 years. The average number of years in the industry was 13.9.

The sample was made up of representatives from several different construction trades. The most commonly represented trade was carpentry with nine of the participants being carpenters by trade and two being apprentice carpenters. However, two carpenters indicated they have a supervisory role in their present jobs and another indicated he was primarily a demolition worker. Supervisory roles were also represented with three participants describing their job as being a builder or site manager and one participant being a site foreman. Three plumbers participated in the study. One participant was a roof tiler and one participant was a landscape gardener. The landscape gardener's apprentice also took part in the study.

22.2.4 Hypotheses

All participants received a standard first-aid training course, the content of which is solely devoted to techniques for the immediate management of injuries and acute illnesses when they occur. This course does not contain any information about construction site risks or their control. Thus, participants received no information that would directly assist them to recognise hazards in their work environment or understand ways in which injuries or illness could be prevented in this context. Thus, in terms of Surry's model, it was expected that first-aid training would not change participants' ability to perceive or recognise workplace hazards. Neither was it expected that first aid training would change their recognition of appropriate avoidance action or ability to avoid a hazard. Thus hypotheses one and two were formulated as follows:

- First-aid training does not improve participants' understanding of acute and delayed effect OHS risks relevant to their workplaces; and
- First-aid training does not improve participants' understanding of OHS risk control for known OHS risks.

However, previous research findings have demonstrated a link between first aid training and an enhanced willingness to take personal responsibility in the avoidance of occupational injury. In the light of the lack of direct preventive information in the first-aid course content, one explanation for this effect is that first-aid training has a positive effect on participants' safety motivation. Thus, in terms of Surry's model, first-aid training changes participants' cognitive response at the 'decision to avoid' stage, making them more likely to decide to avoid a potentially dangerous situation. Subject to their level of pre-existing knowledge of workplace hazards and appropriate risk control responses, this change in decision-making will lead to enhanced OHS behaviour. Thus, hypotheses three and four were formulated as follows:

- First-aid training increases participants' motivation to control known OHS risks; and
- First-aid training improves participants' risk control (safety) behaviour in relation to known OHS risks.

22.2.5 Experimental design

A single-case experiment design was used (Barlow and Hersen, 1984). The design, was organised into two phases and first-aid training was introduced to participants at different timings. The advantage of this experiment design is that it controls bias in the selection of participants with different characteristics in control and experimental groups, since comparisons are made within subjects and not between them. Single-case experiment designs overcome the difficulties of achieving random assignment of subjects into control and experimental groups in real workplace settings (Komaki and Jensen, 1986). Non-random assignment of subjects into control and experimental groups has been a methodological flaw in previous experiments to assess the effect of first-aid training on OHS awareness and performance (see for example McKenna and Hale (1982)).

22.2.6   Data collection

Objective safety performance was measured by a researcher directly observing workers on site once before and once after they received the first-aid training. Two behaviour rating scales were developed for the direct observation of behaviour (Sarvela and McDermott, 1993). First, a literature review was conducted to identify the historical experience of OHS incidents in the small business construction industry context. Secondly, objective OHS performance indicators were identified for pertinent categories of performance. These indicators were then expressed as measurable items in a behaviour rating scale and observers were trained in their content. Items were rated proportionally as proportional rating scales have been found to yield high levels of inter-rater reliability in previous OHS intervention studies (Duff, *et al.*, 1994; Lingard, 1995).

Researchers rated the general level of safety on each site using a Global Safety Measure (GSM) and the safety performance of individual participants using an Individual Safety Measure (ISM). The GSM contained 21 items measuring four different categories of safety performance. These were site 'Housekeeping' (6 items), use of 'Personal Protective Equipment' (6 items), 'Use of Tools' (3 items) and 'Access to Heights' (6 items). The ISM contained 14 items measuring four different categories of safety performance. These were use of 'Personal Protective Equipment' (6 items), 'Use of Tools' (3 items), 'Manual Handling' (1 item) and 'Access to Heights' (4 items).

Inter-rater reliability was assessed three times during the study. This involved two researchers visiting the site and independently measuring safety performance. The percentage agreement between the two raters on items scored and observed was 89%, which constitutes an acceptably high level of inter-rater reliability.

A structured interview theme list was developed to elicit information about participants' cognition (knowledge) of chronic and acute effect OHS risks, the options available for the control of these OHS risks and issues in emergency preparedness. A structured approach was suited to the research purpose in that it is useful in ensuring the comparability of data and enabled the identification of similarities and differences between participants' understandings before and after their receipt of first-aid training.

Interviewers were trained in the use of prompts and probes to reduce the problem of over-simplistic responses to interview items. Interviews were conducted in the workplace, during the course of a normal work day, and were tape-recorded for later transcription and analysis.

22.2.7 Data analysis

Pre-training and post-training scores for each category of safety performance measured during the on-site observation of OHS behaviour, were analysed using Paired Samples T-tests. These tests were carried out for both individual safety performance (ISM) and global safety performance (GSM) measurements. Paired Samples T-tests are well-suited to analysing experiment data in which the same person is observed before and after an intervention (Norusis, 1998).

The interview transcripts were double coded independently by two different researchers for major themes and analysed in more depth using ethnographic content analysis (Tesch, 1990). Double coding of data was undertaken to ensure reliability of the coding framework developed. The coding was an iterative process during which the two independent coders reached agreement on the thematic content of data. The importance of emergent concepts for participants' risk awareness, understanding of risk control, risk-taking tendencies and emergency preparedness was determined on the basis of the frequency with which they were mentioned.

22.3 RESULTS

Key themes that emerged from the pre-training and post-training interview data are presented in Table 22.1.

Table 22.1 Important themes emerging from interview data

Theme	Pre-training	Post training
OHS risk perception and recognition	Immediate effect OHS risks (falls, power tools, trenches) Fatal consequences, permanent damage, dread	Acute effect OHS risks (falls, power tools, trenches) Fatal consequences, permanent damage, dread Infectious diseases
OHS risk sources	Carelessness or complacency (other workers) Inexperience (other workers) Chance (self)	Carelessness or complacency (other workers and self) Inexperience (other workers)
Likelihood of injury/illness	Low probability 'Won't happen to me'	Medium to high probability 'Can happen to me'
OHS risk-taking	Accept risk taking to 'get the job done' Don't think of consequences	Unwilling to take certain risks to 'get the job done' Consideration of the costs/benefits of risk-taking Aware of consequences
OHS risk control	Individual workers' responsibility Current risk controls sufficient	Individual workers' responsibility Current risk controls not sufficient Cost/organisational constraints (no more can be done) Need more enforcement

22.3.1   OHS risk perception and recognition

Before first-aid training, the most commonly cited risks were falling from heights (18 participants), power tools (13 participants) and collapse of trenches (four participants). Chronic or delayed effect risks were mentioned much less frequently. Only two participants mentioned delayed effect health risks without prompting, and, even when prompted, most participants indicated that the delayed effect OHS risks they identified, which included inhalation of dusts (four participants) noise (three participants) and skin disease (two participants), were relatively unimportant. As one participant said *'Illness? Well mainly in what I do you are pretty right. I know there is some propensity when you are cutting certain materials that you have to wear masks, but I don't know.'*

Following first-aid training, this emphasis on immediate effect OHS risks did not change. Out of the twenty-two participants who completed first-aid training, 17 identified falls or working at height to be relevant and 14 identified working with power tools and machinery. Four participants identified the collapse of trenches. The strong emphasis on acute effect compared to delayed effect OHS risks was expressed by one participant as follows: *'I suppose with heights you are more prone to falling. With dust and things like that, you don't realise it until later on in life so you look at things that happen there and then rather than the long-term'.*

Only five participants identified delayed effect OHS risks without prompting. These included skin cancer (two participants), hearing loss (one participant) and dust (two participants). When prompted to consider work-related illnesses, a further six participants mentioned skin cancer, three mentioned asbestos, two mentioned dust, one mentioned noise and another mentioned working with chemicals. As in the pre-training interviews most participants suggested that these risks were not serious because they are easily controlled, especially in the case of asbestos and skin cancer. Another important theme arising from the discussion of occupational health risks was that participants who are aware of the risk, assume it to be minimal because they are uncertain of the credibility of the source of the information they have received. For example one participant said of chemical risks *'A couple of people think there might be a bit of a risk in some primer. But nothing's ever been said so I don't think so'* and a carpenter said of the risk of wood dusts *'They talk about the dangers of working with dust but I don't think it really affects us that much'.*

The only change in participants' awareness of specific OHS risks was the increased awareness of infectious diseases following their receipt of first-aid training. Only one participant mentioned the risk of infectious diseases, such as hepatitis, without prompting but when prompted seven participants identified these illnesses as being relevant to their work. Johnson (1999) identifies the importance of first-aid training in increasing individuals' awareness of the importance of treating small cuts and abrasions immediately to prevent infections from untreated wounds. Before first-aid training, participants thought of minor injuries such as nicks, cuts, scratches as a cost of doing business and indicated that they would often ignore such wounds.

These results are consistent with the results of previous studies of laypersons' perceptions of non-occupational risks. For example, Slovic, Fischoff and Lichtenstein (1981) found that risks associated with a perceived lack of control, fatal consequences, high catastrophic potential, reactions of dread, inequitable distribution of risks and benefits and the belief that risk was increasing and not easily reducible are considered to be greater than risks that are unknown, unobservable, new and delayed in their manifestation.

Thus it appears that, other than raising awareness of the risk of infection, first-aid training had little impact upon participants' understandings of specific OHS risks relevant to their work or the seriousness of these risks. The results generally support hypothesis number one.

22.3.2   OHS risk control

Before training, OHS risks were understood to arise as a result of individual factors. The most commonly cited source of OHS risks affecting other workers was their carelessness or complacency (eleven participants before training). For example, one participant said:

> *You can educate people till the cows come home. Overconfidence causes most of the problems and there is nothing you can do about it...You can put up a handrail and provide someone with a harness, but they have to choose to use the harness and they have to operate within the boundaries of the handrail. If someone decides it is a nuisance, they will take it down.*

However, although before training participants tended to believe accidents to others were attributable to a lack of care or complacency about OHS, only five participants identified carelessness or complacency as being relevant to their personal experience of OHS risk. Eight participants expressed the belief that accidents to themselves were attributable to factors beyond their own control, such as the negligence of others. For example, one participant said: *'There is always the risk of stepping into a puddle and finding out that someone has been negligent and dropped a power cord in there and there is a fault in the leakage switch'*. Another commented: *'Well hopefully I won't but things can happen where it is not your fault either, I mean someone could drop a hammer and it could hit you in the head...there is nothing you can do'*.

Eleven participants also expressed the fatalistic view that their own personal experience of occupational injury or illness was a matter of luck or chance. For example, one participant said *'Put it this way, in ten years I've had one injury that has taken me to hospital, so that is not to be I think'*. Another commented *'I think it is hard to say. It's your own fate'.* The belief that accidents are chance events that 'just happen' is likely to discourage workers from taking appropriate preventive action. Saari (1988) reports a similar result in that 40% of foremen in his survey attributed occupational accidents to chance.

The tendency to attribute events to others to internal causes, and events to oneself to external causes has been observed in previous risk research and is termed 'self-other' bias (DeJoy, 1994).

Following first-aid training, participants still largely attributed OHS risks to individual factors, such as complacency or carelessness. However, the 'self-other bias' that their own personal experience of OHS risks was beyond their control appears to have been reduced. Following first-aid training, 17 participants expressed the importance of taking care and concentrating to avoid occupational injury or illness to themselves. Only four participants mentioned chance or fate and four mentioned the negligence of others as important influences on their own personal experience of OHS risk.

This indicates an increased recognition by participants that their own behaviour is also important in the prevention of occupational injury and disease and is similar to McKenna and Hale's finding that people trained in first aid express a greater willingness to take personal responsibility for safety (McKenna and Hale, 1982).

When asked to identify control measures that should be implemented for the OHS risks they identified as being relevant in their work, participants emphasised individual controls for both themselves and others. Individual controls are those which are intended to act on workers' behaviour whereas technological controls are intended to act on the work environment. This is consistent with previous research findings that, where OHS risks are attributed to factors internal to affected workers, understandings of risk control will be limited to individual controls (Holmes, *et al.*, 1999).

In the technical literature, OHS risk controls should be selected according to a risk control hierarchy. The principle of this hierarchy is that control measures that aim to target hazards at source, and act on the work environment are more effective than controls that aim to change the behaviour of exposed workers. Thus, technological control measures, such as the substitution of hazardous substances or processes and engineering controls are preferable to individual controls, such as the introduction of safe work practices or the use of personal protective equipment. The effectiveness of individual controls for OHS risk is limited because humans are prone to error. The reliance on personal protective equipment is particularly limited since such equipment:

- frequently does not provide the protection claimed;
- is uncomfortable to use;
- often makes working difficult;
- can create a hazard itself; and

The most commonly cited control measures before first-aid training were use of personal protective equipment (seven participants), training and education (six participants), supervision (three participants) and correct work procedures (two participants). In contrast, technological controls, such as the use of scaffolds and handrails to prevent falls or the use of mechanical lifting devices to avoid manual handling were mentioned by only six participants. Following first-aid training participants still predominantly emphasised control measures that individual workers could or should take. For example, one participant said of the risk of falling:

> 'With construction, falls is up to you. If you work off a ladder, you've got to try and tie the ladder up. So it all comes back to you doing the things you're supposed to be. If you fall off, have a harness so that you can't hit the ground'.

Eleven participants mentioned personal protective equipment, 12 participants mentioned the need to take care and concentrate at work, five participants mentioned safety education and training and four participants mentioned safe work procedures. Nine participants also mentioned the need for technological or engineering controls to reduce OHS risks.

Understandings of OHS risk control that focus on individual controls are likely to hinder the adoption of more effective technological solutions to control OHS risks (Lingard and Holmes, 2000). First-aid training does not appear to change participants' emphasis on individual OHS risk controls and they appear resigned to bearing the burden of responsibility for the prevention of injury and illness themselves. Thus, the research results generally support hypothesis number two.

Despite the fact that first-aid training did not substantially change participants' understandings of the different options for controlling OHS risks relevant to their work, interview data do suggest that participants were less willing to accept or tolerate prevailing levels of OHS risk after receiving first-aid training. Before receiving first-aid training, fourteen participants expressed the belief that the OHS risk controls currently adopted in the building industry were sufficient. Following first-aid training, eleven participants expressed the view that OHS risk control measures currently implemented are not sufficient whereas only eight participants believed them to be sufficient. For example, one participant questioned the effectiveness of personal protective equipment, saying:

> 'You have to go up and work. You could have safety harnesses but they become very awkward to work with. You get tangled and that becomes a hassle for workers too and you can probably still get injured if you do fall off with a safety harness'

and another commented:

> 'Sanding, I try to wear a mask at all times. But on a day like today where it is 35 degrees, you can't breathe through a mask so you don't wear it but I try to keep the dust away from my face'.

Thus it appears that after training workers did not believe enough was being done to reduce OHS risks in their workplace. Yet at the same time, many participants were not aware of any feasible alternative risk control measures that could be implemented. As one participant remarked 'There are probably a lot more things we could do but I just don't know about it'.

22.3.3 Responsibility for accident prevention

In the pre-training interviews, many participants expressed the unrealistically optimistic belief that 'it won't happen to me'. In comparison to others in their workplace, the largest proportion of participants (nine) indicated that others were more likely to suffer from an occupational injury or illness than themselves. For example, one participant expressed this by saying:

> 'You make scaffolds that aren't up to scratch I would be the only one to walk on them because I know it's safe for me but I wouldn't want any one else doing it'.

Nine participants also indicated they had a high degree of personal control over OHS risks and therefore they could avoid injury. One participant expressed this by saying 'I think if you have got your wits about yourself, you can deal with anything'. This is consistent with previous research findings that a belief that personal skill is more important in avoiding accidents and a tendency to over-estimate the degree of personal control over events contributes to the belief that 'it won't happen to me' (De Joy, 1994).

Before first-aid training, another group of participants attributed their comparatively low probability of suffering a work-related injury or illness to their experience in their job. One participant expressed this by saying:

> 'I'm probably less likely [to have an accident] because I've been doing it a long time. Not like the young guys running around madly...[they] run into things, fall off the roof and try to carry heavy weights too quickly'.

Weinstein (1980) also discovered that a sense of 'unrealistic optimism' about the probability of being involved in an accident increased with experience.

Following first-aid training, 18 participants indicated that they had a medium to high probability of personally suffering from a work-related injury or illness. Only three participants said the chance of them suffering a work-related injury or illness was low. One of these was an office-based site manager while another had just returned to work on 'light duties' having suffered a work-related back injury. First-aid training seems to reduce workers' sense of 'unrealistic optimism' about their likelihood of experiencing a work-related injury or illness and strengthens the link between risk-taking behaviour and undesirable outcomes in the minds of participants. Thus it seems that first-aid training can help to overcome the motivational problem that workers' direct personal experience of serious negative OHS consequences is rare.

Before participants underwent first-aid training, when asked whether they ever knowingly took unnecessary OHS risks at work, the majority (23) indicated that they did. When asked what types of risks these were, 12 participants indicated that they were associated with working unsafely at height, for example using unsafe scaffolding, using improvised means of gaining access to height or failing to use a safety harness when required. A further five participants indicated that they occasionally took unnecessary risks using power tools and another four said they sometimes failed to use the correct personal protective equipment. There was a strong acceptance of risk-taking behaviour as 'part of the job'. Only two participants suggested that risks should not be taken or that they were concerned about taking risks.

These results suggest that, despite the perceived seriousness of the risks of falling from height and power tools, risk-taking behaviour is widely accepted in these two aspects of construction site safety. This apparent willingness to take OHS risks that are understood to be serious may be due to the prevalent belief before first-aid training, that 'it won't happen to me'.

When asked why they took such risks, the most commonly cited response (nine participants) was 'to get the job done.' This reflects a strong production orientation among construction workers. Four participants also explained risk-taking behaviour in terms of work habits developed over a long period. For example, one participant said people took risks:

'because in the days when they started work it was accepted work practice and they have done it through their entire working career and not had an accident and continue to work in that fashion'.

Following first-aid training, participants did not express such a ready acceptance of risk-taking behaviour. Only eight participants expressed an unreserved willingness to take OHS risks to 'get the job done'. Twelve participants suggested that they would take OHS risks but only under certain circumstances. Five participants indicated that they had taken such risks in the past but that they were less likely to do so now. Three participants said they sometimes took risks that they recognised that they should not take. Four participants indicated that they would consider the costs and benefits before taking an OHS risk and base their behaviour on a 'calculated risk', only taking risks where the benefits outweighed the costs and where they considered the risk to be 'worth it'.

It seems likely that participants' stronger belief that they could personally suffer an occupational injury or illness, following first-aid training, renders them less comfortable about taking unnecessary risks. The recognition that unsafe behaviour can result in injury or illness thus appears to increase participants' motivation to work safely. These results lend support to hypothesis number three.

22.3.4   OHS behaviour

Table 22.2 shows the average ISM and GSM performance scores for the four categories of performance measured before and after participants received first-aid training.

Table 22.2   Pre and post-training safety scores

Global safety measure			Individual safety measure		
Category	Pre-training (% safe)	Post-training (% safe)	Category	Pre-training (% safe)	Post-training (% safe)
Use of tools	97	96	Use of tools	94	98
Access to height	47	78	Access to height	51	93
PPE	60	95	PPE	64	96
Housekeeping	79	85	Manual lifting	85	80

In the ISM, safety performance was generally highest in the 'Use of Tools' category with scores of 94% safe and 98% safe for the pre and post-training measurements respectively. Safety performance was lowest in the 'Access to Heights' category in which the pre-training score was only 51%. The post-training score for 'Access to Heights' rose to 93%. The use of 'Personal Protective Equipment' was found to be 65% safe before and 96% safe after participants received training. 'Manual handling' was found to be 85% safe before training and 80% safe after training.

Paired Samples T-tests were conducted, using the SPSS 8.0 for Windows software package, to assess the statistical significance of these changes. The improvement in the use of 'Personal Protective Equipment' was found to be highly significant (t=3.352, 13df, p=0.005). The improvement in the 'Use of Tools' approached significance (t=2.066, 8df, p=0.073). However, neither the improvement in 'Access to Heights' (t=2.018, 4df, p=0.114) nor the deterioration in 'Manual Handling' safety (t=1.890, 2df, p=0.199) was significant.

As with the individual safety scores, safety performance was generally highest in the 'Use of Tools' category with scores of 97% safe and 96% safe for the pre and post-training measurements respectively. Safety performance was lowest in the 'Access to Heights' category in which the pre-training score was only 47%. The post-training score for 'Access to Heights' rose to 78%. The use of 'Personal Protective Equipment' was found to be 60% safe before and 95% safe after participants received training. Site 'Housekeeping' was found to be 79% safe before training and 85% safe after training.

Paired Samples T-Tests were conducted to assess the statistical significance of these changes. The improvement in the use of 'Personal Protective Equipment' was found to be very highly significant (t=5.325, 19df, p=0.000). The improvement in 'Housekeeping' was also found to be significant (t=2.757, 20df, p=0.012). However, neither the improvement in 'Access to Heights' (t=1.090, 11df, p=0.299) nor the slight deterioration in 'Use of Tools' (t=0.346, 19df, p=0.733) was significant.

In both individual and global performance, safety scores in three of the four categories measured were higher following first-aid training indicating an improvement in safety over pre-training performance. These improvements, some of which were statistically significant, suggest that first-aid training has a positive effect on OHS behaviour. Thus some evidence was found to support hypothesis number four.

The finding that improvements in safety behaviour were not confined to the individuals who had undergone first-aid training was unexpected. It is not clear why this effect occurred but it was possibly due to the fact that many of the participants performed a supervisory role and, through this supervisory influence, their increased motivation to avoid injury brought about a general improvement in site safety.

22.4 CONCLUSIONS AND RECOMMENDATIONS

In combination, the results of the behavioural observations and the interviews yield important information about the effect of first-aid training. Observations at participants' worksites suggested that, for the most part, the first-aid training had a positive effect on the OHS behaviour of participants. However, interview data revealed that, other than raising awareness of the risk of infectious diseases, the first-aid training did not increase participants' understandings of the nature or severity of specific OHS risks relevant to their work. First-aid training appeared to reduce participants' 'self-other' bias, making them more aware that their experience of OHS risks was not beyond their control but that their own behaviour was also an important factor in the avoidance of occupational injury and illness. First aid training also appeared to reduce participants' willingness to accept prevailing levels of OHS risk. Participants' understandings of methods by which OHS risks can be controlled were unchanged by the first aid training and are limited to individual controls. First-aid training did appear to increase participants' perception of the probability that they would suffer a work-related injury or illness and they also expressed greater concern about taking risks at work after receiving first aid training. It therefore appears that first-aid training enhanced participants' motivation to avoid occupational injuries and illnesses. It is possible that the motivational effect of first aid training, coupled with the knowledge, skills and abilities provided by traditional OHS training would yield greater improvements in workers' OHS behaviour than either first-aid training or traditional OHS training on its own. As such, first-aid training may be a valuable tool in facilitating the transfer of safety knowledge, skills and abilities, learned in preventive OHS training, into behaviour at the workplace, ultimately leading to enhanced organisational performance.

Traditional first-aid training does not currently improve participants' understanding of specific OHS risks relevant to their workplace. Neither does it raise general awareness of what can be done to control OHS risks. However, the implication of these findings for training course content and delivery is that it may be beneficial to integrate first-aid training and preventive OHS training to achieve greater overall effectiveness. For example, there may be an opportunity in industry-specific first-aid training to incorporate information about the OHS risk-profile and the nature and severity of OHS issues relevant to the industry. This may be particularly helpful in the case of lesser known, health risks. Similarly, there may be an opportunity to discuss preferred, technological options for OHS risk control, such as the substitution of hazardous chemicals with less hazardous alternatives or the provision of ventilation to prevent dust inhalation, in first-aid training.

The extent to which can be achieved in industry-specific integrated training courses should be investigated. However, before such an approach has been developed and rigorously evaluated, an interim approach could be to link preventive safety and first-aid information in courses designed to complement one another. One solution to this would be to ensure the training programme closely reflected the work environment to ease the transfer (Yamnill and McLean, 2001). For example, first-aid training courses could be designed to encourage people to think about OHS risks in their workplace in using visual imagery so that the first aider can relate the training to their own work environment. This is a similar approach to that used in sports psychology in which the athlete is taught to mentally rehearse performance. Alternatively, training could be conducted at the workplace and injury scenarios created by placing the Resusci-Annie at a location on site. First-aid trainees could be given 'homework' of thinking up three or four commonplace situations they find themselves in at work and asked to imagine emergency situations involving breathing, bleeding, breaks and burns in each of the situations (Cooper, 1995). This would encourage them to think about OHS risks relevant to their work and would probably enhance the motivational effect of first-aid training on their safety behaviour as well as the perceived relevance of the first-aid course to their work. industry-specific first-aid courses could present information about injuries and illnesses and their treatment in the context of how they could occur in the participants' own workplace. Another option would be to relate injuries to the OHS risks relevant to first-aid training participants' work, such as discussing wounds occurring while using power tools, burns from bitumen, or occupational asthma arising from exposure to chemicals. This would serve the dual purpose of providing participants with the skills to treat injuries and illnesses and improving their understanding of specific workplace OHS risks.

22.5 REFERENCES

Australian Bureau of Statistics, 1998, *Business Register data.*

Ajzen, I., 1988, *Attitudes, Personality and Behavior*, Open University Press, Milton Keynes.

Barlow D.H. and Hersen, M., 1984, *Single Case Experimental Designs: Strategies for Studying Behavior Change* (New York: Pergamon Press).

Cooper, I., 1995, First-aid training; a holistic approach. *Occupational Health*, 47, 4, pp. 128-9.

Cox, S. and Tait, R., 1998, *Safety, Reliability and Risk Management: An Integrated Approach* (Oxford: Butterworth-Heinemann).

DeJoy, D.M., 1994, Managing safety in the workplace: An attribution theory analysis and model. *Journal of Safety Research*, 25, pp. 3-17.

Donovan, P., Hannigan, K. and Crowe, D., 2001, The learning transfer system approach to estimating the benefits of training: empirical evidence. *Journal of European Industrial Training*, 25, pp. 221-228.

Duff, A.R., Robertson, I.T., Phillips, R.A. and Cooper, M.D., 1994, Improving safety by the modification of behaviour. *Construction Management and Economics*, 12, pp. 67-78.

Goldstein, I.L., 1993, *Training in Organizations*, 3rd edition, Brookes/Cole, Pacific Grove, CA.

Job, R.F.S., 1990, The application of learning theory to driving confidence: The effect of age and the impact of random breath testing. *Accident Analysis and Prevention*, 22, pp. 97-107.

Johnson, L.F., First aid by numbers, 1999, *Occupational Health and Safety*, 68, 4, pp. 40-43.

Holmes, N., 1995, *Workplace understandings and perceptions of risk in OHS*, Unpublished PhD thesis, Monash University, Australia.

Holmes, N., Lingard, H., Yesilyurt, Z. and DeMunk, F., 1999, An exploratory study of meanings of risk control for long term and acute effect occupational health and safety risks in small business construction firms. *Journal of Safety Research*, **30**, pp. 251-261.

Holton, E.F.III, 1996, The flawed four-level evaluation model. *Human Resource Development Quarterly*, **7**, pp. 5-25.

Komaki, J. and Jensen, M., 1986, Within-group designs: an alternative to traditional control group designs. In *Health and Industry: A Behavioral Medicine Perspective*, edited by M.F. Cataldo and T.J. Coates (New York: John Wiley and Sons).

Lindell, M.K., 1994, *Occupational Medicine: State of the Art Reviews*, 9, **2**, Hanley and Belfus Inc., Philadelphia, pp. 211-240.

Lingard, H., 1995, *Safety in Hong Kong's Construction Industry: Changing Worker Behaviour*, unpublished PhD thesis, The University of Hong Kong, Hong Kong.

Lingard, H. and Holmes, N., 2000, Understandings of occupational health and safety risk control in small business construction firms: Barriers to implementing technological controls, accepted for publication in *Construction Management and Economics* (in press).

Loosemore, M., Dainty, A. and Lingard, H., 2003, Human Resource Management in Construction Projets - Strategic and Operational Aspects (London: Taylor and Francis Ltd.).

Mayhew, C., 1995, *An Evaluation of the Impact of Robens Style Legislation on the OHS Decision-Making of Australian and United Kingdom Builders with Less than Five Employees*, Worksafe Australia Research Grant Report, Sydney.

Mayhew, C., 1997, *Barriers to the Implementation of Known Solutions in Small Business*, Commonwealth of Australia, Canberra.

McKenna, S.P. and Hale, A.R., 1981, The effect of emergency first aid training on the incidence of accidents in factories, *Journal of Occupational Accidents*, 3, pp. 101-114.

McKenna, S.P. and Hale, A.R., 1982, Changing behaviour towards danger: the effect of first aid training. *Journal of Occupational Accidents*, 4, pp. 47-59.

McVittie, D., Banikin H. and Brocklebank, W., 1997, The effect of firm size on injury frequency in construction. *Safety Science*, 27, pp. 19-23.

Miller, G. and Agnew, N., 1973, First-aid training and accidents. *Occupational Psychology*, 47, pp. 209-218.

Norusis, M.J., 1998, *SPSS 8.0 - Guide to Data Analysis Upper Saddle River* (N.J.: Prentice Hall).

Saari, J., 1988, Successful accident prevention: An intervention study in the Nordic countries. *Scandinavian Journal of Work Environment and Health*, 14, pp. 121-123.

Sarvela, P. and McDermott, R., 1993, *Health Education Evaluation and Measurement - A Practitioner's Perspective* (Dubuque, IA: WCB Brown & Benchmark), pp. 265-271.

Slovic, P., Fischoff, B. and Lichtenstein, S., 1981, Facts and fears: understanding perceived risk. *Society Risk Assessment*.

Steers, R.M. and Porter, L.W., 1991, *Motivation and Work Behavior* (New York: McGraw-Hill Inc.).

Surry, J., 1974, *Industrial Accident Research*, University of Toronto, Toronto Canada.

Tesch, R., 1990, *Qualitative Research: Analysis Types and Software Tools* (New York: The Falmer Press).

Tracey, J.B., Hinkin, T.R., Tannenbaum, S. and Mathieu, J.E., 2001, The influence of individual characteristics and the work environment on varying levels of training outcomes. *Human Resource Development Quarterly*, 12, pp. 5-23.

Vojtecky, M.A. and Schmitz, M.F., 1986, Program evaluation and health and safety training. *Journal of Safety Research*, 17, pp. 57-63.

Vroom, V.H., 1964, *Work and Motivation* (New York: Wiley).

Weinstein, N.D., 1980, Unrealistic optimism about future life events. *Journal of Personality and Social Psychology*, 39, pp. 806-820.

Yamnill, S. and McLean, G.N., 2001, Theories supporting transfer of training. *Human Resource Development Quarterly*, 12, pp. 195-208.

22.6 ACKNOWLEDGEMENTS

St John Ambulance, Australia funded the research and St John Ambulance, Victoria provided the first-aid training.

Section 5

Safety Technology

CHAPTER 23

Safety and Technology

Steve Rowlinson

23.1 THE ROLE OF TECHNOLOGY IN SAFETY MANAGEMENT

One way of dealing with the safety issues on site is to provide innovative technological solutions to problems. However, it must be borne in mind that new technologies often bring with them new risks and workers must be thoroughly trained and appraised of new risks before new work processes are implemented. However, many innovations have been made in the ergonomics field and these lead to a reduction in injuries to backs and other parts of the anatomy. This issue is addressed in Davis, Lorenc and Bernold (1999), to the conference in Hawaii:

> *Occupational back injuries are on the rise and seem to be universal. Due to its large impact on the life of people and the related cost, it should be prevented at all costs. This paper presented one approach, technological interventions, namely a simple and advanced engineering solution to the operation of nailguns in building construction. After an extensive period of site observation and analysis, the process of nailing sub-floors was identified as a high-risk operation. Based on the nature of the work, the Ergonomic nailing System (ENS) was designed, prototyped, and tested in the field. The first comparative tests showed that the proposed simple system works well in the field of manufactured housing, is immediately embraced by workers that suffer from constant bending during nailing, and does not result in loss of productivity. Laboratory tests also showed the drastic change of this innovative system on the sagittal position of the human, who is not required to bend anymore* (Davis, Lorenc and Bernold, 1999).

This issue was also addressed by the Health and Safety Executive (1998) as follows:

Plan all material handling to avoid the risk of injury. Where possible, avoid people having to lift materials at all. Provide mechanical handling aids wherever possible to avoid manual handling injuries. Make sure that all equipment used for lifting is in good condition and used by trained and competent workers.

Plant for material handling:

- before the job starts, decide what sort of material handling is going to take place and what equipment will be needed;
- avoid double handling - it increases risks and is inefficient;
- make sure that any equipment is delivered to the site in good time and that the site has been prepared for it;
- ensure the equipment is set up and operated only by trained and experienced workers;
- co-ordinate site activities so that those involved in lifting operations do not endanger other workers and vice versa;
- arrange for the equipment to be regularly inspected and where necessary examined and tested by someone who fully understands the safety matters. Make sure records are kept. The Lifting Plant and Equipment (Record of Test and Examination, etc.) Regulations 1992 give details of what has to be covered in the tests and examinations.

23.2   VIRTUAL REALITY

Taking technological solutions to the ultimate limits, at present, the use of virtual reality or visualisation is becoming very important as a future mechanism for improving construction site safety. Hadikusumo and Rowlinson (2003) discuss the use of the VR systems to assist in construction site layout and safety analysis. They address the issue as follows:

Virtual product and process data supports the users goal of embedding the Design-for-Safety-Process by using theories of accident causation as well as axioms derived from safety best practice and regulation. This gives information related to the visual data as follows:

- The possibility for the safety engineer to do a virtual walkthrough in order to analyse the proper design, for example, of bamboo scaffolding, safety nets and other temporary safety protection.
- The collision detection facility of virtual reality can be used to evaluate the product and process model from, *inter alia*, 'inadequate aisle space' horizontally and vertically in a permanent or temporary construction.
- An illumination facility in the virtual model can be used to design proper lighting during the construction process.

26.3   SITE LAYOUT

Hadikusumo and Rowlinson (2003) go on to discuss the use of visualisation in what they term the design for safety process (DFSP) approach:

> Visualisation allows the user to interact with the data by a virtual walk through which is very useful for presentation of product data. The user can see the product from any position as well as distance intended. The user can also walk inside the virtual product (building) which can not be done by physical product data such as rapid prototyping of a miniature of building. Therefore, many members from different organisations can walk through in the virtual product model and discuss the product in the design for safety process mode - DFSP.

23.4   ROBOTICS

Robotics is an important area of health and safety improvement. When dangerous substances have to be sprayed, a robot can be used rather than exposing a human being to the risk. Where processes are tediously repetitive, and the problem of boredom and fatigue come into play, a robot will perform much better than a human. There are many advances in robotics but most of these have been specifically directed at manufacturing industries where the process is much more static than in the construction industry. However, as robotics develop and artificial intelligence systems improves the scope for use of robotics on construction sites to deal with dangerous and repetitive tasks becomes more and more a reality.

23.5   ACCESS TO HEIGHT - SCAFFOLDING SYSTEMS AND BAMBOO SCAFFOLDING

Access to heights has always been a dangerous issue on construction sites. Preventing falls, even from small distances, is essential. Any fall can be fatal or lead to serious injury. Rules to prevent falls from Scaffolding Systems, be they temporary work systems or access platforms, are indicated in the following abstract from the UK Health and Safety Executive:

26.5.1   Summary of steps to take before working at height

- check there is a safe method of getting to and from the work area;
- decide what particular equipment will be suitable for the job and the conditions on site;

- make sure work platforms and any edges from which people are likely to fall have guard rails and toe boards or other barriers;
- make sure that the equipment needed is delivered to site in good time and that the site has been prepared for it;
- check that the equipment is in good condition and make sure that whoever puts the equipment together is trained and knows what they are doing;
- make sure those who use equipment are supervised so that they use it properly. The more specialised the equipment (for example, boatswain's chairs and rope access equipment), the greater the degree of training and supervision required to ensure safety;
- check any equipment provided by another company is safe on site before using it;
- find out who to tell if any defects need to be remedied or modifications need to be made and keep them informed.

When selecting a means of access, remember:

- only when it is not practicable to provide a work platform with guard rails should other means of access (for example, boatswain's chairs or rope access techniques) be used;
- only when no other method is practicable, or when work platforms cannot comply with all requirements for safe work (e.g. a guard rail has to be removed to land materials), should a way of arresting falls (for example, a harness and lines or nets) be relied upon;
- only use harnesses and lines or nets to provide safety as a last resort they only provide protection for the person using the harness in the event of a fall - they do not prevent falls;
- harnesses or nets may also be needed to protect those working to put guard rails or other protection in place;
- ladders should only be used as workplaces for short periods and then only if it is safe to do so. It is generally safer to use a tower scaffold or MEWP even for short-term work;
- before any work at height, check that there is adequate clearance for equipment. For example, overhead power lines can be a risk when erecting scaffolds or using MEWPs; there can be a risk of crushing against nearby structures when mobile access platforms are manoeuvred.

Another area, surprisingly, where falls from heights are common is the use of ladders. Few workers, it seems, have a thorough understanding of safety and the use of ladders.

Cameron, Gillan and Duff report that the UK construction industry suffered 84 fatal accidents during the year 2000. Almost half of these fatalities were due to 'falls from heights' and around half of these were due to 'falls from roof edge' and 'falls through roof'. The statistics also suggest that new build (e.g. profiled aluminium roofs) are just as dangerous as refurbishment (e.g. asbestos cement roofs). These figures demonstrate that falling accounts for the majority of fatal accidents in the UK construction industry.

Despite the amount of regulations and laws which aim to force upon contractors the planning of roofwork and introduction of technical measures to prevent falls in the UK the situation has not improved. For instance, Section 2 of the Health and Safety at Work Act 1974 requires contractors to provide a safe place of work and this includes roofwork. Other regulations require the risks associated with working at heights to be assessed and controls introduced to reduce this risk to as low a level as is reasonably practicable and places an emphasis on measures which protect the whole workforce and thus yield the greatest benefit. This means that collective protection (e.g. safety net) is favoured over individual protection (e.g. safety harness) and general preference is given to methods of fall protection which take advantage of technological progress.

Cameron, Gillan and Duff report that in accordance with the principles of risk prevention and protection hierarchy, there are five key areas which their research programme is evaluating because they represent the current state of 'best-practice' when working at heights in construction:

- The benefits and limitations of three different safety systems in roofwork.
- The benefits and limitations of cable based fall arrest systems as a means of protection when working at heights and/or near a leading edge.
- The benefits and limitations of the use of currently specified fall arrest systems.
- when erecting and dismantling scaffold.
- The practical usefulness of these methods of fall protection were investigated through a combination of expert opinion and experiences drawn from live field trials on case study sites and the results of the study are reported.

Fang, *et al.* discuss and compare the use of steel and bamboo scaffold systems in Hong Kong. They report that it is because of its light weight, a large pool of skilled labour and exceptional low cost, that bamboo scaffolding enjoys advantages in the construction bidding process. However, the disadvantage is its low strength, inflammability, irregular shape, changeable nature and uncertain quality that adversely influences construction safety and quality as well as the corporate image of many building contractors. Metal scaffolding overcomes the shortcomings of bamboo scaffolding, in addition to its advantages of ease of erection, safety and reliability, and meeting the demands for standardisation and automation in the construction industry. Now, more and more contractors are considering its use, although the cost is higher than bamboo scaffolding. Since bamboo and metal scaffolding have their inherent advantages and disadvantages, both have been touted as being more appropriate for the Hong Kong construction industry.

It is recorded that three quarters of the construction fatalities in Hong Kong are falls from height. Mak (1998) pointed out that among 41 fatal accidents occurring on construction sites in 1997, 20 were related to falls, in which 6 (30%) were workers falling from bamboo scaffolding. Wong (1998) studied the management of bamboo scaffolding and provided material information for carrying out supervision and inspection of works associated with bamboo scaffolding and to improve the safety of bamboo scaffolding in Hong Kong. Fu (1993) indicated that bamboo scaffolding provides a flexible, practical and economical means of support. He also pointed out that there was no specific regulations concerning design, calculation and drawing of bamboo scaffolding in Hong Kong and that qualification of structural design and construction of bamboo scaffolding would be solely dependent on the experience of scaffolding specialist contractors. However, there are many aspects of bamboo scaffolding safety that need to be addressed and improved. Chan, *et al.* (1998) suggested a computerised method to design and analyse bamboo scaffolding. However, there is limited research on comparing safety and future application of the two types of scaffolding. This chapter reviews the situation of bamboo *vis a vis* steel scaffolding. The objectives of this chapter are to synthetically and systematically evaluate the application of the two types of scaffolding by investigating and comparing their safety performance, economical effectiveness and influences on other major factors, and to explore the development trend of scaffolding in Hong Kong and the countermeasures the government and contractors should consider.

Lew, *et al.* discuss excavation and trenching safety, highlighting existing standards and challenges. Trenching fatalities and injuries continue to plague the construction industry. While complete and accurate records of the actual number of fatalities occurring in trenching incidents are not maintained, 'the estimate of 100 fatalities per year due to cave-ins and other excavation accidents (Hinze and Bren, 1996)', and 7000 injuries, is perhaps a reasonable approximation of the magnitude of the problem. In addition to the possibility of trench cave-ins, workers in trenches can 'be harmed or killed by engulfment in water or sewage, exposure to hazardous gases or reduced oxygen, falls, falling equipment or materials, contact with severed electrical cables or improper rescue (ELCOSH, website).'

Findings of a mail survey conducted by Equipment World (1998) show the following alarming statistics:

Nearly 41% of all respondents said they experienced a trench collapse on one of their jobs. Out of this group, 29.4% said that someone was injured or killed in the collapse. Of the nearly 41% who had experienced a trench collapse on a job, 76.5% said that the trench collapse was due to unstable soil, 29.4% said it was due to human error, and 11.8% said it was due to insufficient shoring/shielding.

In addition to fatalities, injuries caused due to unsafe trenching practices are costly in terms of direct and indirect costs to the construction industry. The direct costs include medical and workers' compensation payouts. In 1995, construction accounts for 15% of all workers' compensation spending while construction workers are only about 6% of the labour force. The employers and society also pay large indirect costs. Hinze (1991) estimated that the ratio of indirect to direct costs for injuries resulting in lost work time was 20 to 1. The indirect costs range from lost productivity among co-workers and management, and lawsuits, to reduced worker morale, especially when fatalities occurred.

In September 2001, Purdue University received a grant from the National Institutes of Occupational Safety and Health (NIOSH) to develop strategies for safer trenching operation. This chapter discusses the findings of this study and focuses on the following issues:

- Role of the competent person in excavation safety
- OSHA regulations related to excavation safety
- Causes of accidents
- Characteristics of accidents.

23.6   REFERENCES

Chan Siu-Lai, Wong, Francis K.W., So, Francis Y.S. and Poon. S.W., 1998, Empirical design and structural performance of bamboo scaffolding.   In *Proceedings for the Symposium on Bamboo and Metal Scaffolding*, pp. 5-15.

Davis, M.L., Lorenc, S.J. and Bernold, L.E., 1999, Technological interventions to reduce the incidents of back injury in building construction. In *Proceedings of the Second International Conference of CIB Working Commission W99 Implementation of Safety and Health on Construction Sites*, edited by A. Singh, J. Hinze and R.J. Coble, Honolulu, Hawaii, 24-27 March, p. 388.

Equipment World, 1998, *Digging Deeper: Survey on Trench Shoring Practices.*

Fu, W.Y., 1993, Bamboo scaffolding in Hong Kong. *The Structural Engineer*, 71, **11**, pp. 202-204, June.

Hadikusumo, B.H.W. and Rowlinson, S. 2003, Visualisation: An aid to safety management. *International Journal of Internet and Enterprise Management*, Special Issue on 'Product and Process Modelling in the building and related industries', Inderscience Enterprises Limited, 1, **2**, April.

Health and Safety Executive, 1998, *Health and Safety in Construction*, HSE Books.

Hinze, J., 1991, *Indirect Costs of Construction Accidents*, The Construction Industry Institute (CII), Austin, Texas, *Source Document*, 67.

Hinze, J., and Bern, K., 1996, Identifying OSHA paragraphs of particular interest, ASCE. *Journal of Construction Engineering and Management*, 122, **1**, March, pp. 98-100.

Mak, Dominic Hung-kae, 1998, Legislative control regime for ensuring safe use of scaffolding. In *Proceedings of Symposium on Bamboo and Metal Scaffolding*, Hong Kong.

Wong, F.K.W., 1998, *Bamboo Scaffolding-Safety Management for the Building Industry in Hong Kong*, Department of Building and Real Estate, Hong Kong Polytechnic University.

CHAPTER 24

A Report on Research Investigating the Practical Usefulness of Current Fall Prevention and Protection Methods when Working at Heights

Iain Cameron, Gary Gillan and Roy Duff

INTRODUCTION

The UK Construction Industry suffered 84 fatal accidents during the year 2000. Almost half of these fatalities were due to 'falls from heights' and around half of these were due to 'falls from roof edge' and 'falls through roof'. The statistics also suggest that new build (e.g. profiled aluminium roofs) are just as dangerous as refurbishment (e.g. asbestos cement roofs). These figures demonstrate that falling accounts for the majority of fatal accidents in the UK Construction Industry.

There are numerous pieces of statute which aim to plan roofwork and introduce technical measures to prevent falls in the UK. Section 2 of the Health and Safety at Work Act 1974 requires contractors to provide a safe place of work and this includes roofwork. Regulation 3 of The Management of Health and Safety at Work Regulations 1999 require the risks associated with working at heights to be assessed and controls introduced to reduce this risk to as low a level as is reasonably practicable. Regulation 4 of these regulations places an emphasis on measures which protect the whole workforce and thus yield the greatest benefit. This means that collective protection (e.g. safety net) is favoured over individual protection (e.g. safety harness) and general preference is given to methods of fall protection which take advantage of technological progress.

Further, Regulation 6 of The Construction (Health, Safety, Welfare) Regulations 96 place an absolute duty to prevent falls when working at a height of over 2 metres. This is achieved by a requirement for guardrails and toeboards and where this is not possible by the introduction of methods which will arrest a fall as a last line of defence. Also, the planning of a safe system of work and risk controls will form an important section of a Principal Contractor's Health and Safety Plan issued under the Construction (Design and Management) Regulations 94.

In accordance with the principles of risk prevention and protection hierarchy, there are five key areas which this research programme intends to evaluate because they represent the current state of 'best-practice' when working at heights in construction:

1. The benefits and limitations of safety nets during roofwork.
2. The benefits and limitations of purlin trolley systems during industrial roofwork.
3. The benefits and limitations of air inflated safety bag systems when working at heights and/or near a leading edge.
4. The benefits and limitations of cable based fall arrest systems as a means of protection when working at heights and/or near a leading edge.
5. The benefits and limitations of the use of currently specified fall arrest systems when erecting and dismantling scaffold (SG4).
6. The benefits and limitations of lightweight plastic and aluminum decking systems.
7. The benefits and limitations of fall prevention during maintenance and refurbishment work.

The practical usefulness of these methods of fall protection are currently being investigated through a combination of expert opinion and experiences drawn from live field trials on case study sites.

24.1   RESEARCH OBJECTIVES

This research will investigate the following objectives:

1. To evaluate the benefits and limitations of safety nets during roofwork.
2. To evaluate the benefits and limitations of trolley systems during industrial roofwork.
3. To evaluate the benefits and limitations of air inflated safety bag systems when working at heights and/or near leading edges.
4. To evaluate the benefits and limitations of cable based fall arrest systems as a means of protection when working at heights and/or near a leading edge.
5. To evaluate the benefits and limitations of the use of currently specified fall arrest systems when erecting and dismantling scaffold (SG4).

24.2   OBJECTIVES – BACKGROUND INFORMATION

24.2.1   The benefits and limitations of safety nets during roofwork

These systems are growing in popularity within the UK Construction Industry as a direct result of Regulation 6 of the Construction Regulations 96. The systems are becoming easier to use and install as new suppliers and rigging companies enter the marketplace. Further, the systems have been widely championed in the UK through HSE. The use of safety nets during industrial roofwork is a published 'HSE Enforcement Priority for the Year 2001'. It is also possible that safety nets can be used on some refurbishment work for example, to protect falls through fragile rooflights in circumstances where the primary protection (the solid roof light cover) is carelessly removed.

It is the writers' belief that there are some problems with the use of safety nets which prevent there use on all occasions, and although they are an excellent method of fall protection, they have limitations and these need to be investigated.

For instance:

a) How feasible are nets during refurbishment work with highly serviced roof spaces?
b) How feasible are nets to install if the site encounters poor ground conditions?
c) How feasible are nets to erect during mezzanine or swimming pools construction?
d) How feasible are nets during the construction of tall buildings beyond the reach of normal access equipment?
e) What changes do designers need to make to allow easier installation of nets and their border ropes and how can this be done so as not to impede contraction for example, gutters?
    Furthermore, are some contractors negating the purpose of safety nets by the systems of work they design. For instance, some contractors are insisting on the use of safety harnesses in addition to the use of nets. They argue that nets gives people a false sense of security and this encourages workers to take risks. These harnesses place a massive shock load on the body and this introduces other problems ('suspension trauma' and rescue practicalities and damage to internal organs etc.). This introduces another research question in relation to nets:

f) How prevalent is this practice and how can contractors be persuaded to abandon this practice and let the net serve its purpose?

24.2.2 The benefits and limitations of purlin trolley systems during industrial roofwork

These systems have been around for a long time and are usually used in conjunction with safety harnesses. This is because, traditionally, the purlin trolley has a double handrail on the 'Leading Edge' (the opposite side to that being worked on) which provides protection. But, has an open 'Working Edge' (the side where the sheets will be progressively installed) and thus requires a harnesses attached to the trolley to prevent falls at the working edge.

However, a number of technological advantages have been made in this area in recent years. For instance, patented systems (e.g. Rossway) now mange to protect the 'Working Edge' by means of a trolley which limits the open area by the provision of a horizontal barrier (attached to the trolley) which rests about six inches below the location where the roof sheet will be fixed to. This means that if someone accidentally stands on the unfixed sheet, the sheet will be caught (and therefore the worker) by the horizontal barrier. The system eliminates the need for harnesses, etc. and has been endorsed by HSE as a safe system of roof work which provides an alternative to the use of nets and/or harnesses (www.rosswaydowd.co.uk).

This system may well be an alternative to nets, for use on occasions where nets are limited, in order to be an alternative to nets and possibly limit the risk from net installation. The system is suitable for simple roof designs that do not feature curved surfaces or intricate plan shapes and cost effective systems for industrial warehouses.

This programme of research reported in this paper will investigate the kinds of building designs where this system can be used and the kind of design changes that will required to be made to allow this systems to function effectively. This may include the specification of roof panels of the same modular width and the location of anti sag bars, etc.

24.2.3 The benefits and limitations of air inflated safety bag systems when working at heights and/or near a leading edge

There are situations in construction where nets are impractical and the alternative means of fall protection has normally been a harness and line. These include the installation of precast slabs where there is always a leading edge at each floor of the building. The concept of air inflated safety bags is growing in recognition within the industry. These systems are being trialed by the Precast Flooring Federation (PFF) at present. Also, it is possible that these systems could augment traditional scaffold 'crash decks' by overlaying the air bags on the scaffold deck and thus provide a softer landing during atrium construction, etc.

Also, perhaps the greatest opportunity for these systems is in domestic housing, this sector of the industry has always struggled with the concept of safety nets and it is probably fair to say that nets have their limitations when used during low level construction. Further, harnesses are also problematic during this type of work because it is difficult to find attachment points and workers have to attach/detach frequently. Therefore, safety air bags may have much to offer the housing sector.

The opportunities for this systems of work will be investigated by this research. Further, problems of perception will also be viewed which may inhibit the industry's acceptance of these methods, for example, air bags being viewed as a child's amusement park toy, etc.

24.2.4 The benefits and limitations of cable based fall arrest systems as a means of protection when working at heights and/or near a leading edge

These systems can be installed as either a last line of defence during new construction work or as permanent maintenance systems installed during new building. Also, the systems are useful during refurbishment work. The facts are that there are some occasions when it is possible that safety nets could be impractical. This could be the case during unusual new build structures – parabolic and ellipse type roofs. These roof shapes also render purlin trolley systems useless.

Therefore it is possible that steel cable systems which operate with a harness and inertia reel attached to the running line may offer practical and cost advantages over nets. Also, even if cable based systems prove to be of only limited value during new build, it is conceivable that occasions will exist during refurbishment work where cable systems are preferred over nets.

This is particularly the case during the renovation of historic buildings with dome roofs and other unusual features. These buildings may make nets difficult and dangerous to install and may mean that the net is well over 2m below the apex of the roof. Also, it is conceivable that cable systems may need to be specified as part of the maintenance system because they are unobtrusive and thus sympathetic to the architectural needs of historic buildings.

The programme of research reported in this paper will evaluate the types of projects where cable systems should be considered on the grounds of practical usefulness.

24.2.5 The benefits and limitations of the use of currently specified fall arrest systems when erecting and dismantling scaffold (SG4)

The National Federation of Scaffold Contractors (NFSC) published Guidance Note: SG 4 - *'The Use of Fall Arrest Equipment when Erecting, Altering and Dismantling Scaffold'* during 2000. The guide is endorsed by the Construction Confederation, HSE, several major contractors, NFSC member companies. The guide is in direct response to Regulation 6 of the Construction Regulations 96.

The system appears to be well presented but on close inspection there are a number of limitations that the researchers believe require to be addressed. It is well documented that the guide does not comply with Regulation 6 of CHSWR 96 and is in fact a compromise. The guide only applies when working at a height of over 4m. The reason for this is that if someone wears a 2m lanyard attached at foot level (=worst case position), then they will fall a minimum of 2 metres with a shock load to the body of 12KN (circa 12,000Kg). This is unacceptable and has to be reduced to half this load under EU Standards. This requires the introduction of a 1.75m shock absorbing lanyard. This lanyard will 'tear' for a distance of 2m when worn as part of a fall prevention strategy. This means that the scaffolder (if clipped on at the worst case foot position) would fall 1.75m plus the 2.0m of the lanyard tear - a total of approximately 4m with the shock load to the body being reduced to 6KN.

It is clear that this does not fully comply with Reg. 6 of the Construction Regulations 96 and the possibility of secondary injuries during this 4m fall need to be considered. This point has been forcefully demonstrated by the UK Construction Industry Training Board (CITB) who found that the skull of a mannequin was literally smashed due to striking a scaffold transom during a 4m fall whilst wearing the shock absorbing lanyard. The CITB's spokesperson believed that the strength of the mannequin's skulls was at least equal to that of a human skull.

A safer systems would be to introduce an inertia reel as part of the installation. These reels are very lightweight and compact and are similar in construction, operation, and looks to a car seat belt. And, if used as part of the fall prevention installation, would drastically reduce the free fall distance of the scaffolder - to around 2m if attached at the worst (foot height) case, and about 0.5m if attached at the best (head height) case. This amended system would comply with the Construction Regulations 96 (Reg. 6) and would reduce the risk from secondary injuries. Alternatively, a fixed anchorage point attached to scaffold uprights (standards) would reduce falls to only a short length. For example, 'The Jordan Clamp', which ensures that all scaffold lanyard hooks are always attached at above head height.

This programme of research reported in this paper will evaluate the feasibility of this systems and the views of the key stakeholders, and some major contractors, on the worth of this revised methodology.

24.2.6 The benefits and limitations of lightweight plastic and aluminium decking systems.

A range of exciting new height safety innovations have entered the marketplace in relation to working at heights on tasks such as the installation of block and beam floors, etc., lightweight safety decking systems have emerged. These systems act as a means of preventing falls and thus they are a more fundamental risk control solution than systems that aim to mitigate the consequences of a fall (e.g. nets). These systems are lightweight plastic or aluminum and offer cost/time savings over the traditional tube and fitting scaffold 'crash' deck solution.

There are at least three exciting decking systems currently vigorously marketing their benefits: Trellis systems ('Oxford'), ABS plastic decking ('Graceland'/'Tarmac'), HL Planking ('Suredeck') for use when working at heights, each system with its benefits and limitations. It is clear that these systems can add value to the safety of the construction process by acting as temporary platforms. Therefore, their practical utility needs to be evaluated critically.

24.2.7 The benefits and limitations of fall prevention during maintenance and refurbishment work

This category looks at the applications of all of the above systems in refurbishment and maintenance works. This area of work is often under-planned and under-resourced and it is believed that new applications can be derived for fall prevention and fall arrest equipment for non-standard works.

24.3 RELATED ISSUES OF CONCERN – TRAINING OF ROOF WORKERS

The training of roof workers and the equipment they hold has always been an issue of concern. Roofwork is much more dangerous, at least statistically than steel erection. It is fair to say that the modern day steel erector is better trained, older, and more psychologically aware of the danger than a roof worker. The steel worker operates along open steel and is spatially aware. The roof worker on the other hand walks across a platform that appears to be solid and is thus subject to a false sense of security and therefore less aware of the dangers. The roof worker tends to be younger, had little previous training, and works for a small company with limited access and safety equipment.

The Roofing Industry Alliance Hallmark Scheme in the United Kingdom - a kind of 'Quality Mark' or 'Kite Mark' - hopes to overcome some of these problems by ensuring the necessary training of experienced construction workers and assessing the quality of companies via a rating scheme. The scheme intends to weed out the 'cowboys' and restore credibility to the roofing industry.

The research reported may well investigate the merits of this scheme and the ways in which Principal Contractors can be persuaded to endorse the scheme and therefore reduce the opportunities for non-skilled roof workers to operate in the construction industry. Presently, however, this 'competency' objective is less certain because it seems HSE have allocated this research to others.

24.4 RESEARCH METHODOLOGY

The research reported here will be delivered through a combination of a desk top review of the suitability of current technical guidance and standards for fall prevention when working at heights. This will be followed by a series 'focus group' meetings of experts and vested interest groups. This group will identify a series of live case study sites of each different method of fall protection which the research team will visit and observe. Also, a research 'steering group' maintains the direction of the research. Finally, interviews with experienced site managers, suppliers, and operatives are to be conducted. These interviews will determine the experiences of site managers and operatives in relation to the identified fall prevention methods.

The 'focus group' of interested parties may well include: The National Federation of Roofing Contractors (NFRC), The National Federation of Scaffold Contractors (NFSC), The Scottish Building Employers Federation (SBEF) and/or The Construction Confederation (CC), The Health and Safety Executive (HSE), The Major Contractors Group (MCG), and other interested stakeholders, e.g. designers, suppliers, trade unions, trade contractors.

The focus group will identify case study sites and industrial partners who will provide access for the research team to observe the use of safety nets, purlin trolleys, safety air bags, cable base systems, harnesses during scaffold operations, etc. This will allow the research team to evaluate the practical usefulness of the different methods of fall prevention methods for working at heights and the ways in which each of these methods can be improved. This may well involve interviews with construction managers, construction workers, and construction suppliers in order to investigate the optimum arrangements required for the implementation of these methods of fall prevention.

The research will culminate in a 'best-practice' publication which will be derived from the desk top literature review, the focus group findings, and the live case study sites identified from the steering group and interviews.

24.5   CONCLUSION

This paper has presented a planned programme of study which is currently being negotiated with the UK's Health and Safety Executive (HSE). The initial findings of the study and early collaborator consultation has suggested that there are a range of potential fall prevention and protection methods available for use when working at heights, each system with benefits and limitations. For example, some are suited to new build, others to refurbishment, and some are dependent on other circumstances such as plan shape of the building. It is clear that all methods can add value to the construction process, however, a Technical Guide is required in order to outline to designers and contractors the optimum circumstances of specification and/or use for each respective system. This technical guide is currently lacking but this research hopes to deliver this technical standard and further details will be reported at the conference.

24.6   REFERENCES

CITB, 2000, *If You Should Fall*, Construction Industry Training Board, CITB Video 9VID066), CITB, Bircham Newton, Norfolk.

HSE, 2000, *Health and Safety in Roofwork*, HSG33, Health and Safety Executive (HSE Books).

HSE, 2002, Proposed draft of *The Temporary Work at Height Regulations 2004*, (HSE Books).

NASC, 2000, *The Use of Fall Arrest Equipment whilst Erecting, Altering and Dismantling Scaffold*, Guidance Note SG4, The National Association of Scaffold Contractors, UK.

CHAPTER 25

A Comparative Study on Safety and Use of Bamboo and Metal Scaffolding in Hong Kong

Dongping Fang, Shenghou Wu,
Francis K.W. Wong and Q.P. Shen

25.1   INTRODUCTION

Finance, real estate, tourism and manufacturing are the four major economic sectors in Hong Kong. The Construction industry, which is intimately linked to the real estate industry, has been played an important role in the economy of Hong Kong. Being an essential trade for construction, scaffolding provides workers with working platforms, safe access and protection.

It is because of its light weight, trained skillful workers and especially its comparatively low price that bamboo scaffolding has the advantages in the construction bidding process. However, the disadvantage is its low strength, inflammable nature, irregular shape, changeable nature and uncertain quality that adversely influence construction safety and quality as well as companies' image. Metal scaffolding overcomes these disadvantages and has such advantages as ease of erection, is safe and reliable, and meets the demand for standardisation and automation in the construction industry. Now, more and more contractors are considering to use it, although it is much more expensive than bamboo scaffolding. However, metal scaffolding remains in an inferior position to it and is being adopted very slowly in Hong Kong.

Three quarters of the total number of construction fatal accidents occurring in Hong Kong are due to falls from height. Mak (1998) pointed out that among 41 fatal accidents occurring on construction sites in 1997, 20 were related to falls, in which 6 (30 %) were workers falling from bamboo scaffolding. F.K.W. Wong (1998) studied the management of bamboo scaffolding and provided useful information for carrying out supervision and inspection of works associated with bamboo scaffolding and to improve the safety of bamboo scaffolding in Hong Kong. Fu (1993) indicated that bamboo scaffolding provides a flexible, practical and economical means of support. He also pointed out that there were no specific regulations concerning design, and calculation for structural safety of bamboo scaffolding in Hong Kong and that structural design and construction of bamboo scaffolding would depend on the experience of scaffolding workers. However, in many aspects of safety, it needs to be improved. Chan and Wong (1998) suggested a computerised method to design and analyse bamboo scaffolding. Y.Y. Wong (1998) and Chow (1998) discussed the requirements of laws and regulations for inspection and supervision of bamboo and metal scaffolding and they suggested a method to improve the current situation. However, there are few research studies comparing safety and the application of these two kinds of scaffoldings.

The objectives of this study are to synthetically and systematically evaluate the application of these two kinds of scaffolding by investigating and comparing their safety performance, economic effectiveness and influences on other factors, and to explore the development trend in Hong Kong and the measures the government and contractors should consider.

25.2   RESEARCH METHOD

Figure 25.1 shows the research framework of this project. This research focused on safety performance and economic effectiveness. As a temporary structure, scaffolding is provided by suppliers, rented by contractors, and erected and dismantled by workers. Workers and the scaffolding form a human-environment system in which safety is emphasised. Contractors and suppliers form a market system in which economic issues are the emphasis.

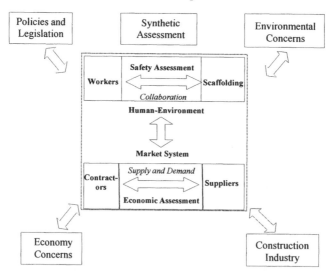

Figure 25.1   Research framework

25.2.1   Comparison of Safety Performance

Scaffolding is erected by scaffolders and its quality is directly influenced by scaffolders' skills. The scaffolding structure is the working platform for workers and this influences the safety and quality of workers' performance. Contractors are responsible for the management of erecting, dismantling and carrying out daily maintenance of scaffoldings. In this human-environment system, the safety of the two kinds of scaffolding has been assessed by hazard analysis, analytic hierarchy process and experimental psychology based methods.

Hazard analysis originated from a method adopted by the American Atomic Energy Commission (Xiao, 1982). Its principle is to evaluate the risk of an event by comprehensively analysing the probability of occurrence and the consequences of this event. The process is to decompose a complex safety assessment problem into several sub-problems that can be evaluated independently, and then to assess these sub-problems, and finally to sum up the results of sub-problems to an overall conclusion of the original problem. Thus, the risk index of an event can be obtained through the following functions:

$Hi=L \times E \times C$
Hi – risk index of an factor
L – failure probability of an factor
E – accident probability of a failed factor
C – possible outcome of a failed factor

$$H = \sum_{i=1}^{n} H_i$$

N – number of factors

Analytic Hierarchy Process (AHP) has been used to analyse complex decision-making problems in social, political, economical and technological fields (Xu, 1988). Its solution process is as follows. 1) A complex problem is structured by decomposing it into a hierarchy with enough levels to include all factors to reflect the goals and concerns of the decision maker ; 2) Elements are compared in a systematic manner using the same scale to measure their relative importance, and the overall priorities among the elements within the hierarchy are established ; 3) The relative standing of each alternative with respect to each criterion element in the hierarchy is determined using the same scale and ; 4) The overall score for each alternative can then be aggregated, and a sensitivity analysis can be performed/to see the effect of change in the initial priority setting, while the consistency of comparison can be measured.

The experimental psychology based test uses multi-channel physiology equipment to measure the influence of scaffolding conditions on human beings by measuring mental pressure and the level of nervousness caused by different working conditions; bamboo scaffolding and metal scaffolding (Yang, 1990). The authors developed this method to provide an objective tool for assessing working conditions' impact on workers' behaviors, which are influenced by the level of nervousness.

25.2.2   Comparison of Economic Effectiveness

The unit rate for scaffolding is determined by its cost and volume of work, which in turn is influenced by the cost of a construction project as well as contractors' profit. The lifetime analysis of net present value and the payback period of the two kinds of scaffoldings is conducted from both suppliers' and contractors' points of view. After major economic indexes were defined based on a survey in the summer of 2000, a questionnaire survey was conducted, and the static cost and dynamic cash flow were analysed.

25.2.3   Synthetic Comparison

Safety and economy are two correlated factors that determine contractor's choice of scaffoldings. The study takes the two major aspects and other factors such as construction quality and company's image into consideration to assess these two kinds of scaffolding synthetically, so as to provide fundamental information for the government to enact relevant policies and for contractors to select their scaffolding.

25.3   QUESTIONNAIRE DESIGN AND INVESTIGATION

Four sets of questionnaires and a survey on economic issues were developed. 20 questionnaires were distributed to expert, and 17 responses were received. 400 questionnaires for workers were distributed, and 180 useful replies were received. 28 questionnaires were distributed to safety officers, all of which replied. 8 suppliers were surveyed, 10 construction sites were visited, and 30 project managers were interviewed. Based on the outcome of these questionnaires and surveys, the safety performance and economic effectiveness of bamboo and metal scaffolding were compared and analysed.

25.3.1   Safety Performance

By hazard analysis, the risk of accidents in metal scaffolding is found to be 0.32 and that in bamboo scaffolding is found to be 0.68. By AHP, the risk of accident in metal scaffolding is found to be 0.39 and that in bamboo scaffolding is found to be 0.61. The safety assessment by the two methods results consistent in conclusions. Figure 25.2 shows the risk index of every factor by hazard analysis. The result of the experimental psychology based test indicates that the workers using bamboo scaffolding became nervous, tired, and act erratically more easily, all of which are likely to cause accidents.

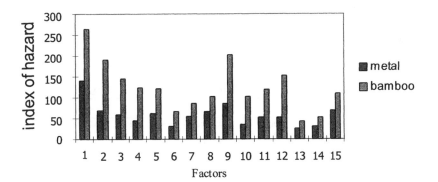

KEY:
1. broken putlog or no putlog;
2. damaged or broken node;
3. rotten member of scaffolding;
4. broken member of scaffolding;
5. without or broken working platform
6. without or broken toeboard;
7. without or broken guardrail;
8. broken bracing;
9. overload;
10. hit by falling object;
11. without safety belt;
12. without anchoring safety belt;
13. without safety helmet;
14. tired and distracted;
15. violate operation regulation

Figure 25.2 Hazard indexes of factors

Table 25.1 show the safety risk ranking by AHP. The results of the questionnaire survey also indicate that safety knowledge is very important. The professional skill and safety knowledge provides workers with the capability to identify hazards and to decrease unsafe behaviors. It is also found that the safety training should pay more attention to the putlog, bracing, and so on to ensure workers understand the consequences of destroying these parts of the scaffolding. It is also very important to improve the working conditions of scaffolding because the spacious and neat working platforms could decrease workers' nervous emotions and unsafe behaviors. Providing a better working platform, safety belt and places for fastening the safety belt could improve workers' safety performance.

25.3.2   Economic Effectiveness

The cost of bamboo scaffolding is low and the initial investment in metal scaffolding is relatively high, the rent of metal scaffolding is generally 2-2.5 times of that of bamboo scaffolding. The payback period of bamboo scaffolding is shorter and the risk is low, while the payback period of metal scaffolding is longer and the risk is higher. For metal scaffolding, the rent could be decreased if there is a stable market demand. The integral elevated scaffolding system, which has been widely used in China, has such advantages as short payback period, low risk and less material consumption.

Table 25.1   Safety risk ranking

Factors	Hierarchy	Factor and weight					
	B	Human (B1)			Condition (B2)		
		0.656			0.344		
	C	Physic and mentality (C1)	Knowledge (C2)	Material (C3)	Erecting and dismantling (C4)	Destroyed by outside action (C5)	Weight
		0.485	0.515	0.382	0.383	0.235	
	B*C	0.318	0.338	0.131	0.132	0.081	
Human	Degree of fatigue (D1)	0.601					0.19
	Working space (D2)	0.399					0.13
	Professional skill (D3)		0.498				0.17
	Safety knowledge (D4)		0.502				0.17
Condition	Selecting material (D5)			0.500			0.07
	Maintenance (D6)			0.500			0.07
	Design (D7)				0.594		0.08
	Experience (D8)				0.406		0.05
	Environment (D9)					0.455	0.04
	Human (D10)					0.545	0.04

25.3.3   Synthetic Comparison

Metal scaffolding is better than bamboo scaffolding in terms of safety performance, economic effectiveness, as well as quality, company images, etc. The synthetic assessment value is 0.87 for metal scaffolding and is 0.67 for bamboo scaffolding. Metal scaffolding is higher in price, and the cost difference between the two kinds of scaffolding will increase dramatically with the increase in the height of building. The rent of metal scaffolding will not decrease largely on the basis of information from suppliers. The integral elevated scaffolding system provides excellent synthetic assessment results and its cost is relatively low. At the same time, the unit cost will decrease with an increase in the height of building. The assessment, synthetically considering safety, economy and other factors indicates that metal scaffolding has advantages over bamboo scaffolding and that integral elevated scaffolding would have a bright prospect in Hong Kong due to its safety performance and reasonable cost.

25.4   PHYSIOLOGICAL REACTIONS TO BAMBOO AND METAL SCAFFOLDING

With economic development, the demand for workplace safety has been increasing. Employers have come to recognise that improving workplace safety can reduce losses due to accidents, enhance employees' loyalty, and increase employees' working efficiencies.

To promote safety on construction sites, workplace safety evaluation plays a very important role. At present, there are different kinds of evaluation methods, such as safety checklist, hazard analysis, event tree analysis, fault tree analysis. These methods, however, emphasise the physical environment but neglect human physiological reactions.

As an example of comparing physiological reactions to different working conditions, this chapter introduces a comparative study on pulse, pectoral breath, ventral breath and electrodermal reactions to bamboo and metal scaffolding. These four indexes are chosen because they are easy to measure in construction sites and a lot of research has been done to these indices. Yang (1990) has studied the change of the human pulse frequencies in response to physical stimulation, and proved that pulse frequencies raise under tension. Tian and Li (1996) has recommended a change of 3 beats per minute in pulse as the critical value for detecting unusual change in the pulse frequencies under environment stimulation by studying more than 1,000 cases. An experiment to compare physiological reactions was conducted in Hong Kong to validate the safety evaluation results obtained by other research methods.

25.5   DESCRIPTION OF THE EXPERIMENT

25.5.1   Experiment Preparation

Volunteers were randomly divided into 3 groups with around 10 examinates in each group. The first group worked on bamboo scaffolding, the second group worked on metal scaffolding, and the third group worked in a normal in-service building for benchmarking. It was required that the examinates be male and similar in basic characteristics, including age, height, weight, and experience.

Three buildings were chosen: two were under construction (one using bamboo scaffolding, the other using metal scaffolding), and one was in service. The experiments of the three groups were under taken at almost the same time of day with similar weather conditions. The condition of the area of the examinates' movement, including the height of scaffolding, the distance of movement, and presence of safeguards, was similar except for the type of scaffolding.

The experiment was made in the following sequence:

1. Preliminary remarks were made and measuring equipment was shown to the examinates;
2. Serial numbers were given to each examinate for identification;
3. Physiological indexes(i.e. pulse, pectoral breath, ventral breath and electrodermal response) of the examinates, were recorded in quiet, steady and isolated conditions when they gathered under the buildings;
4. Examinates were asked to go upstairs and took a rest of half an hour in the test area;
5. Examinates were asked to walk through the test area on the scaffolding at a prescribed speed (Normal speed for a worker to pass it) one by one;
6. Physiological indexes of the examinates were recorded immediately after they walked on the scaffolding.

25.5.2   DATA COLLECTION

Three experiments were conducted as shown in Table 25.2.

Table 25.2   Information of the experiments

Group	Number of Examinates	Location	Time	Condition
Benchmarking	14	Industrial Center of Hong Kong Polytechnic University	PM, January 15, 2001	The weather of those three days was similar. It was noisy on sites.
Bamboo Scaffolding	11	Students' Dormitory of Hong Kong Polytechnic University	PM, January 16, 2001	
Metal Scaffolding	8	The Elementary School in Junk Bay, Hong Kong	PM, January 17, 2001	

The examinates' four physiological indexes were measured with a polygraph for 70-120 seconds first on the ground, and then on the scaffolding or equivalent altitude in the benchmarking building.

The polygraph used in this experiment could record only the relative values of the signals and not the absolute value. Thus from the collected signals, the absolute value of the pulse and the peak value of pectoral breath could not be identified.

25.6   ANALYSIS AND DISCUSSION

25.6.1   Analysis of the Pulse

An 80-order Auto Regressive (AR) (Zhao and Zhou, 1996) and Burg method (Sun and Luo, 1996) was used to calculate the power spectrum of the pulse signals. The statistical analysis of the first prime peak of the power spectrum gave the following results:

- The inter-group independent T test shows that the difference in the pulse frequencies of the samples between groups on the ground is not significant at the 95% level.
- The results of the intra-group paired T test of the pulse frequencies are listed in Table 25.3. The difference between the pulse frequencies measured on the ground and those measured on the scaffolding (or equivalent altitude in the benchmarking building) of the bamboo group is significant at the 70% level, but not significant at the same level for the metal group and the benchmarking group.

Table 25.3 Intra-group paired T test of the pulse

Paired Samples Test

| | Paired Differences | | | | | | | |
| | | | | 70% Confidence Interval of the Difference | | | | |
	Mean	Std. Deviation	Std. Error Mean	Lower	Upper	t	df	Sig. (2-tailed)
Pair 1[a]	5.68E-02	.15496	4.6724E-02	5.68E-03	.10783	1.215	10	.252
Pair 2[b]	-3.3E-03	9.1714E-02	3.0571E-02	-3.7E-02	3.06E-02	-.107	8	.917
Pair 3[c]	1.71E-02	5.2941E-02	1.5283E-02	4.94E-04	3.37E-02	1.120	11	.287

a. The pulse between on the ground and on the altitude in the bamboo scaffoldings group

b. The pulse between on the ground and on the altitude in the bamboo scaffoldings group

c. The pulse between on the ground and on the altitude in the benchmark group

- The results of the inter-group one-way analysis of variance (ANOVA) of the absolute values of the difference between the pulse frequencies measured on the ground and that measured on the scaffolding (or equivalent altitude on the benchmarking building) are shown in the Figures 25.3, Table 25.4 and Table 25.5. Table 25.4 shows that variances are not homogeneous, thus the results by Tamhane's T2 method should be chosen in Table 25.5. Considering the absolute values of the difference between the pulse frequencies measured on the ground and that measured on the scaffolding, Table 25.5 shows that the difference between the absolute values in the bamboo group and those in the benchmark group (or that in the metal group) is significant at the 90% level. But at the same level the difference between the absolute values in the metal group and those in the benchmark group is not significant.
- The absolute value of the change of the pulse frequencies in the bamboo group equals 0.131Hz (7.9 times per minute), while 0.072Hz (4.3 times per minute) in the metal group, and 0.046Hz (2.8 times per minute) in the benchmarking group.

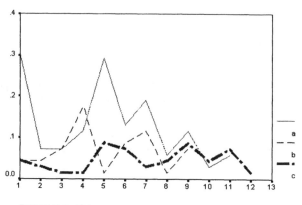

Sequence number

a. The Absolute Difference of the Pulse of the Bamboo Group

b. Metal Group   c. Benchmark Group

Figure 25.3 Absolute difference between the pulse on the ground and on the scaffolding

Table 25.4   Test of homeogeneity of variances by noe-way ANOVA

Levene Statistic	df1	df2	Sig.
4.861	2	29	.015

Table 25.5   Multiple comparisons of the absolute difference between the pulse on the ground and on the scaffolding, by LSD and Tamhane's T2 of one-way ANOVA

	(I) Type	(J) Type	Mean Difference (I-J)	Std. Error	Sig.	90% Confidence Interval[a]	
						Lower Bound	Upper Bound
LSD	Bamboo	Metal	5.9486E-02*	.028	.045	1.1134E-02	.10784
		Benchmark	8.4697E-02*	.026	.003	3.9792E-02	.12960
	Metal	Bamboo	-5.949E-02*	.028	.045	-.10784	-1.113E-02
		Benchmark	2.5211E-02	.028	.374	-2.223E-02	7.2648E-02
	Benchmark	Bamboo	-8.470E-02*	.026	.003	-.12960	-3.979E-02
		Metal	-2.521E-02	.028	.374	-7.265E-02	2.2226E-02
Tamhane	Bamboo	Metal	5.9486E-02	.028	.244	-1.646E-02	.13544
		Benchmark	8.4697E-02*	.026	.040	1.4881E-02	.15451
	Metal	Bamboo	-5.949E-02	.028	.244	-.13544	1.6463E-02
		Benchmark	2.5211E-02	.028	.506	-2.024E-02	7.0667E-02
	Benchmark	Bamboo	-8.470E-02*	.026	.040	-.15451	-1.488E-02
		Metal	-2.521E-02	.028	.506	-7.067E-02	2.0244E-02

*. The mean difference is significant at the .1 level.

a. Dependeng Variables. Absolute Difference

25.6.2   Analysis of the Breath

The pectoral breath and ventral breath signals were filtered by a Butterworth band-pass filter. A 20-order AR model and Yule-Walker method were used in the power spectrum estimation. The inter-group independent T test shows that the difference of the breath frequencies measured on the ground among groups is not significant at the 95% level.

Figures 25.4, 25.5 and 25.6 show the difference of the pectoral breath frequencies measured on the ground with those measured on the scaffolding (or equivalent altitude in the benchmarking building). It can be concluded that the difference between groups is not significant, although the average value of the bamboo group are highest. The results of the analysis of ventral breath are similar to those of pectoral breath.

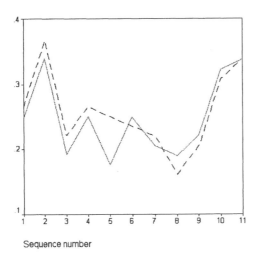

Sequence number

*   Solid line. The data on the ground;
    Dashed line. The data on the scaffolding

Figure 25.4   Pectoral breath of the bamboo group

25.6.3   Analysis of the Electrodermal Response

Because the experiment was conducted in summer and the electrodermal response changes with perspiration, the electrodermal response is not suitable for analsysis.

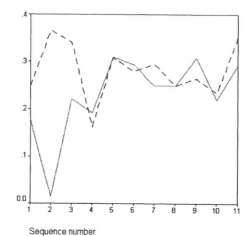

Sequence number

*   Solid line. The data on the ground;
     Dashed line. The data on the scaffolding

Figure 25.5   Pectoral breath of the metal group

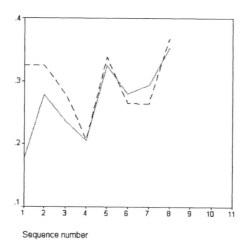

Sequence number

*   Solid line. The data on the ground;
     Dashed line. The data on the scaffolding

Figure 25.6   Pectoral breathing of the benchmarking group

25.7   LIMITATION OF THE STUDY

The following are some limitations of this study:

- Due to the complexity and sensitivity of physiological reactions, any study such as this requires many replications to be valid.
- The polygraph in this experiment couldn't record the absolute values of the signals.
- Although some research has been reported, there is no widely accepted model to relate the physiological changes with the emotion changes, or to relate the emotion changes with the workplace safety condition. Thus it requires further research to determine whether the results of this comparative study of physiological reactions means that metal scaffolding is safer than bamboo scaffolding. Despite this limitation, this research is the first step to conduct a safety evaluation by studying physiological reactions.

25.8 CONCLUSIONS

The statistical results of the pulse frequencies show that the pulse reaction to bamboo scaffolding is higher than that to metal scaffolding, and the pulse reaction to metal scaffolding is not different from that to the benchmarking group.

Taking a change of 3 beats per minute in pulse as the critical value for the normal condition, the pulse reaction to the bamboo scaffolding is greatly out of the normal change range, while the pulse reaction to the metal scaffolding is in the normal change range. This implies that people are usually more nervous on bamboo scaffolding than on metal scaffolding.

25.9   ACKNOWLEDGEMENTS

Appreciation is given to the Works Bureau of Hong Kong, Gammon Construction Limited., Wei Loong Scaffolding Works Company Limited, and AES Scaffolding Engineering Limited for their great support to this research. Professor Albert Kwok of the Hong Kong Polytechnic University and Dr. S.W. Poon of the Hong Kong University contributed their knowledge and experience in this study. This research was also supported by the National Natural Science Foundation of China (Grant number 70172005) and the Beijing Natural Science Foundation (Grant number 8002013).

25.10 REFERENCES

Chan, S.L. and Wong, F.K.W., 1998, Empirical design and structural performance of bamboo scaffolding. In *Proceedings of Symposium on Bamboo and Metal Scaffolding.*

Chow, J.M.K. and Ng, H.K., 1998, Independent checking and control of the design and erection of metal scaffolding. In *Proceedings of Symposium on Bamboo and Metal Scaffolding.*

Fu, W.Y., 1993, Bamboo scaffolding in Hong Kong. *The Structural Engineer*, 71, **11**, pp. 202-204.

Mak, H.K., 1998, Legislative control regime for ensuring safe use of scaffolding, In *Proceedings of Symposium on Bamboo and Metal Scaffolding.*

Sun, D.M. and Luo, Z.C., 1996, Parametric modeling method of pulse pressure signal and its application in the evaluation of cardiovalcolar function. *Journal of Beijing Polytechnic University*, 22, **1**, pp. 39-47, Mar.

Tian, S.X. and Li, J.X., 1996, The effects of infrasound on blood pressure and pPulse in humans. *ACTA Acustica*, 21, **3**, pp. 254-258, May.

Wang, Y.H., 1999, *Safety System Engineering* (Tianjin: Tianjin University Publishing House).

Wong, C.K., 1998, *Identification of the Key Factors involved in Bamboo-Scaffolding-related Accidents on Construction Sites in Hong Kong.* A report submitted as partial fulfilment of the requirements for master of applied science (safety management).

Wong, F.K.W., 1998, Bamboo scaffolding-safety management for the building industry in Hong Kong. In *Proceedings of Symposium on Bamboo and Metal Scaffolding.*

Wong, Y.Y., 1998, Maintenance and inspection of bamboo and metal scaffolding. In *Proceedings of Symposium on Bamboo and Metal Scaffolding.*

Xiao, A.M., 1982, *Safety System Engineering*, China Labour Press.

Xie, J.Y, and Li, X.J., 1999, *Employee's Safety and Health* (Beijing: Economy Management Publishing House).

Xu, S.B., 1988, *Practical Decision-making Method: The Theory of Analytic Hierarchy Process* (Tianjin: Tianjin University Press).

Yang, Z.L., 1990, *Experimental Psychology* (Shanghai: East China Normal University Press).

Zhao, E.Y. and Zhou L.Q., 1996, *Fundamental Digital Signal Processing* (Beijing: People's Posts and Telecommunications Publishing House).

CHAPTER 26

Trenching Accidents and Fatalities: Identifying Causes and Implementing Changes

D. Abraham, J. Lew, J. Irizarry, C. Arboleda and R. Wirahadikusumah

INTRODUCTION

Trenching accidents can be caused by failure of the soil in which the trench was excavated, or the lack of protection of the trench walls by means of protective structures, or contacts with equipment or materials near the trench area, etc. While complete and accurate records of the actual number of fatalities occurring in trenching incidents are not maintained, 'the estimate of 100 fatalities per year in the USA due to cave-ins and other excavation accidents (Hinze and Bern, 1996)', and 7,000 injuries, is perhaps a reasonable approximation of the magnitude of the problem. In addition to the possibility of trench cave-ins, workers in trenches can 'be harmed or killed by engulfment in water or sewage, exposure to hazardous gases or reduced oxygen, falls, falling equipment or materials, contact with severed electrical cables or improper rescue (ELCOSH website)'.

The financial impact of poor excavation safety practices results in economic losses incurred from cave-ins and other accidents that are phenomenal in magnitude. When an accident occurs or when a contractor is cited for violations of safety codes and standards, the contractor incurs sizable, unnecessary expenses. Due to the risks of excavation work and the unknown nature of subsurface conditions, contractors' insurance costs to contractors can be excessive. These factors combine to increase excavation construction costs. Thompson (1982) reported that the cost of excavation failures resulting from not following safe excavation procedures adds about seven to eight percent to the cost of construction. Establishing and following safe excavation procedures makes good economic sense.

A worker performing excavation and trenching operations is protected by OSHA (Occupational Safety and Health Administration) standards, which involve most safety aspects: knowledge, training, and experience of the people responsible under the codes. Unfortunately, the codes are often misunderstood or ignored. OSHA inspections are limited to establishing the cause of the trenching related accident and verifying if the required measures, according to their standards, were used. If the required measures to ensure safety were not taken, then OSHA can impose fines to the company performing the work. However, these inspections may not indicate the reasons why the OSHA trench safety standards were not used or why they failed to protect the workers.

Many studies have analysed accident reports of agencies such as the Occupational Safety and Health Administration (OSHA) and the Bureau of Labor Statistics and determined the various reasons for trenching related accidents (Hinze and Bern, 1996, Suruda, Smith and Baker, 1988). These accidents can be caused by soil failure, the lack of protection of the trench walls by means of structures such as trench boxes, etc. A study conducted by OSHA in 1990 that analysed construction fatalities from 1985 to 1989 determined that 79% of trenching related fatalities occurred in trenches less than 15 feet (4.6 metres) deep and 38% percent occurred in trenches less than 10 feet (3.0 metres) deep.

Findings of a mail survey conducted by Equipment World (1998) show the following alarming statistics:

- Nearly 41% of all respondents said they experienced a trench collapse on one of their jobs. Out of this group, 29.4% said that someone was injured or killed in the collapse.
- Of the nearly 41% who had experienced a trench collapse on a job, 76.5% said that the trench collapse was due to unstable soil, 29.4% said it was due to human error, and 11.8% said it was due to insufficient shoring/shielding.

In addition to fatalities, injuries caused due to unsafe trenching practices are costly in terms of direct and indirect costs to the construction industry. The direct costs include medical and workers' compensation payouts. In 1995, construction accounted for 15% of all workers' compensation spending while construction workers accounted for only about 6% of the labour force. The employers and society also pay large indirect costs. Hinze (1991) estimated that the ratio of indirect to direct costs for injuries resulting in lost work time was 20 to 1. The indirect costs range from lost productivity among co-workers and management, and lawsuits, to reduced worker morale, especially when fatalities occurred.

In order to determine possible intervention strategies, it is important to learn more about the circumstances under which the accidents occur, and to investigate the relationship between the accidents and the adherence/non-adherence to OSHA Safety Standards. Such an endeavor is pivotal in understanding why applicable standards were not used, or if they were used, why they were unsuccessful in ensuring safety in the trenching operation.

This chapter presents a discussion of recent efforts on investigating the causes of trenching-related fatalities, and determining the characteristics of these fatalities. It also discusses the existing OSHA excavation and trenching standards, specifically describing the requirements and the roles of a competent person, and other issues in OSHA Standard 1926 Subpart P. The research study underlying this chapter is funded by a grant from the National Institutes of Occupational Safety and Health (NIOSH) to the Construction Safety Alliance (CSA).

26.1   ROLE OF THE COMPETENT PERSON IN EXCAVATION SAFETY

To function as a competent person at an excavation site a competent person must be:

- Thoroughly knowledgeable with excavation safety standards including soil classification.
- Capable of identifying existing and predictable and hazards and unsafe conditions.
- Knowledgeable in the proper use of protective systems and trench safety equipment.
- Designated to have the authority to stop work when unsafe conditions exist.

A person must have documented experience and training in the first three requirements, and be designated as the competent person by the employer with the authority indicated in requirement four. Construction management must be aware of these requirements, and that the responsibility to comply with these requirements rests with the managers or owners of construction companies.

The role of the competent person must be completely understood by owners, contractors, and engineers. Managers of construction work need to be knowledgeable of the general and common legal applications of the OSHA Standard 1926 Subpart P. Some examples are presented in the following:

1. Training of the competent person must be done in an acceptable manner that complies with the standard. Training must cover the requirements listed in the section on Requirements of a Competent Person. Training can be either formal, or on-the-job, so long as it is documented and meets the competent person requirements. In as far as OSHA is concerned, the compliance officer will look at and inspect the conditions of the jobsite, and then question the competent person to determine his or her qualifications. If necessary, the compliance officer will conduct interviews with company management and employees to determine if the competent person has the ability, knowledge, and authority to ascertain and correct hazards at the jobsite.
   The competent person is the representative of the employer at the jobsite. However, as far as OSHA is concerned, the employer, not the competent person, is responsible for the safety of the employees.

2.   The designated competent person may not always be competent in all areas of safety, but can still be considered competent. For example, the recognition of hazardous atmosphere may not be within the area of expertise of the site competent person. However, if an employer knows a person capable of recognising and evaluating hazardous atmosphere conditions, then that person can be called upon to assist the competent person. That person must be available to provide the competent person with the information needed to make decisions relative to protective systems and abatement measures. This has to be done only when the possibility of hazardous atmosphere conditions can be reasonably expected to exist at a construction site.

3.   If an employer should override a decision or action of the competent person, then the employer is negating the authority of the competent person at a jobsite, and thus would be in violation of the standard. In any case, OSHA always holds the employer responsible for protection of employees.

4.   Geographic location of the jobsite is not the determining factor for jobsite control by a competent person. Employee safety control at the jobsite is the determining factor. Each excavation with employees at risk or exposed must have a competent person at the site and must comply with the standard.

5.   The competent person must perform daily inspections and other inspections as needed to ensure the safety of exposed employees. This means that if site conditions change, inspections and tests such as soil classification tests, must be taken to reevaluate and reclassify as necessary.

6.   If an employer always slopes all excavations as if the soil were type C, then, technically, no competent person would be required at the site. However, this is not a practical approach to safety. When evidence of a condition arises indicating an existing or changed condition, such as hazardous atmospheres or the presence of water, then a competent person would be required.

7.   Prior to the completion of an OSHA inspection, plans or manufacturer's tabulated data must be made available to the OSHA compliance officer. The plans or manufacturer's tabulated data must be signed and stamped by a registered professional engineer.

8.   The standard requires that vibration producing machinery be kept at least two feet away from the edge of an excavation. However, distance is not as critical as are the effects of vibration on the ability of a support system to function safely. If a competent person notes that vibrations from traffic or heavy equipment causes visible distresses in a trench or excavation, the competent person must take corrective measures to ensure worker safety. A possible action would be to keep equipment further away than two feet.

26.2   LEGAL APPLICATIONS

The effects of 'competent person' training with respect to excavation and trenching safety are demonstrated by the following cases.   They explain the concept of 'willful trenching charge':

1. A citation issued to a company charging it with willful failure to shore and protect the vertical faces of an eight foot deep trench in previously disturbed soil was appealed to an OSHA review board. The company's defence was isolated employee misconduct. A portion of the trench had been properly shored, but two employees were observed by an OSHA compliance officer in an unshored section of the trench during an inspection.   The competent person had left the worksite to secure equipment and materials before the approaching storm struck.   The workers were hurrying to complete work because of the threat of the storm and entered the unshored portions of the trench. The review board upheld the citation, because the competent person at the jobsite should have discovered the workers' presence in the unshored portion of the trench and should have removed the workers.

2. A prime or general contractor hired an excavation subcontractor to dig and protect a trench 12 feet deep at a large building construction site. The electrical subcontractor required to work in this trench and was cited for a willful excavation charge.   During a site visit the president of the electrical subcontractor saw that the trench was too deep and unprotected. He then instructed his site foreman to stop working in the trench and, to instead fabricate pipe outside the trench until the excavation subcontractor made the trench safe. The site superintendent for the general contractor was instructed to and agreed to contact the excavation subcontractor and make the trench safe.   The electrical subcontractor's president then left the site.   The electrical foreman ordered the crew to reenter the trench and resume work after the president left the site. This action was against the president's specific instructions and outside of the scope of the foreman's authority. This electrical construction company had a well communicated, enforced, and documented safety programme which included weekly meetings and safety training covering 'competent person' requirements with the use of safety equipment. The review board dismissed wilful excavation charges on grounds of employee misconduct.

OSHA citations that are classified as wilful violations are based on the principle of intentional disregard of OSHA standards. When a wilful violation is upheld by the review board, the offending company must pay the determined fine. On the other hand, if a wilful charge is dismissed on the grounds of employee misconduct, then no fine is paid, as in the case above. At present in the USA, no individual (employee) can be required to pay a fine. Fines are only paid by companies. If a review board dismisses charges, the citation no longer exists against the company or the individual. However, the offending employee may be subject to company discipline.

26.3   DANGEROUS TRENCHING SITUATIONS

Some of the more dangerous situations in trenching operations are: crossing under another pipeline; a broken or leaking water or sewer line; and trenching parallel to adjacent or intersecting pipelines. Other situations are: intersecting open trenches or manhole excavations; excavation in an old backfilled trench or *previously distributed* soil; sudden change in the type or condition of a soil; and excavation in a fissured soil.

An explanation of particularly dangerous situations must be included as a key part of excavation safety training. For example, the term *previously disturbed* refers to more than the actual excavated area itself. When an excavation is made the soil is disturbed a distance back from the face of the excavation an amount equal to one or two times the depth of the excavation. This explains the severity of the situations of crossing pipelines and trenches, and trenching parallel to existing pipelines, both of which occur in previously distributed soil.

The competent person at the jobsite must be able to recognize these and other dangerous situations and take prompt corrective measures. These are practical examples of how the standard requires the competent person to be 'capable of identifying existing and predictable hazards'. Analysis of numerous fatal trench accidents indicates that in most cases, the hazard was a dangerous trenching situation that was not recognised by jobsite supervisors or employees due to lack of experience or training.

26.4   OSHA REGULATIONS RELATED TO EXCAVATION SAFETY: OSHA STANDARD 1926 SUBPART P

The OSHA (Occupational Safety and Health Act) standard consists of three main sections with six appendices. The first section contains definitions clearly explaining the terms used in the excavation standard. It is important that these definitions be understood before reading the standard and applying the rules of the standard to the worksite. The competent person can use the standard to a maximum depth of 20 feet. Excavations deeper than 20 feet require the approval of a registered professional engineer

The second section contains the general requirements. The General Requirements of the OSHA standard can be used as the basis of a trench safety programme. All underground and above ground installations must be located before approaching them with a trench. These installations must be supported so as to safeguard employees so that an existing utility line will not fail, causing an accident. Installations must be located prior to the start of excavation work to prevent accident, injuries to workers, and damage to the utility line itself. Access and egress must be provided for employees in excavations over 4 feet in depth to prevent falls when entering or exiting excavations. Employees working in trenches shall be protected from cave-ins, loose rock and soil, and from falling loads. Employees shall be protected from hazardous atmospheres. Both surface and subsurface water must be controlled with water removal equipment or diversion ditches and this process must be supervised by a competent person. Adjacent structures must be underpinned before start of excavation work. All required inspections shall be conducted by a competent person on a daily basis. Also, inspections must be conducted while employees are in the excavation, as needed to identify any signs of cave-ins or unsafe conditions. Fall protection must be provided where appropriate in excavations and over trenches.

The third section specifies the actual Requirements for Protective Systems that must be provided by the employer to protect workers who enter excavations. The standard requires that employees entering excavations which are five feet or greater in depth be protected from cave-ins. The requirements for protective systems are divided into two categories, *sloping and benching* and *support systems*, each of which contains four options, thus eight OSHA options are available to the competent person. *Support systems* include *shoring systems* and *shielding systems*.

These eight OSHA options are the methods that must be used to protect the employees and must be understood by the competent person.

Sloping is defined in the OSHA standard as *a method of protecting employees from cave-ins by excavating to form sides of an excavation that are inclined away from the excavation so as to prevent cave-ins'*. Workers must be instructed to slope both sides of an excavation. The competent person must classify the soil, and then select the correct angle of inclination. Fatalities have been reported in excavations sloped at 45 degrees.

Shoring is defined in the OSHA standard as '*a structure such as a metal hydraulic, mechanical or timber shoring system that supports the sides of an excavation and which is designed to prevent cave-ins'*. The use of aluminum hydraulic shoring requires the use of manufacturer's tabulated data and extensive training in the use of the equipment and a complete understanding of manufacturer's recommendations. The advantage of aluminum hydraulic shoring is that workers can install and remove them quickly outside the trench.

A *shield* is defined in the OSHA standard as '*a structure that is able to withstand the forces imposed on it by a cave-in and thereby protect employees within the structure'*. Trench shields are commonly called trench boxes. Shields are commercially available and can be fabricated from aluminum or steel. Shields are easy to assembly and install with the properly sized piece of equipment, and they provide safe work areas. The use of shields requires the use of manufacturer's tabulated data and extensive training in their use and the accompanying manufacture's recommendations.

The primary appendices of the standard are: Appendix A - Soil Classification; Appendix B: Sloping and Benching; Appendix C: Timber Shoring; and Appendix D: Aluminum Hydraulic Shoring.

26.5   CAUSES OF ACCIDENTS

Data on occupational accidents can be obtained from several agencies: the Bureau of Labor and Statistics (BLS), the Occupational Safety and Health Administration (OSHA), and the National Institute for Occupational Safety and Health (NIOSH). Each agency maintains its database for different specific objectives; thus, the types of information included in the database and the focus of the investigation varies from agency to agency.

In 1991 OSHA conducted a study to analyse the most cited standards in the construction industry in order to identify the causes of accidents and provide suggestions on how to eliminate, control or mitigate such hazards. This study showed that 4 of the top 25 standards cited were related to trenching as shown in Table 26.1.

Table 26.1 Most cited trenching related standards (adapted from OSHA, 1991)

RANK	DESCRIPTION OF THE STANDARD		Standard (1926.
5	Trenching/Excavation	Protective Systems for trenching/excavation	652(a)(1)
11	Trenching/Excavation	Daily inspection of physical components of trench and protection systems	651(k)(1)
16	Trenching/Excavation	Spoil pile protection	651(j)(2)
22	Trenching/Excavation	Access/Egress from trench/excavation	651(c)(2)

The National Safety Council (NSC) has adopted the BLS figures (beginning with the 1992 data year), as the authoritative count for work related deaths in the United States. However, the categories in the BLS system on fatal injuries do not isolate 'trench-related' injuries. While most injuries would be classified as 'caught in or crushed in collapsing materials' or 'excavation or trenching cave-in', trench-related injuries could also be categorised as 'falls', 'contact with electric current', etc. Using the BLS data alone, the death or injury counts can be misleading and they are in most cases, understated. Therefore, to measure the trenching hazard and to study the causes of trench-related accidents, the database from all three agencies should be used concurrently.

In order to understand the fatalities causes associated with trenching operations, and to develop the intervention strategies, it is necessary to access the National Institutes for Occupational Safety and Health (NIOSH), specifically, the Fatality Assessment and Control Evaluation (FACE) programme. This is a research programme designed to identify and study fatal occupational injuries. The goal of the FACE program is to prevent occupational fatalities across U.S by identifying and investigating work situations at high risk for injury and then formulating and disseminating prevention strategies to those who can intervene in the workplace.

The research project funded by NIOSH has considered and studied 52 (48 out 52 construction operations) reports associated with Trenching and Excavation Operations. All reports were extracted from NIOSH web site. The reports covered the period from 1985-2000.

The preliminary results show a similarity with previous research studies (e.g. Hinze, 1998, Hinze and Bern, 1997). Sewer systems (35%) and water supply systems (15%) are areas with the highest trenching related fatalities. The analysis also indicated that electrocutions in trenching accidents are increasing.

An analysis of the *type of accidents*, cave-in was cited as the main cause in seventeen cases. In cases (94%) that involved cave-ins, the walls were not protected by shoring, shielding or sloping. Eighteen equipment related accidents were included in the FACE reports. These accidents were due to improper equipment operations, equipment working near trenching areas, lack of signals, inexperienced operators and mechanical deficiencies.

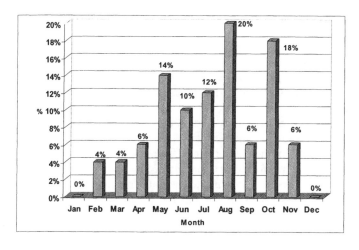

Figure 26.1 Month of occurrence vs. fatalities

The *month of occurrence* of the accident was examined. Figure 26.1 identifies August and October as the months with the highest percentages of fatalities. *Company Safety Programs* were also analysed. In 50% of the cases, the company had an Official Safety Programme, but in 60 of the cases, a competent person did not conduct a safety site evaluation prior to the accident.

Another characteristic that was analysed was the victim's age. In 70% of the cases, the victims were younger than eighteen. Figure 26.2 shows the age range for all cases in the FACE reports. It is noted that 51% of the workers were younger than thirty-five.

26.6   CHARACTERISTICS OF ACCIDENTS

Another source of important information related to trenching related accidents are the OSHA investigation reports, which make up the largest single source for this type of information. To analyse this information a total of fifty fatal and non-fatal cases were identified from 1996 to 1997. The data was obtained from the OSHA Database System. The following parameters were studied, and the observations of this study are discussed in this section:

- Month of event
- Accident outcome (injury or fatality)
  - Gender of workers affected
  - Classification by SIC code
  - Time of day of accident
  - Union status of workers
  - Trench characteristics.

Month of event: In 1996, 21% of the accidents occurred during the month of October. In 1997, 18% of the accidents occurred during the month of December. Overall the month with the highest incidence of accidents during the period of investigation (1996-1997) was October (16%). Figure 26.3 shows this information in a graphical format.

Accident Outcome: Of all the cases studied more than half (65%) resulted in fatalities and only 35% resulted in injuries.

Gender of workers: From the data obtained from the OSHA reports, it was observed that all of the workers involved in trenching accidents were male.

Classification by SIC code: According to the data from the OSHA reports 40% of the accidents reported involved workers for water, sewer, and pipeline contractors, i.e. SIC (Standard Industrial Classification) Code - 1623.

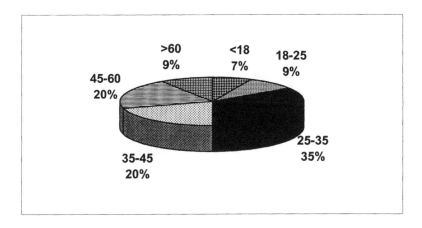

Figure 26.2 Age vs. percentage of trenching accidents

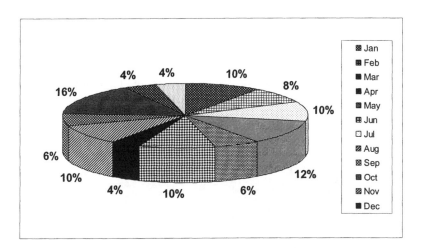

Figure 26.3 Total accident occurrence by month

Time of day: From the data available a comprehensive analysis of the time of day of occurrence of the accidents was not possible. The time of day of the accidents was reported on only seven (14%) of the 50 cases analysed.

Union status: The majority of the workers involved in trenching related accidents were non-union workers (98%). This gives us an indication that workers who are not union members are more likely to have accidents due to lack of proper training.

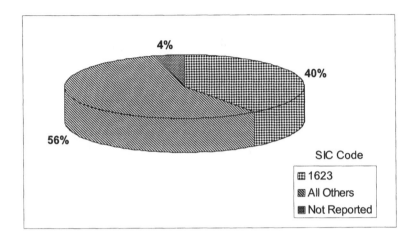

Figure 26.4 Total accident occurrence by SIC code

Trench characteristics: Of the 50 cases studied, 27 reported information related to the depth of the trenches in which the accidents took place. The depth of the trenches varied from 0 to 20 ft with ten instances (37%) in the range from 0 to 5 ft. This indicates that even in shallow trenches the possibility of accidents still exists.

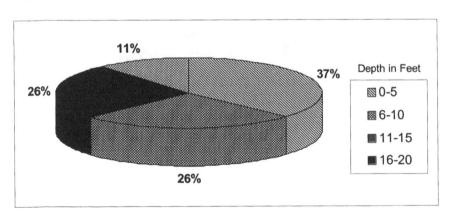

Figure 26..5 Total accident occurrence by trench depth

In five cases studied the presence of excavation support structures was reported. In eight cases there was no trench support structure present and in 37 of the 50 cases (74% of the cases analysed) the presence of excavation support structures could not be determined from the OSHA accident reports.

The type of excavation protection structures or methods depends on the characteristics of the trench. Depth and soil condition are the predominant factor when deciding if and what type of protection will be used. Of the five cases where excavation support structures were present, shoring was used to protect two of the trenches, trench boxes were used in two cases and sloping was used in one case.

Risk Factors: Various risk factors that contribute to trenching accidents were identified from the OSHA accident reports. Misjudgment of hazardous situations was identified in 39% of the instances, making it the most common risk factor. This reinforces the need to have a 'Competent Person' capable of correctly identifying risks so that actions can be taken to reduce the probability as well as severity of accidents.

26.7 CONCLUSIONS

This chapter discussed the role of the competent person in excavation safety and analysed characteristics of accidents based on FACE and BLS records. Based on these initial findings, continued site visits and interviews with craftspeople, and front-line supervisors, potential intervention strategies can be identified. Two key observations from the initial review and analysis of FACE records point to the need for a competent person at the excavation work site and effective worker training prior to the commencement of construction operations. Other intervention strategies that are being investigated include recommendations to OSHA regarding the existing standards for trench safety, engineering controls, and safety management issues in construction.

26.8 REFERENCES

Boom, J., 1999, Trenching is a dangerous and dirty business. *OSHA's Job Safety and Health Quarterly Magazine*, 11, **1**, Fall.

Bureau of Labor Statistics, 2000, *National Census of Fatal Occupational Injuries, 1999*, United States Department of Labor, Washington, D.C.

Equipment World, 1998, *Digging Deeper: Survey on Trench Shoring Practices*.

Hinze, J., 1991, *Indirect Costs of Construction Accidents*, The Construction Industry Institute (CII), Austin, Texas. *Source Document, 67.*

Hinze, J. and Bern, K., 1996, Identifying OSHA paragraphs of particular interest, ASCE. *Journal of Construction Engineering and Management*, 122, **1**, March, pp. 98-100.

Hinze, J. and Bern, K., 1997, The causes of trenching related fatalities and injuries. In *Proceedings of Construction Congress V: Managing Engineered*

Construction in Expanding Global Markets, edited by Stuart D. Anderson, ASCE, pp 389-398.

Hislop, R.D., 1999, *Construction Site Safety - A Guide for Managing Contractors* (Florida: Lewis Publishers), Boca Raton.

Lew, J.J., 1994, Trenching and excavation safety. In *Proceedings of the Fifth Annual Rinker International Conference Focusing on Construction Safety and Loss Control*, October, pp. 43-52.

National Safety Council, 1991, *Accident Facts*.

Occupational Safety and Health Administration, 1990 (Revised). *Excavations - OSHA Publication*, 2226, US Department of Labor.

Occupational Safety and Health Administration, 1992, *OSHA Standard*, 1926, Subpart, P., US Department of Labor, July.

Occupational Safety and Health Administration, 1994, *OSHA Standard*, 1926 Subpart, P., US Department of Labor, October.

Rekus, J.F., 1992, Safety in the Trenches. *Occupational Health & Safety*, February pp. 26-37.

Stanevich, R.L. and Middleton, D.C., 1988, An exploratory analysis of excavation cave-in fatalities. *Professional Safety*, 33, pp. 24-28.

Suruda, A., Smith, G. and Baker, S., 1988, Deaths from trench cave-in in the construction industry. *Journal of Occupational Medicine*, 30, 7, July, pp. 552-555.

Thompson, L.J. and Tannenbaum, R.J., 1975, Excavations, trenching and shoring: The responsibility for design and safety. *A Report from the Texas A&M Research Foundation Project Number RF-3177*, College Station, TX, September.

Thompson, L.J. and Tannenbaum, R.J., 1977, Responsibility for trenching and excavation design. *Journal of the Geotechnical Engineering Division,* ASCE, 103, April, pp. 327-338.

Thompson, L.J., 1982, Trenching and excavation safety. *ASCE Specialty Conference on Construction Equipment and Techniques for the Eighties*, March 28-31, Purdue University, West Lafayette, Indiana, pp. 302-307.

US Department of Labor: Occupational Safety and Health Administration, 1990, *Analysis of Construction Fatalities – The OSHA Data Base 1985-1989*, November, Washington, D.C.

US Department of Labor: Occupational Safety and Health Administration, 1995, 29 CFR 1926/1910. *Construction Industry-OSHA Safety and Health Standards*. Revised July 1, Washington, D.C.

US Department of Labor: Occupational Safety and Health Administration, 1995, The most frequently cited OSHA construction standards in 1991. *A Guide for Abatement of the Top 25 Associated Physical Hazards*, March, Washington, D.C.

26.9 ACKNOWLEDGEMENTS

This work was supported by the National Institutes for Occupational Safety and Health (NIOSH). The contents of this paper reflect the views of the authors, who are responsible for the facts and the accuracy of the data presented herein. The contents do not necessarily reflect the official views or policies of National Institutes for Occupational Safety and Health (NIOSH).

Section 6

Accident Analysis

CHAPTER 27

The Need for Accident Reporting Systems: Legal versus Company Needs

Steve Rowlinson

Accident reporting systems stem from two sources. There is normally a legal requirement based on the employees' compensation system which forces employers to report on all accidents. However, there is also a company need to monitor accidents. This company need is based on the reporting of accidents for insurance premium calculation purposes and also as part of the safety management system to reduce the occurrence of accidents by analysing and reporting on accident causation and highlighting areas where action is needed.

27.1 ACCIDENT CAUSATION

One of the first theories of accident causation was developed by Heinrich, Petersonand Roos (1980) in the 1920s. Heinrich developed his 'axioms of industrial safety'. These axioms dealt with areas such as accident causation, the interface between worker and machine, the relationship between accident frequency and accident severity, the underlying reasons for unsafe acts, the relationship between management functions and accident control, organisational responsibility and authority, the costs of accidents and the relationship between efficiency and safety.

Heinrich's axioms are summarised below:

- Injuries result from a sequence of factors, the last of which is the accident itself.
- Unsafe acts of persons cause the majority of accidents.
- Three hundred narrow escapes from serious accidents will have occurred before a person suffers injury caused by an unsafe act.
- The severity of injury incurred is largely a matter of chance.
- Four basic methods are available for accident prevention.
- Accident prevention methods are similar to those methods used for cost control, quality control and production management. Management is in the best position to initiate accident prevention strategies.

- The art of supervision is the most important influence on successful accident prevention.
- The humanist view of accident prevention can be usefully enhanced by a consideration of economic forces.

    Bird (1974) extended the scope of Heinrich's work by drawing on other work done in the field of management in order to produce a model known as the Domino Sequence.

27.2   THEORIES OF CAUSATION

Hinze (1997) discusses accident causation theories in detail and the reader is referred to this comprehensive text. However, of the many types of causal models presented a couple have a significant impact on the design of accident reporting systems. These are discussed below:

27.2.1   Report system philosophy

One approach which is important to bear in mind when designing reporting systems is the concept of multi-causality. Peterson (1971) queried the basis of the domino theories. He asked the question, when an act or condition that caused an accident is identified, how many other causes are overlooked? Hence, his view is that many causes may come together to cause an accident and Peterson used the example of a man falling from a defective step ladder to illustrate his point. Under the domino theory one act or condition would be identified as the underlying cause of the accident:

- The unsafe act is climbing a defective ladder.
- The unsafe condition is the defective ladder.
- The corrective action is to remove the defective ladder from use.

    However, if one was to view the accident from a multiple causality viewpoint then the contributing factors to the accidents could be probed by asking questions such as:

- Why was the defective ladder not identified during inspections?
- Why did the supervisor allow the ladder to be used?
- Why did the employee use the ladder when he should have known that it was defective?
- Had the employee been properly trained?
- Did anybody consider the safety infrastructure required when assigning this task?

- Answering the questions above might lead to the following actions:
  - Improvements in inspection procedures;
  - Improved training courses;
  - A review of supervisors' responsibilities and roles.

The underlying principles which can the gleaned from this brief review of accident causation models is that most accidents occur due to a sequence of events and the underlying root causes of accidents may well be multiple. Hence, one must identify these root causes during accident investigation, not just the unsafe acts or conditions which are the symptoms of a problem. Such a philosophy leads to the adoption of an epidemiological approach to accident analysis (see below).

27.2.2   Epidemiological

Once it is accepted that accidents are 'caused', then analysis of the causal factors for the purposes of directing prevention efforts becomes a worthwhile exercise. In order to analyse the factors involved in accidents, some working model of the accident process is necessary. The epidemiological approach outlined by Gordon (1949) provides a useful conceptual basis for descriptive accident analysis. As explained by Lingard (1995), under the epidemiological model, accidents are viewed as the result of a combination of forces from three different sources, host, environment and agent. The epidemiological approach, as outlined by Gordon (1949) and Suchman (1961) is based upon the idea that accidents, including occupational accidents, bear similar characteristics to infectious and non-infectious diseases which has led to a better understanding of their causes and possibilities for their control can equally be applied to accidents.

Accidents are often described as 'unanticipated' events which manifest 'undesirable' consequences. In the same way, diseases may strike an individual without warning and are certainly 'undesirable'.

The public health approach to disease concern itself with, not only treating the consequences of the disease but in trying to find out why certain individuals are more susceptible to catching an illness and what environmental conditions are conductive to the spread and risk of disease.

Suchman points out that this approach is in sharp contrast to the medical approach to accidents. From the medical perspective, the consequences of accidents are considered but rarely is attention given to their etiology. This is put down to the problem that accidents are regarded as chance events, which are not suitable for etiological enquiry (Lingard, 1995). This approach thus concurs with Bird's and Heinrich's axioms and is described by Lingard in relation to studies undertaken for the Hong Kong Housing Authority by the author and Lingard.

27.2.3   Objectives of a reporting system

The objective of an accident reporting system is:

- To monitor accident rates.
- To identify accident causes.
- To monitor the effect of site safety initiatives.
- To estimate the costs of accidents.

In Chapter 28 Tang reports on Hong Kong's construction industry's very poor safety record (Lee, 1991) accounting for more than one third of all industrial accidents and most fatalities in industrial accidents occurred in the construction industry. There is a general consensus that construction contractors should increase their safety investment in their construction projects. The assumption is that the higher the safety investment, the better the safety performance. However, the extent of the investment is always a major concern. Recent research has revealed that in Hong Kong, most contractors set aside an amount of less than 0.5%, and some even less than 0.25% of the contract sum for investing in safety for their contracts. But is that enough?

Safety investments cannot be limitless. A methodology has been developed in a study (Tang, Lee and Wong, 1997) to quantify the minimum amount of safety investment for a building project. In that research, only financial costs of construction accidents have been considered. The social costs have not been included. This chapter will discuss both the financial costs and describe the social costs of construction accidents.

27.2.4   Accident Causality

Ahmad and Gibb discuss accident causality and describe a series of issues which affect the ability of organisations to effectively analyse accidents, as detailed below:

Issues relating to the studies and methodology
Difficulty in obtaining studies and accessing   Minor nature of the incident
involved persons   consequences
Defensive attitude by many parties

Key aspects identified from the interviews
Accident investigation   Time, cost and work pressure
Management/supervision   Method statements/risk assessments
Communication, language and instructions   Abilities, skills transfer and training
Tools, equipment, materials, PPE and task   Working environment, ergonomics and
execution   health factors

In each case the issue is discussed, drawing on comments made by participants.

Chua and Goh discuss the identification of the accident sequence and causal factors of accidents, in particular the underlying factors, forms the first step in the learning process. The acquired information is seen to serve as invaluable inputs for preventive measures. However, there had been very few comprehensive studies on how and why construction accidents happen. Whittington, Livingston and Lucas (1992) attempted to analyse the management and organisational factors of construction accidents, but it was realised that the accident data available within most companies were insufficiently detailed to permit a comprehensive analysis. Most other studies on construction accidents focuses on immediate causes, characteristics of accident victims or accident sequence (Kartam and Bouz, 1998; Cattledge, Hendricks and Stanevich, 1996; Jeong, 1998; Hinze, Pedersen and Fredley, 1998). Information like this is important, but they will not be complete without compilations of the frequency of occurrence of underlying factors and Safety Management System failures.

Thus, this chapter attempts to identify how (accident sequence) and why (immediate factors, underlying factors and safety management system failures) construction accidents occur and to make appropriate recommendations to improve construction safety.

Poon, Wang and Zhang discuss accident statistics in China. They note that there has been an increasing number of construction projects undertaken in China since the beginning of the Deng Xiaopeng era. Engineering quality and construction safety, however, are the two main problems affecting the construction industry. The Chinese government has been making continuous efforts to improve the performance in these areas by issuing and enforcing laws and regulations, strengthening education and training, and learning from the experience of other industrialised markets. The Construction Law of Peoples Republic of China promulgated on March 1, 1998 is a milestone. It clearly demands that all construction activities must meet stringent quality and safety requirements found in Item 3 of Chapter 1. Although the performance of construction site safety has improved since then, there are still many outstanding site safety issues that need to be addressed. The authors have collected 307 pieces of casualty information on construction sites in southern China. They report on the main causes of accidents, activities with high occurrence, dangerous types of work, etc., by statistical analysis. Based on that, a number of improvements are recommended to improve safety management in the future.

27.3   REFERENCES

Bird, F., 1974, *Management Guide to Loss Control*, Institute Press, Atlanta.

Cattledge, G.H., Hendricks, S. and Stanevich, R., 1996, Fatal occupational falls in the U. construction industry, 1980-1989. *Accident Analysis and Prevention*, 28, 5, pp. 647-654.

Gordon, J.E., 1949, The epidemiology of accidents. *American Journal of Public Health*, 39, pp. 505–17.

Heinrich, H.W., Peterson, D. and Roos, N., 1980, *Industrial Accident Prevention* (New York: McGraw Hill Book Company).

Hinze, W.J., 1997, *Construction Safety* (New Jersey: Prentice Hall), USA.

Hinze, J., Pedersen, C. and Fredley, J., 1998, Identifying root causes of construction injuries. *Journal of Construction Engineering and Management*, 124, 1, pp. 67-71.

Jeong, B.Y., 1998, Occupational deaths and injuries in the construction industry. *Applied Ergonomics*, 29, 5, pp. 355-360.

Kartam, N.A. and Bouz, R.G., 1998, Fatalities and injuries in the Kuwaiti construction industry. *Accident Analysis and Prevention*, 30, 6, pp. 805-814.

Lee, H.K., 1991, *Safety Management: Hong Kong Experience*, Lorrainelo Concept Design, Hong Kong.

Lingard, H., 1995, *Safety in Hong Kong's Construction Industry: Changing Worker Behaviour*, Ph.D. Thesis, Department of Surveying, The University of Hong Kong.

Petersen, D., 1971, *Techniques of Safety Management* (New York: McGraw-Hill).

Suchman, E.A., 1961, A conceptual analysis of the accident phenomenon. In *Behavioural Approaches to Accident Research*, Association for the Aid of Crippled Children, New York, 1961, The Risk Management Handbook, 1995.

Tang, S.L., Lee, H.K. and Wong, K., 1997, Safety cost optimization of building projects in Hong Kong. *Construction Management and Economics*, 15, 2, pp.177-186.

Whittington, C., Livingston, A. and Lucas, D.A., 1992, Research into management, organizational and human factors in the construction industry. *Health and Safety Executive Contract Research Report No. 45/1992*, Sudbury (Suffolk: HSE Books).

CHAPTER 28

Financial and Social Costs of Construction Accidents

S.L. Tang

INTRODUCTION

In Hong Kong, the construction industry, especially for building projects, has a very poor safety record (Lee, 1991). The number of construction accidents is consistently at a very high level (Hong Kong Government, 1995). It accounted for more than one third of all industrial accidents and most industrial fatalities occurred in the construction industry. There is a general consensus that construction contractors should increase the safety investment in the construction projects. The higher the safety investment is, the better the safety performance will be. However, the extent of the investment is always a major concern. Recent research has revealed that, in Hong Kong, most contractors set aside an amount of less than 0.5%, and some even less than 0.25% of the contract sum for investing in safety on their contracts.

28.1   SAFETY ISSUES

The question addressed here is as follows: 'What is an optimal level of safety investment?'

Safety investments cannot be limitless. A methodology has been developed in a study (Tang, Lee and Wong, 1997) to quantify the minimum amount of safety investment for a building project, in which only financial costs of construction accidents have been considered. The social costs have not been included. The following section will discuss only the financial costs and the results of previous research and subsequently the author deals with the social costs of construction accidents.

28.2   FINANCIAL COSTS OF CONSTRUCTION ACCIDENTS

28.2.1   What Are Financial Costs?

Financial costs of construction accidents represent the losses incurred by the private investors, such as contractors, due to the occurrence of construction site accidents. In financial analysis, market prices are always used to represent benefits and costs (Tang, 1996). There are a number of studies concerning accident costs (e.g. Heinrich, Peterson and Ross, 1980; Lee, 1991; Levitt and Samelson, 1993). The following financial losses were used in the research carried out by Tang, Lee and Wong 1997):

a) Loss due to the injured person
    -   The compensation paid to the injured worker by the contractor is 2/3 of the wage of the injured person for each day of absence from work
    -   Disability compensation, which depends on the percentage of disability (determined by the Employees Compensation Board by a registered doctor) that the injured worker suffers
b) Loss due to the inefficiency of the worker who has just recovered from injury upon resuming work
    -   When the injured worker returns to work, he cannot initially work with 100% efficiency
    -   A formula to calculate the loss is given below:
    Loss = Wage of injured worker × (Days lost × 1/10 + % of disability)
c) Loss due to medical expenses
    -   Medical expenses of the injured worker, including the cost of transport to hospital
d) Loss due to fines and legal expenses
    -   If the contractor faces prosecution, he may have to pay solicitor's fees and fines imposed by the court
e) Loss of productivity of other employees
    -   The safety officer, site agent, site engineer and the foremen may be involved in assisting the injured and carrying out works relating to the accident such as accident investigation and accident report writing
    -   Other workers may have to stop work immediately after the occurrence of the accident
    -   Loss assumptions (based on the experience of site safety staff interviewed):
        (1) Site agent: 0.05 day
        (2) Site engineer: 0.05 day
        (3) Foreman: 0.25 day

(4) Other workers: 0.25 day for each worker and on average four other workers are involved in each accident
f)   Loss due to damaged equipment or plant
g)   Loss due to damaged material or finished work
h)   Loss due to idle machinery or equipment
-    After the accident has occurred, the workers may stop work temporarily and hence there will be idle machinery or equipment
-    Loss formula is based on the assumption that 20% of the contract sum is attributable to plant and equipment and that 2% of the plant and equipment will be idle on the day of accident:

$$Loss = \frac{Contract\ sum \times 20\% \times 2\%}{Number\ of\ working\ days\ of\ the\ contract}$$

Figure 28.1 was the questionnaire the researchers used to acquire the financial costs of accident from contractors. Readers should note that insurance premium, particularly the additional premium paid to the insurance company by a contractor when his safety record is poor, represents financial cost. This is, however, extremely difficult to quantify and therefore is not considered in the research. This is obviously a drawback. The insurance payment, together with fines imposed by the court, however, are financial costs only and not social costs.

28.2.2   Accident costs and safety performance

The total costs of accidents on a construction site depend greatly on project safety performance. If the safety performance is good, the accident costs will be low and vice versa. In order to compare site accident costs of projects of different contract sums and carried out at different time (so that no inflation adjustment is necessary), the Accident Loss Ratio (ALR), a dimensionless quantity, is defined as follows:

$$ALR = \frac{TC}{Contract\ Sum} \times 100\%$$

where TC is the total costs of site accidents in a project.

The assumed general shape of an ALR versus safety performance curve is shown in Figure 28.2.

Cost items arising from each accident

1. Injured person   (Job nature: _____)
   - Day loss   _____ days
   - Amount of compensation   HK$_____
   - % of disability   _____ %
   - Disability compensation   HK$_____
2. Loss from injured person (after resuming work)
   - Day loss ($\times 1/10 + \%$ of disability $\times 100$)   _____ days
   - Equivalent loss   HK$_____
3. Medical services and expenses
   - Hospitalisation/medical expenses   HK$_____
   - Others   HK$_____
4. Fines and legal expenses
   - Fine by court and solicitor fees   HK$_____
   - Others (e.g. transportation costs, etc.) HK$_____
5. Lost time of other employees
   (time taken by other employees in assisting the injured person)

Post	Monthly wages	Time incurred	Amount
Site Agent	HK$_____	_____days	HK$_____
Site Engineer	HK$_____	_____days	HK$_____
Site Foreman	HK$_____	_____days	HK$_____
* Other Labourers			HK$_____

6. Equipment or plant loss
   - Damaged/replacement cost   HK$_____
   - Repairing cost   HK$_____
   - Others   HK$_____
7. Damaged material or finished work
   - Cost of damaged material   HK$_____
   - Cost of damaged finished work   HK$_____
   - Others   HK$_____
8. Idle machinery/equipment
   ** Idle machinery/equipment   HK$_____
9. Other costs
   Items
   _____   HK$_____

                           Total=   HK$_____

* (4 workers) $\times$ (1/4 day) $\times$ (Daily wage of injured person)

** (Contract sum $\times$ 20%)/(No. of working days) $\times$ 2%

Figure 28.1 Financial losses of each site accident

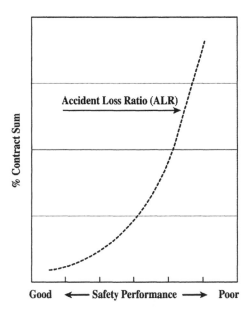

Figure 28.2 Accident loss ratio (ALR) versus safety performance

28.2.3 Safety investment

Safety investment is aimed at protecting the health and physical integrity of workers and the material assets of the contractor. Safety investment consists of the following three components:

(a) Safety administration personnel
   - Site staff and head office staff: according to Hong Kong law, a contractor has to employ safety officers and safety supervisors on site to monitor safety-related matters
   - Some large contractors will also employ safety managers/senior safety officers to direct and coordinate site safety staff
   - The salary of these personnel and their supporting staff (e.g. clerks, typists) are part of the safety investment
(b) Safety equipment
   - Purchasing of safety boots, goggles, helmets, safety fences, first-aid facilities, etc. which are related to the provision of safety on site
(c) Safety training and promotion
   - Safety training courses are organised by contractors for their employees

- Safety promotion includes the printing of pamphlets and posters, the production of safety advertising banners and boards, organiszation of safety campaign and monetary rewarding of individual workers who achieve a good safety standard of work, etc.

Figure 28.3 was the questionnaire used by the researchers to acquire information on safety investment.

Safety investment on each project

1. Investment on safety administration personnel
   1.1 On-site module

Post	Number	Monthly wages
- Safety supervisor	( )	HK$_____
- Safety officer	( )	HK$_____
- Secretary/typist/clerk	( )	HK$_____
- Others	( )	HK$_____

   1.2 Head office module
   (please fill in monthly wages on pro rata according to no. of projects supervised in the same period)

Post	Number	Monthly wages
- Safety manager	( )	HK$_____
- Chief safety officer	( )	HK$_____
- Senior safety officer	( )	HK$_____
- Secretary/typist/clerk	( )	HK$_____
- Others	( )	HK$_____

2. Safety equipment investment on the project
   Safety equipment investment   HK$_____

3. Safety training cost
   Safety training cost   HK$_____

4. Safety promotion cost
   Safety promotion cost   HK$_____

5. Other costs
   i.   _____   HK$_____

   ii.   _____   HK$_____

   iii.   _____   HK$_____

   Total =   HK$_____

Figure 28.3 Safety investment on each site/project

28.2.4   Safety investment and performance

The safety performance of a construction site varies with the amount of safety investment in the project. The higher the safety investment, the better the safety performance should be, and vice versa. As for ALR, safety investments on projects at different sizes and of different times can be compared if a dimensionless quantity, the Safety Investment Ratio (SIR) is used. SIR is defined as follows:

$$\text{SIR} = \frac{\text{TSI}}{\text{Contract Sum}} \times 100\%$$

where TSI is the total safety investment in a project.

Figure 28.4 shows the assumed shape of an SIR versus safety performance curve.

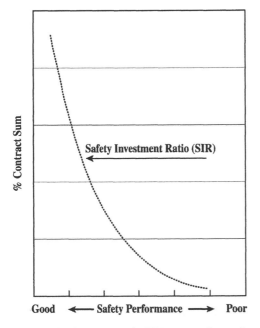

Figure 28.4 Safety investment ratio (SIR) versus safety performance

28.2.5   ALR and SIR curves combined

If the two curves in Figures 28.2 and 28.4 are combined, a third curve of total costs ratio (i.e. ALR + SIR) versus safety performance can be obtained. This curve will have a minimum point, as shown in Figure 28.5, which corresponds to the optimal safety investment of a construction project.

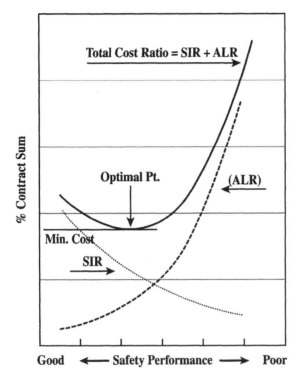

Figure 28.5 Total costs ratio (ALR + SIR) versus safety performance

28.2.6   Accident occurrence index

Accident Occurrence Index (AOI) is used to measure safety performance. It is defined as follows:

$$AOI = \frac{\text{Total equivalent day loss}}{\text{Total man-days required for the project}}$$

where   Total equivalent day loss $= \sum_{i=1}^{n}$ Equivalent days lost of an accident,

where n is the total number of accidents in a project, and
Equivalent day loss of an accident $=$ Days lost $+ (20 \times \%$ disability$)$

For example, if a worker is injured in an accident, has been absent for 30 days, and is certified by a doctor to be of 2% permanent disability as a result of the accident, then the equivalent days lost will be 70 days (30+20×2). The maximum percentage of disability is 100. Twenty times 100 is 2,000, and this means that the equivalent day loss for a case of 100% disability is 2,000 days. A worker in Hong Kong earns an average of HK$600 a day, and the compensation for a fatality is HK$1,200,000. This is consistent with the current compensation practice if a factor 20 is assumed. The factor 20 will be changed for calculating equivalent days lost if the compensation policy is changed or if the average daily wage of labourers in Hong Kong is changed. Note that the equivalent days lost of a very serious injury may be higher than that of a fatal accident.

To calculate the accident occurrence index, the total equivalent days lost should be divided by the total man-days required of the project. For example, a contractor recorded that there were 12 accidents when a building project was completed. There were a total of 100 days lost for these 12 accidents and a total of 10% disability of the injured workers. There were 10,000 man-days of labour force recorded for completing the project. Then the accident occurrence index would be (100 + 20 × 10)/10,000 = 0.03.

Figure 28.6 shows the questionnaire used to acquire information about AOI.

Contractor name	: _____
	_____
Contract title	:_____
	_____
Contract sum	: HK$_____
Contract period	: _____ to _____
Number of working days	: _____days
Total man-days employed (include subcontractors)	: _____
Number of accidents	: _____
Result of injury: _____	Result of death: _____

Figure 28.6 Data for calculating accident occurrence index of each site/project

28.2.7   Result of research and optimal safety investment

Altogether 576 accidents from 18 building projects were investigated (Tang, Lee and Wong, 1997). The data obtained were used to plot a curve as shown in Figure 28.5. A minimum point can be obtained from the curve. It was found that in Hong Kong, the optimal safety investment on a building project was about 0.8% of the contract sum. The total cost to the contractors (accident loss + safety investment) was found to be 1.2% of the contract sum. In fact, the 0.8% should be regarded as a minimum amount of safety investment. An investment greater than 0.8% will result in intangible benefits, such as greater peace of mind of workers, better reputation of the company, greater job satisfaction and so on, which, although not considered in this mathematical model, will definitely be valuable assets to the contractor.

28.3   SOCIAL COSTS OF CONSTRUCTION ACCIDENTS

28.3.1   What are social costs?

Social costs represent losses incurred by society due to the occurrence of construction site accidents. Social costs are defined as any items that will result in the utilisation of national resources. Social costs are not based on the contractor's point of view as discussed in the previous sections, but are based on society's point of view. However, most of the financial costs (but not all) are also social costs, and costs incurred by society are of a wider perspective than those incurred by private investors (contractors) and some individual items costs incurred by society are higher than costs incurred by contractors. For example, hospital fees paid by a contractor (financial loss) to the injured worker are HK$68 per day, but the actual cost incurred (social loss) is about HK$1,800 per day in 2,000. That is to say, society (the Hong Kong Government) subsidises the injured worker by HK$1,732 per day for his stay in a hospital. This point will be further discussed below. The following are examples of social costs (Ngai and Tang, 1999):

a) The productive years of the injured worker. To evaluate the loss of the productive years of a worker, the method stated in the 'Employees' Compensation Ordinance' published by the Hong Kong Government should be adopted. The ordinance establishes the compensation of an injured worker for the case of permanent total incapacity and the case of permanent partial incapacity, with reference to earnings, age and the extent of loss of earning capacity of the injured worker.

b) Families and relatives losses. This refers to the opportunity costs of housewives' work and relatives' work to take care of the injured workers.

c) Fire Department and rescuer services. Costs are incurred by society to provide rescue services such as the ambulance transportation and first-aid services. Besides, fire-engines services and the wages of the related staff are also social costs. It is found that the average costs per accident, including the human resources and the operation costs of equipment of the Fire Department in 1996 and 1997 were HK$9,014.00 and HK$9,885.00 respectively.

d) Losses due to the medical expenses and hospitalisation. In Hong Kong, the hospitalisation fees of local residents in public hospitals were HK$60.00 and HK$68.00 per day per person in 1996 and 1997 respectively. However, the actual expenses that the hospitals incurred were HK$1,570.00 and HK$1,770.00 per day per person in 1996 and 1997 respectively. The losses incurred by the society are the actual expenses.

e) The Hong Kong Police Force. When a construction site accident is reported to the police, the latter will tackle the case and take immediate action. The police force also maintain discipline on site and assists factory inspectors from the Labour Department in investigating accidents. This is also a cost to society.

f) The Social Welfare Department. This includes the administration/personnel costs of the Social Welfare Department to provide assistance to the injured worker.

g) The Labour Department. This includes the costs for regular site inspection for prevention of accidents and the costs for investigation and reporting if accidents occur.

h) The Court. When a serious or a fatal accident happens, the Court will carryout an investigation to find out the reasons for the injury or the death of the worker, especially when there is any argument between the employer and the family of the employee. This is another cost to society.

Items (a) through (h) described above, when added to the financial costs described in Section 28.2.1, represent total social costs of construction accidents. Note that financial losses such as fines and insurance premiums should be excluded from the social costs consideration because they represent internal transfers (Tang, 1996) rather than the costs of the society.

28.3.2   Safety investment by society

Under the social costs of construction accidents, safety investment consists of resources invested both by the contractors and the government. Safety investments by contractors have been discussed in Section 28.2.3. Safety investments made by the Hong Kong Government include:

(a) Safety administration personnel from government departments
  -    Safety inspectors, safety advisors, senior safety advisors of the
       Labour Department
  -    Other supporting staff in the Labour Department and the
       Occupational Safety and Health Council (OSHC) who are
       responsible for safety in the construction industry
(b) Safety training and promotion organised by government departments
  -    Printing of pamphlets and posters, the making of safety advertising
       boards and banners, the organising of safety campaigns, etc.
  -    The costs of safety training courses offered by government
       education institutions [e.g. The Hong Kong Polytechnic University
       (HKPU) and the Construction Industry Training Authority (CITA)]

28.3   IMPACT OF SAFETY INVESTMENT ON CONSTRUCTION SITE SAFETY

In view of the present unsatisfactory situation of construction accidents in Hong
Kong, there is no doubt that the government, promoters and contractors should
increase their safety investment in construction projects to improve site safety.
More measures should be implemented to improve site safety. The increase in
safety investment will result in higher social benefits.

Social benefits are defined as any items that will result in a saving of
national resources. Social benefits refer to the resources saved owing to the
reduction in the number of fatal and/or injured workers. For instance, the number
of fatal and injury cases per thousand workers in the construction industry in 1995
were 0.96 and 232.7 respectively and in 1996 were 0.681 and 219.86 respectively.
Therefore, the social benefits gained by society are the resources saved owing to
the reduction of 0.279 fatalities per thousand workers and the 12.84 injuries per
thousand workers. As the saved resources can be used elsewhere in society, these
represent benefits to the society.

Using the statistical data, the social costs incurred by and the safety
resources invested by contractors and the government in 1999 and 2000 can be
evaluated. The difference of social costs of 1999 and 2000 represents the social
benefits (the social costs in 1999 is expected to be higher than that in 2000 due to a
higher safety investment in 2000 than in 1999). As a result, the impact of safety
investment on social benefits for construction projects can be established.

28.4   REFERENCES

Heinrich, H.W., Peterson, D. and Ross, N., 1980, *Industrial Accident Prevention: A Safety Management Approach*, 5th edition (New York: McGraw-Hill).

Hong Kong Government, 1995, *Consultation Paper on the Review of Industrial Safety in Hong Kong*, Education and Manpower Branch, Hong Kong Government.

Lee, H.K., 1991, *Safety Management: Hong Kong Experience*, Lorrainelo Concept Design, Hong Kong.

Levitt, R.E. and Samelson, N.M., 1993, *Construction Safety Management*, 2nd edition (New York: John Wiley & Sons)

Ngai, K.L. and Tang S.L., 1999, Social costs of construction accidents in Hong Kong. In *Proceedings of the 2nd International Conference of CIB Working Commission W99 on Implementation of Safety and Health on Const4ruction Sites*, Honolulu, Hawaii, USA, March, pp. 525-529.

Tang, S.L., 1996, *Economic Feasibility of Projects*, revised edition (McGraw-Hill).

Tang, S.L., Lee, H.K. and Wong, K., 1997, Safety cost optimization of building projects in Hong Kong. *Construction Management and Economics*, 15, **2**, pp. 177-186.

CHAPTER 29

Towards Effective Safety Performance Measurement - Evaluation of Existing Techniques and Proposals for the Future

Kunju Ahmad and Alastair Gibb

INTRODUCTION

The problem of measuring safety performance has existed since the very beginning of organised attempts to control accidents and their consequences. The level of safety performance within an organisation reflects the loss that the organisation will face. This loss may be due to either an accident or incident resulting in injury or property damage each time it occurs. The loss is not just monetary, the biggest expense is that of human life and it is therefore essential that organisations place specific emphasis on maximising safety performance.

The construction industry, in particular, needs to have a different outlook on safety. Safety must have equal status to other primary business priorities within construction. Accident prevention forms good business practice in that a safe operation is usually an efficient one. In order to reduce the accident or incident level and therefore cut losses, it is important to ensure that safe working practice is being observed. The only way of knowing if safety really exists is to measure it and as the saying goes 'if you don't keep score, you are only practising'. Measurement is a pre-requisite to identifying the factors that contribute to accident potential and need control. The current reactive approach adopted in most construction organisations does not reveal either how safe a site is or what the safety culture is. Merely relying on post-accident data will not reveal sufficient information in order to improve the safety level.

This paper presents some of the findings of a three-year research programme at Loughborough University to develop a proactive safety performance measurement tool (SPMT).

29.1 COLLATION OF EXISTING SAFETY PERFORMANCE MEASURES

Review of available literature indicates a strong safety culture in the petrochemical engineering industry compared with that of the general construction industry. Du Pont, a world leader in safety, claimed that 96% of lost workday and restricted workday cases are caused by unsafe acts of people who created unsafe conditions (Hubler, 1995). The overall construction industry is still looking at positive ways to change to a safer working environment with many researchers including Hinze and Gambates (1996), Gibb *et al.* (2001 and 2002) and Suraji and Duff (2001) trying to understand the causes of accidents. In general the objective of these studies of accident causation is to prevent accidents. These accident causation theories have gone through various changes based on the foundation of the domino theory. Over the years the domino theory has been updated with an emphasis on management as a primary cause in accidents and the development of a multi-causal approach to accident causality.

Many factors help to activate the concern for safety such as trade unions, consumerism, technology and others. With the influence of safety activism factors, safety is becoming everyone's concern - not just the worker or individual. Safety is looking beyond accidents and more towards human behaviour and culture. Measurement will enable comparison, benchmark performance and track progress over time. Once the principle and the practice of measurement become the norm, this will transform motivations, attitudes and choices in construction companies. However, traditionally only reactive performance measurement, based on accident frequency, has achieved widespread use.

The existing safety measurement strategies presented in this paper come from extensive literature reviews and contact with industry. From the literature reviews, the author has identified at least thirty different measurement techniques, all of which are in-house systems. Typically companies design their own safety manual and procedures. These safety assessments were mainly carried out within the petrochemical construction sector where safety is the highest priority. In addition a few assessments are presented that focused on the overall construction industry. A summary of existing safety performance approaches is discussed below while a detailed explanation of the assessment can be found in Kunju Ahmad, Gibb and McCaffer (2001, 1999).

The following discussion describes the existing safety performance measurement approaches adopted by different organisations. These approaches also include research undertaken to improve safety performance. The discussion is organised as follows, based on the literature shown in Table 29.1:

- Reactive performance measurement
- Periodic audits
- Behavioural approach
- Culture/climate evaluation

- Benchmarking
- Proactive performance measurement

Table 29.1 Existing safety performance assessment

Types of assessment	Authors	Year	Name of assessment
Reactive performance measures	Laufer	1986	Safety performance measurement
	Haines amd Kian	1991	Group Unified Accident Reporting Database (GUARD) reporting system
	Azambre	1991	Occupational Accident Analysis & Reporting System (OCCAR)
	Wagenaar	1997	TRIPOD
	Anon	Various	Government accident statistics
Periodic safety audits	Magyar	1983	Performance rating
	Bond	1985	International Safety Rating System (ISRS)
	Waldram	1991	Elements Loss Prevention Management
	Cote and Rochet	1991	TOTAL
	Byrne	1996	Three levels of audits
	Hurst and Donald	1996	Process Safety Management (PRIMA)
	EPSC	1996	REALM
	EPSC	1996	Operating system
	HASTAM	1999	CHASE
Behavioural safety approach	Ramsey, Buford and Beshir	1986	Classification of unsafe behaviour
	Duff, et al.	1993	Goal-setting & feedback technique
	Hubler	1995	Behaviour Accident Prevention Process (BAPP)
	EPSC	1996	Measurement of behaviour
	HSE	1997	HSE Climate Survey Tool
	Loughborough University	1998	Offshore Safety Climate Assessment Technique
Benchmarking safety performance	CII/ECI	1999	Benchmarking initiatives
	Respect for People	2002	Key Performance Indicators
Proactive safety performance measurement	Fitts	1994	Safety Management System (SMS)
	Asheidu and Okoene	1996	Safety performance improvement
	Jaselskis and Anderson	1996	Successful safety management
	Cameron and Duff	1999	Safety Performance Model
	NOHSC	1999	OHS Performance Measurement

29.2   REACTIVE PERFORMANCE MEASUREMENT

The most common form of safety performance measures are reactive and they are the basis of most governmental measurement systems. Laufer (1986) suggested that safety measuring methods are characterised primarily by the manner in which they relate to the criteria of safety effectiveness, the events measured and the method of data collection. The frequency element of the undesirable event usually splits up into four categories:

- Lost day cases – cases which bring absence from work;

- Doctor's cases – non-lost workday cases that are attended by a doctor;
- First aid cases – non-lost workday cases requiring only first aid treatment; and
- No-injury cases – accidents not resulting in personal injury but including property damage or productivity disruption.

Despite the backward-looking nature of reactive measures, several efforts have been made to use them as learning tools to affect future performance. Sarawak Shell looked at the possibilities for using the information gathered when accidents and near misses are investigated to generate not only a useful measure of safety performance but also a target that would discourage undesirable behaviour with regard to reporting. Two pieces of information related to every accident or near miss were considered; the potential for injury and the standard underlying cause category. An estimate is made on a standard scale of how severe the accident or near miss could have been and how many people might have been affected. Standards are based on research work done for Shell on accident causes and were introduced with the new Group Unified Accident Reporting Database (GUARD) incident reporting system (Haines and Kian, 1991).

The Occupational Accident Analysis and Reporting System (OCCAR) is a computerised system used by those responsible for any activity on sites to identify the origin of an accident/incident when it occurs and the chain of events leading to the event. Azambre (1991) divides the origins of accidents into the following three factors:

- human factors;
- procedural factors; and
- technical factors.

The TRIPOD accident causation model (Wagenaar, 1997) aims to produce a profile of the extent to which certain generic failure types are present in the organisation and typically comprises the following four stages:

- What people did?
- How this cause disturbance?
- How the barriers gave way?
- How it all ended in accident?

Even the latest initiatives are still heavily reliant on some form of reactive measure. For example, the UK's Respect for People Initiative (2002) observes that 'many companies are adopting a measure of working hours since the last 'lost-time' accident as a pro-active safety indicator', despite the fact that it is intrinsically a backward-looking measure.

29.3   PERIODIC SAFETY AUDITS

Performance rating provides an objective means for evaluation of the elements which are essential to injury. In this case the elements identified are safe/work practice, housekeeping standards, storage practice, machinery equipments and injury experience. Elements of Loss Prevention (Waldram, 1991) is an assessment of monitoring managerial aspects of the safety policy including the performance of contractors. The International Safety Rating System (ISRS) was an audit system that measured 20 elements including visible management involvement with site visits, comprehensive job analysis procedures, safety training at all levels, structured inspections and follow-ups (Bond, 1985).

TOTAL developed and refined a specific methodology to assess the operating and safety conditions of hydrocarbon process installations (Cote and Rochet, 1991). This exercise assesses plant conditions, site organisation and personnel behaviour. Shell Expo use a three-level audit system comprising: Level 1: the Corporate audit; level 2: the filed unit safety case audit; and, Level 3: the location audit (Byrne, 1996). Process Safety Management (PRIMA) measures eight key audit areas and for each area comparisons are made with the strength of the control and monitoring loop found during the audit (Hurst and Donald, 1996).

Resource Efficient Auditing for Line Management (REALM) is a 'one-stop' system which delivers a means of auditing, a tool to help managers set Health, Safety and Environment (HSE) plans, targets and a mechanism to facilitate sharing and comparing between business units (EPSC, 1996). This is a system of self-audit against a range of environmental health and safety standards and objectives. Dow Chemicals' Operating Discipline management system (EPSC, 1996) evaluates 13 areas and each area is given substance and complete descriptions which will allow the people operating the safety management system to evaluate performance. Complete Health and Safety Evaluation (CHASE) is a unique management tool designed for both monitoring by line managers and auditing by safety professionals (HASTAM, 1999).

Taylor Woodrow, a major UK-based property and construction organisation, has developed a proactive measurement approach based on periodic audits as part of its overall safety management strategy. Paul Turrell, Taywood's Safety Manager described their approach at the conference to launch Rethinking Construction's new Respect for People toolkit in London in October 2002.

Taywood, along with much of the UK construction industry previously judged a project's health and safety performance using reactive measures, based on accident and incident statistics. Their new approach centres on periodic audits of project health and safety practice. The audits are completed by Turrell's centrally-based staff to ensure a measure of consistency in measurement. Turrell explains that their system is designed for continuous improvement, in that scores are given in each audit, but the standards required to achieve the scores increase each year as the practice improves. The audit is split into a number of areas, including:

- Administration
- Fire protection/prevention
- Working platforms
- Work places
- Mobile plant and equipment
- Other requirements/regulations
- PPE
- Access
- Health and welfare
- Housekeeping
- Public protection
- Training and information
- Emergency/permit procedures
- Environmental

Each area is given a score where 7 is the industry average for good practice; 8 is the company norm; 9 is a model to follow and 10 is outstanding and unique. An example of how the scoring system changes with time can be best explained by the following example: a leading project starts to provide hand-care centres across the site and as a result scores a '10' for this healthcare innovation. Other projects get to hear about this and also implement the innovation. When 25% of the company's projects use the innovation the score drops to 9; when 50% comply the score drops to 8. This has advantages in driving standards up, but of course makes it difficult to directly compare performance year on year. Project scores are available throughout the group via the intranet, Tayweb.

29.4   BEHAVIOURAL SAFETY APPROACH

The behavioural safety approach identifies, emphasises, measures and promotes 'safe' behaviours rather than the punishment of 'at risk' behaviours. It works best when the people at the work place believe that 'safe' behaviour is the acceptable norm. There are four steps of implementation:

- identify and define critical behaviour;
- train observers to recognise and measure the occurrence of the behaviours in the workplace;
- establish a system of ongoing observation and feedback; and
- use gathered data to identify corrective actions and plans for continuous improvement.

The measurement of behaviour is, in principle, directed at two groups:

- people in leadership positions (managers, supervisors and relevant staff); and
- people in operating positions.

The classification system of unsafe behaviour is a tool for identifying casual factors and for evaluating safety related performance. The resultant categorisation is divided into three major types of unsafe worker behaviour: those related to the worker, those related to tools, equipment or materials; and those related to material handling equipment.

The Behaviour Accident Prevention Process (BAPP) was developed by Behavioural Science Technology (BST), California (Hubler, 1995). BAPP takes DuPont's STOP programme further upstream by identifying, defining and scientifically measuring critical behaviours which are essential to performing a specific job safely. The objective of BAPP is to increase the proportion of time that crews perform certain critical behaviour safely, and therefore eliminate the likelihood of employee injury.

Duff *et al.* (1993) developed an objective and quantifiable method of safety measurement by identifying contributory factors in the chain of events which causes accidents by:

a. sorting the accident data by department;
b. identifying the different types of accident within each department; and,
c. classifying on the basis of whether or not the individual's behaviour
    or situation had contributed to the accident.

29.5   SAFETY CULTURE/CLIMATE EVALUATION

A number of organisations have developed a safety culture assessment tool to give organisations a means of identifying the status of their own safety culture (Lee and Harrison, 2000; Guldenmund, 2000). After a wide ranging technical review, the parameters were organised in one of the nine key sub-groups and three overall groups as shown in Table 29.2.

The UK HSE's health and safety climate Survey Tool aims to promote employee involvement in health and safety by seeking people's views on key aspects of health and safety in their organisation using a 71 statement questionnaire and then involving them in improvement strategies based on the information which emerges (HSE, 1997). The offshore Safety Climate assessment technique is based on the use of multiple methods (Loughborough, 1998). This technique was derived from literature, organisational culture and climate, as well as previous studies in the offshore sector. The technique includes three methods for assessing safety climate offshore and seeks to build on current industry initiatives. The three methods are:

- attitude assessment and questionnaires;
- interviews and focus groups; and
- behavioural observational assessment.

Table 29.2 The safety culture framework (EPSR, 1996)

Management and organisational factors	Enabling activities	Individual factors
• Positive organisational attributes	• Reinforcement and incentives	• Individual ownership
• Management commitment to safety	• Communication	• Individual perceptions
• Strategic flexibility • Participation and empowerment		• Training

29.6   PROACTIVE SAFETY PERFORMANCE MEASUREMENT

Kvaerner Construction UK (now Skanska) developed a proactive method of assessment called the Site Safety Performance System (SSPS) (Kvaerner, 1998). Its purpose is to assist site management in the reduction of accidents on construction sites by encouraging the participation of the workforce in a system of measuring and improving site safety performance and promoting safe behaviour at work. SPSS's measurement categories are PPE; access to heights; scaffolding; signing and guarding; mobile plant/equipment; environmental aspects; documentation; mobile access scaffolds; and lifting operations. SSPS uses the following procedure to measure safety performance:

- formal observations;
- score proportion of unsafe situations in each category;
- calculate raw score;
- calculate safety performance level (SPL) using the SSPS equation;
- calculate weekly average SPL; and,
- compare SPL with target.

The North Sea Chapter's safety system (1999) uses a Performance Indicators matrix with the leading indicators shown on Table 29.3 being assessed.

Table 29.3 Leading performance indicators (North Sea Chapter)

Leading indicators	Areas
Management commitment	Compliance with Personal Safety Contract
Health, Safety and Environmental Plans	Tasks achieved verses targets
Safety meetings and Representatives	Safety meetings attended
	Representative trained
Risk awareness	Risk assessment completed
Training and competence	Completion of training
Occupational health	Occupational health plan
Audits and follow up	Achieved audit review
	Corrective actions closed
Technical integrity	Operational availability of safe critical
	equipment

Asheidu and Okoene (1988) indicated that improving safety performance consists of three areas namely: development of a management system; a hazard management process; and rigorous auditing of work sites. Whereas Jaselskis and Anderson (1996) identified the following eight factors necessary to achieve outstanding project safety performance:

- upper management attitude;
- project management team turnover;
- safety representative;
- safety meetings with supervisors;
- specialist-contractors;
- informal meetings with supervisors;
- site safety inspections; and
- worker safety performance fines.

Cameron and Duff (1999) developed a safety performance model where individuals are influenced by a continuous interaction between their attitude, behaviour and the situation in which they find themselves. The model represents a theoretical framework, which requires development to form a practical framework for the measurement of attitudinal, behavioural and organisation factors, which influence safety. Fitts (1994) suggested that the objective of a safety system should be to provide a healthy working environment where risk of injury and illness is as low as reasonably practical.

Australia's National Occupational Health and Safety Commission (NOHSC, 1999) developed positive performance indicators for construction covering the following:

- leadership in Occupational Health and Safety (OHS);

- design and planning initiatives;
- methods for consultation, communication and participation;
- management of sub-contractors;
- system and process to manage OHS;
- training and education initiatives;
- risk management and control of hazards; and
- auditing procedures.

29.7   BENCHMARKING

The European Construction Institute (ECI) benchmark safety performance to assist participants to:

- objectively compare their performance against others;
- provide an indication of how to improve their performance;
- quantify the use and value identified good practices; and
- identify industry norms and trends.

Benchmarking adopts the following key safety elements in its approach:

- adopt a 'zero injury' philosophy,
- establish specific contract safety requirements,
- recognise quality of effort is more important than time spent,
- understand the high cost of worker injury,
- recognise the benefits from considering constructability,
- conduct a safety self assessment,
- ensure owner is a participant,
- ensure all parties employ relevant competent staff,
- ensure subcontractors are active participants, and
- success in eliminating accidents is not guaranteed.

The UK construction sector is wedded to a benchmarking approach for many business measures using KPIs (Key Performance Indicators). These have been driven by the 'Egan' agenda and the Movement for Innovation (now re-badged as Constructing Excellence). There are three separate checklists in the latest version of this benchmarking tool covering the following aspects (Respect for People, 2002):

- Conception, design and planning
    - Conception and procurement
    - Design
    - Planning and site start-up

- • Health and safety plan
- • Site health
  - • Construction phase
  - • Exposure to hazardous substances
  - • Musculo-skeletal hazards
  - • Noise and vibration
  - • Other factors
- • Site safety
  - • Site layout and tidiness
  - • Working places - preventing falls
  - • Plant and machinery (including hand-held tools)
  - • Transport
  - • PPE
  - • Public safety
  - • Health and safety file
  - • Review and audit

These checklists have a 0-5 scoring system and scores are entered onto a radar chart to track results.

29.8   LIMITATION OF EXISTING SAFETY PERFORMANCE MEASURES

From the discussion above, it can be concluded that each company is likely to adopt or create their own measurement technique with each technique being successful in its own way. Each approach will only focus on specific aspects of safety in the organisation. The approach and methods of measurement are based on the organisation's preference and priorities. However, there are limitations to all of these individual approaches. Table 29.4 identifies the drawbacks of some of the measurement approaches:

The latest UK report on the subject, by Respect for People (2002), admits that 'an effective predictive measure is still not available and this needs to be identified and suitable data collected'. They add that 'this is now an urgent need to drive the safety agenda forwards'.

Table 29.4 Limitation of existing safety measurement approach

Measurement approach	Limitations
Accident reporting system	Relying on accident reports is totally reactive and liable to lead to under-reporting to falsely enhance reported performance.
Safety audit	Most safety audits measure only the presence of the safety system. They do not measure the effectiveness of the system on site. After a safety audit has been carried out, an organisation may achieve excellent results on paper. But in practice accidents still keep happening.
Safety behaviour	Measurement of safety behaviour depends on the observer's competency to recognise and measure the acceptable and unacceptable behaviour.
	Individuals often feel threatened when they are being observed.
	Measuring unsafe behaviour is seen as an example of perpetuating the blame culture.
Safety climate	The safety climate measures are mostly carried out on petrochemical sites and changes need to be made to adopt the approach for the whole construction industry.
	The questions are not standard. They can be changed according to the needs of each site and organisation. An example of this approach is the HSE Safety Climate Tool (HSE, 1997).

29.9   THE CASE FOR A SINGLE STANDARD MEASUREMENT TECHNIQUE

Based on the above, there exists a need to develop a single standard measurement technique for construction. By having a single measurement tool, the level of safety could be determined and a benchmark towards measuring safety performance could be formed. Jacobs (1970) agrees that the benefit of having a single measurement is that it must be able to evaluate the magnitude of changes over time or in comparative evaluation of two similar situations.

The development of a single measurement tool must be able to:

- reveal the safety performance level;
- measure safety effectiveness that will enable identification of accident problems;
- provide continuous information concerning change in the safety state within an organisation;
- be sensitive to the fundamental behaviour and condition malfunctions;

- define where remedial actions are required; and
- continuously generate observable improvement in the way people work and thus will definitely lead to a good safety culture.

There is a necessity for a technique to measure safety performance in construction that will enhance the ability to predict and control accident losses. The generic technique must be able to be applied in the ever-changing construction industry. Other considerations are the relative cost involved in using it, the clarity of the system under study, the desired output, its compatibility with programmed activities and its meaningfulness to managers and those who will benefit from it.

What is important throughout the review is that many researchers agree that management involvement is essential. Factors such as safety training, orientation, safety awards, and safety committees are also important. Simon and Piquard (1991) et al suggest that communication is also important so as to encourage employees to suggest safety improvements and report near misses as well as unsafe conditions and practices.

The above discussion urges a strong need for a proactive measurement tool. Hence, the thought of developing a proactive Safety Performance Measurement Tool (SPMT) emerged. Kunju Ahmad, Gibb and McCaffer (1999, 2001) discuss SPMT in detail. SPMT was designed to be a generic measurement tool that is applicable to any construction site at any phase. The development of SPMT as a proactive measurement tool reduces the dependency to measure safety against reactive data. SPMT also involves the participation of various levels of personnel from HQ management to operatives combining interviews, document checks and observations. Thirty Safety Control Measures (SCMs) were established from literature and expert opinion as parameters for SPMT. These measures include safety audits, training, documentation, risk assessments, reporting procedures meetings and communication. These SCMs were verified as the proactive factors that are most likely to influence the safety culture and environment on site. The feedback of results from carrying out SPMT are informative and able to highlight the detail performance of each SCM. Each assessment will be able to highlight the SCMs that performed poorly hence enabling management to plan remedial actions. Overall, SPMT has the potential to be a generic measurement tool to enhance safety culture on a project. The SCMs covered are shown in Table 29.4.

Table 29.5: Safety control measures from SPMT

Safety Control Measures	
Safety audit	Method statements
Up-to-date safety documentation	Permit-to-work system
Pre-tender risk assessment	Machinery and equipment in safe working condition
Procedures for reporting accidents/incidents	Good housekeeping
Procedures for reporting near misses	Material Safety Health Data Sheet
Up-to-date safety policy	Emergency response system
Safety meeting with supervisors	Procedures for recommendation, suggestion or improvement
Safety meeting with specialist-contractors	Procedures for conveying instructions
Selection of sub-contractors based on safety issues	Safety poster, signs and notice boards
Health and Safety Committee	Training
Safety officer	Safe behaviour
Induction training	Safe working environment
Site inspection	Effective health care
Tool-box talks	Motivation to safe behaviour
Construction risk analysis	Right person for the right job

29.10 THE CASE FOR LINKING SAFETY PERFORMANCE TO OTHER PERFORMANCE MEASURES

Experts are divided regarding the benefits and disbenefits of treating health and safety as an integrated part of overall project management. Supporters claim that it ensures that these issues are given equal weighting, but detractors are fearful of the essential nature of the safety of individuals being somehow traded off against other business drivers.

An additional thrust of the Taylor Woodrow approach discussed earlier is to link performance-related bonuses directly to health and safety audit scores. For many years, the organisation has awarded annual performance bonuses to staff. Since 2000, to achieve this bonus, the project where the individual is based must score above the company benchmark of 80%. This is also applied to Directors and other non-project-based staff, where all the projects that they are responsible for must score above the benchmark.

For 2001, Taywood also compared the health and safety audit scores with the profitability of each project. Figure 29.1 shows the results with the main vertical scale being the audit score and the colours of the columns representing varying degrees of profitability (more than 2% above the expected return; within +/- 2% of expected return; more than 2% below expected return). The graph demonstrates that many of the poorer performing projects from a health and safety viewpoint are also performing badly financially (shown as black columns which are groupd towards the lower end of the safety performance spectrum). Clearly the sample would need to be enlarged to argue this point more strongly, however, there is a clear indication of a link that is not going un-noticed within Taylor Woodrow's senior management team.

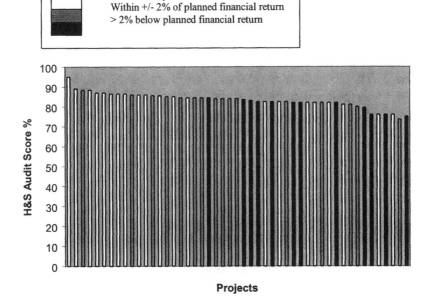

Figure 29.1 Comparison of health and safety audit scores with financial performance

Turrell stresses that this measurement approach is only one part of an overall strategy to improve health and safety: 'There is no *one pound* solution for health and safety – there are only *100 pennies* – but each penny makes a difference.'

As mentioned earlier, the UK's Rethinking Construction drive Respect for People is also striving to see health and safety issues, along with other 'people' issues such as diversity and life-long learning, being used as benchmark measures alongside the harder performance measures such as time, cost and quality (Respect for People, 2002).

29.11   CONCLUSIONS

The over-riding principle is that proactive measures should be used to evaluate safety performance rather than backward-looking techniques.

An industry-wide technique is a potential vision for the future, however, difficulties in applying a single tool to construction remain and, for the foreseeable future, individual organisations are likely to continue to develop their own systems. These individual organisations can derive considerable benefits internally despite being unable to accurately compare their performance with others.

The issue of continual improvement and its effect on performance measurement comparison still requires more work.

29.12   REFERENCES

Asheidu, R.I. and Okoene, A.C., 1988, Improving, construction safety performance in high risk multi-project. In *Proceedings of First International Conference on Health, Safety & Environmental*, pp. 855-861.

Azambre, J., 1991, Accident analysis: A tool for safety management. In *Proceedings* of *International Conference on Health Safety & Environment*, pp. 243-252.

Bond, J., 1985, International safety rating system. *Loss Prevention Bulletin*, pp. 23-29.

Byrne, P.B., 1996, Construction management in regulated safety case regime. In *Proceedings of International Conference on Health, Safety & Environmental*, pp. 863-870.

Cameron, I. and Duff, A.R., 1999, Constructing a construction safety culture: A human approach, Chartered Institute of Building, Ascot. *Construction Paper*, No. 105, CIOB.

Cote, B. and Rochet, B., 1991, Operability and safety audits: A tool for safety management. In *Proceedings of 1st International Conference on Health, Safety & Environment*, pp. 753-762.

Duff, A.R., Robertson, I.T., Cooper, M.D. and Phillips, R.A., 1993, *Improving Safety on Construction Sites by Changing Personnel Behaviour* (London: HMSO Publications).

EPSC, 1996, *Safety Performance Measurement*, edited by Jacques van Steen, (UK: European Process Safety Centre).

Fitts, T.G., 1994, A quality approach to safety management. In *Proceedings of IADC/SPE Drilling Conference*, pp. 419-425.

Gibb, A.G.F, Hide, S, Haslam, R.A. and Hastings, S., 2001, Identifying root causes of construction accidents. *Journal of Construction Engineering & Management*, American Society of Civil Engineers, July/August, 127, **4**, pp. 348, ISSN 0733 9364.

Gibb, A.G.F., Haslam, R.A., Gyi, D.E., Hide, S., Hastings, S. and Duff, R., 2002, ConCA - Preliminary results from a study of accident causality. In *Proceedings of the Triennial International Conference of CIB Working Commission W99*, edited by S.M. Rowlinson, CIB Publication 274, Hong Kong, ISBN 9627757047, pp. 61-68.

Guldenmund, FW., 2000, The nature of safety culture: A review of theory and research. *Safety Science*, 34, **1-3**, pp. 215-257.

Haines, M.R. and Kian, D.V.S., 1991, Assessing safety performance after the era of LTI. In *Proceedings of First International Conference on Health, Safety & Environment*, pp. 235-242.

HASTAM, 1999, *CHASE for Windows*, (UK: HASTAM).

HSE, 1997, *Climate Survey Tool* (UK: Health and Safety Executive).

Hubler W.G., 1995, Behaviour based approach to creating a strong safety culture. In *Proceedings of SPE/IADC Drilling Conference*, pp. 361-373.

Hurst, N.W. and Donald, I. 1996, Measures of safety performance measurement and attitudes at major hazard sites. *Journal of Loss Prevention Process Industry*, 9, **2**, pp. 161-172.

Jacobs, H.H., 1970, Towards more effective safety measurement system. *Journal of Safety Research*, 2, **3**, pp. 161-175.

Jaselskis, E.J. and Anderson, S.D., 1996, Strategies for achieving excellence in contractor safety performance. *Journal of Construction Management & Economic*, 122, **1**, pp. 61-70.

Hinze, J. and Gambatese, J., 1996, *Construction Industry Institute: Addressing Construction Worker Safety in the Project Design* (Austin: Construction Industry Institute, University of Texas).

Kunju Ahmad, R, Gibb, A.G.F. and McCaffer, R., 2001, SPMT an interactive safety performance measurement tool for construction sites, invited paper. *International Journal of Computer Integrated Design and Construction (CIDAC)*, Special issue on computerised safety management, February, 3, **1**, ISSN 1466 5115, pp. 3-15.

Kunju Ahmad, R., Gibb, A.G.F. and McCaffer, R., 1999, Methodology to develop a proactive safety performance measurement technique. In *Proceedings of the Second International Conference of CIB working Commission W99*, Hawaii, pp. 507-514.

Laufer, A., 1986, Assessment of safety performance measurement at construction sites. *Journal Construction Management & Economic*, 112, **4**, pp. 530-542.

Lee, T. and Harrison, K., 2000, Assessing safety culture in nuclear power stations. *Safety Science*, 30, pp. 61-97.

Loughborough University, 1998, Safety climate assessment,

www.lboro.ac.uk/departments/ec/JIP/B3-HTM.

Magyar, S.V., 1983, Measuring safety performance. *Professional Safety*, pp. 18-24.

National Occupational Health & Safety, 1999, *OHS Performance Measurement in the Construction Industry - Development of Positive Performance Indicators* (NSW: NOHSC).

Ramsey, J.D., Burford, C.L. and Beshir, M.Y., 1986, Systematic classification of unsafe worker behaviour. *International Journal of Industrial Ergonomics*, 1, pp. 21-28.

Respect for People, 2002, *A Framework for Action - Report of Rethinking Construction's Respect for People Working Group*, 41 pp.
http://www.rethinkingconstruction.org/documents/Framework%20for%20Acti on.pdf
Simon, J.M. and Piquard, P., 1991, Construction. safety performance significantly improves. In *Proceedings of First International Conference on Health, Safety and Environment*. Netherlands. pp 465-472.

Suraji, A. and Duff, R., 2001, Development of a causal model of accident causation. *Journal of Construction Engineering & Management*, American Society of Civil Engineers, July/August, 127, 4, ISSN 0733 9364, pp. 337-345.

Wagenaar, W.A., 1997, Accident analysis the goal, and how to get there. *Safety Science*, 26, 1, pp. 25-33.

Waldram, I.M., 1991, The use of safety audits: A UK sector operators experience. In *Proceedings of First International Conference on Health, Safety & Environment*, pp. 735-744.

CHAPTER 30

Utilising the Modified Loss Causation Model for the Codification and Analysis of Accident Data

D.K.H. Chua and Y.M. Goh

INTRODUCTION

Various studies have shown that the construction industry needs to improve its safety performance. In order to do so, learning from past incidents is one of the most basic processes in which the industry should engage in. However, currently there is a lack of models and systems to facilitate such a learning process. Thus, this chapter presents an incident causation model, the Modified Loss Causation Model (MLCM), which provides the framework for the codification and analysis of incident investigation information. Detailed categorisations of causal factors and incident sequence variables, based on the MLCM, were developed. The categorisations are hierarchical; allowing the user to zoom in to obtain the necessary details or "zoom out" to obtain sufficient data for statistical analysis. The categorisation framework can accommodate an international construction incident database and incident investigation.

The MLCM and its taxonomy was used to carry out an in-depth study on 140 fatal construction accidents between 1993 and 1999. The study was conducted by the National University of Singapore (NUS), Department of Civil Engineering, in conjunction with the Ministry of Manpower (MOM), Occupational Safety Department (OSD). In Singapore, the OSD investigates all fatal accidents and at the end of each investigation a report is produced for internal usage. This huge pool of accident investigation reports holds great potential for learning from accidents. However, prior to this study there had been no deliberate attempt to explore the wealth of knowledge and learning that is stored in this accident database. The accident reports proved to be useful in identifying the situational variables, incident sequence, immediate causes and underlying job factors of construction accidents. Furthermore, a risk assessment tree constructed based on the 140 accident cases is also used to illustrate how the MLCM-codified information can be applied to improve the quality of risk assessment.

In the Construction 21 report (MOM and MND, 1999), it was identified that despite efforts in improving construction safety, accident rates remain high. This problem is not a localised one. In the United States, the construction industry had consistently been among those industries with the highest injury and fatality rates (Hinze, 1997). In the United Kingdom, the construction industry accounted for 30% of all employee deaths in 1995/96 (Brabrazon, Tipping and Jones, 2000). In Kuwait, the industry accounts for 42% of all occupational fatalities (Kartam and Bouz, 1998) and in Hong Kong the industry accounts for more than a third of all industrial accidents over the last ten years (Tam and Fung, 1998). These studies are among many others that show that the industry has a very poor safety performance record.

In order to ensure that the construction industry learns from accidents or incidents[1], deliberate feedback procedures and systems have to be in place. As can be seen in Figure 30.1, the learning process should be inclusive of four inter-dependent sub-processes, (1) collection of in-depth information during incident investigation, (2) proper storage of the information in a safety knowledge base, (3) detailed data analysis of information in the knowledge base, and (4) systematic utilisation of the stored and derived knowledge. Consequently, this process will lead to an improved Safety Management System[2] (SMS).

With reference to Figure 30.1, when an incident occurs it implies that there has been a failure in the SMS. Thus, during incident investigation the key aim would be to discover information that will be useful to improve the SMS that had failed. At the same time, this information should also be carefully codified based on an incident causation model (such as the Modified Loss Causation Model (MLCM) presented in this chapter), and stored in a safety knowledge base so that it can be used for data analysis and future safety planning. With the knowledge base, various statistical studies can then be carried out to monitor safety performance and to facilitate determination of necessary safety measures during safety planning. In order to fully leverage on the safety knowledge base to improve the quality of construction safety planning, Case Based Reasoning, which is a relatively new branch of artificial intelligence, can be utilised to facilitate retrieval and adaptation of relevant safety knowledge needed during safety planning (Chua and Goh, 2002).

---

[1]   Broadly defined as unintended and undesirable events that may or may not lead to loss or injury.
[2]   From the definition given in BS 8800 (BSI, 1996), SMS can be thought of as an interdependent set of preventive measures, which is targeted at maintaining and improving safety performance of an organisation.

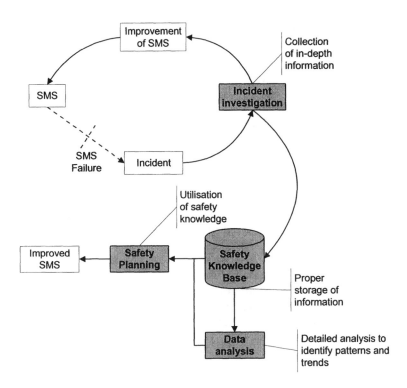

Figure 30.1 Learning process necessary to improve safety performance

However, based the research by the Human Reliability Association (Henderson, Whittington and Wright, 2001), 38% of UK companies (across industries) do not have any documented structure or support for incident investigation and 24% have minimal formal support with the focus of the system on identifying immediate accident causes. Furthermore, there have been very few comprehensive studies on how and why construction accidents happen. Whittington, Livingston and Lucas (1992) attempted to analyse the management and organisational factors of construction accidents, but it was realised that the accident data available within most companies were insufficiently detailed to permit a comprehensive analysis. Most other studies on construction accidents focus on immediate causes, characteristics of accident victims or accident sequence (Kartam and Bouz, 1998; Cattledge, Hendricks and Stanevich, 1998; Hinze, Pederson and Fredley, 1998) Information like these are important, but they will not be complete without knowledge of the key underlying factors and Safety Management System failures that led to incidents.

In view of the current deficiencies, this chapter seeks to demonstrate how the MLCM and its accompanying taxonomy can be utilised to facilitate the learning process through the codification and analysis of incident data. The discussions are based on a 140 fatal construction accident cases obtained from Singapore's Ministry of Manpower (MOM), Occupational Safety Department (OSD). The set of accident cases were codified using the MLCM, and relevant statistical results are discussed here. The applications of the codified information in risk assessment will also be explored.

30.2   THE MODIFIED LOSS CAUSATION MODEL

The MLCM (Figure 30.2) was developed based on an extensive literature review and applications on actual accident analyses. It is meant to guide incident investigation, safety planning and codification of safety information. As the name implies, the MLCM is a modified version of the Loss Causation Model (LCM) (Bird and Germain, 1996), which is employed due to its proactive approach (Covey, 1989), and its clear identification of immediate causes and underlying factors.

As can be seen in Figure 30.2, the MLCM describes six key components of incident causation, (1) situational variables, (2) loss, (3) incident sequence, (4) immediate causes, (5) SMS failures, and (6) underlying factors. Situational variables refer to a set of critical characteristics of the context or situation in which the incident occurred. Through the comparison of situational variables, similarity between construction scenarios of future projects and past incident investigation can then be assessed. In this way, relevant knowledge gained from incident investigations can be retrieved and utilised during the safety planning of new construction projects. Furthermore, the situational variables can also serve as stratification variables during data analysis. Stratification variables separate the data set into smaller sets so as to facilitate more detailed study within or across sub-sets. For example, to study the differences in terms of immediate causes for each type of work (a type of situational variable), the data set is stratified based on the types of work and then the statistics for the immediate causes of each type of work are obtained for comparison. With the right choice of stratification variables new knowledge can be derived.

In the MLCM, loss refers to the undesirable consequences of an incident and it can be in various forms, for example monetary loss, human loss or property loss. Losses due to incidents need to be carefully recorded so as to understand the detrimental impact of incidents. In addition, loss data is also very useful during safety planning, because it facilitates the estimation of the potential losses due to incidents in different types of situations and feeds these data into the risk assessment process.

Incident sequence is defined as the sequence of events that leads to unintentional and undesirable consequences or losses. In the MLCM, the incident sequence is described by two main events, namely, the breakdown event and contact event. A breakdown event is the initiating point of loss of control of a source of energy that, without an intervening event, will lead to the occurrence of a contact event. The contact event is the event where the subject comes into contact with the uncontrolled source of energy that eventually leads to some loss. The breakdown event needs to be identified because it is the starting point of the incident, and through its identification, preventive measures can then be implemented to prevent incidents more effectively. Similarly, clear identification of contact events is equally important because it allows appropriate measures to be implemented to prevent the uncontrolled source of energy from coming into contact with the subject.

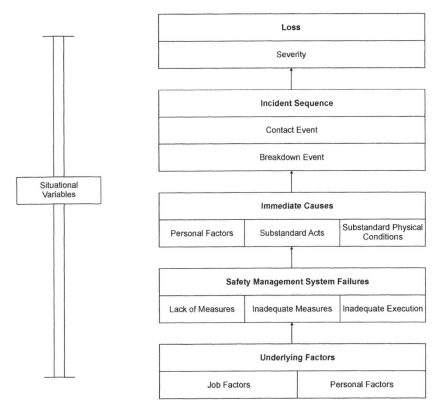

Figure 30.2 The modified loss causation model

Immediate causes are direct causes of incidents, and they are usually related to front-line workers. Immediate causes can be broadly separated into substandard acts, substandard physical conditions and personal factors. Substandard acts and substandard physical conditions refer to observable behaviours and physical conditions (or objects) respectively, that do not fulfil the required safety standards of the organisation or industrial norms. Personal factors refer to factors that are related to an individual's physical, mental and psychological states, which can lead to substandard acts. As compared to substandard acts and physical conditions, personal factors are usually harder to detect, but the potential benefits of their determination can be enormous.

From a proactive paradigm, the occurrence of an incident and the identification of immediate causes should lead to an assessment of the SMS, so that specific SMS procedures or components that had failed are isolated. In this way, the SMS can then be improved and will be able to prevent the recurrence of a similar incident. SMS failure can be broadly categorised into three types: lack of measures; inadequate measures and inadequate execution. If there are no relevant measures in the SMS that could have prevented the immediate causes and incident sequence, it is a 'lack of measures' type of SMS failure. If there are relevant measures, but the measures are inadequately designed or planned to prevent the immediate causes and incident sequence, then the failure belongs to the 'inadequate measures' type. If there are relevant and adequate measures, but the measures are inadequately implemented, the failure belongs to the 'inadequate execution' type.

To improve the SMS, the underlying factors causing the failure of the SMS need to be determined. Underlying factors are factors that contributed to the occurrence of the SMS failures. In the MLCM, underlying factors are separated into two main types, namely, job-related, and personal-related. Job factors are factors related to job functions of the construction project. The personal factors classified under underlying factors are similar to personal factors classified under immediate causes. The only difference is that the underlying personal factors influence job factors, whereas immediate personal factors influence substandard acts. The underlying factors, in particular the personal factors, are usually hidden and are hard to detect. They are also contributory in nature and its determination may have to depend on experts' subjective judgement. The knowledge of underlying factors would certainly facilitate the improvement of safety performance, but due to the difficulties in their determination, these underlying factors are often not available because they are not fully recorded for a number of reasons, such as their consequences in terms of identifying legal liability and the lack of formal procedures for accident reporting.

30.3 THE MODIFIED LOSS CAUSATION MODEL TAXONOMY

To facilitate the analysis of the large amount of safety information it is necessary to develop a codification taxonomy to summarise the information. The development of a taxonomy is an iterative process and the taxonomy should be tailored to the purpose of the analysis. The presented MLCM taxonomy is designed to be generic, so as to cover various types of construction situations and factors.

A thorough review of existing incident variables taxonomies and classifications was conducted with the aim of identifying suitable categorisations for each of the components in the MLCM. Most of the available taxonomies lack a strong underlying incident causation model (Hinze, Pederson and Fredley, 1998; Kartam and Bouz, 1998; Feyer and Williamson, 1991; Sawacha, Naoum and Fong, 1999), making the logic structure of these taxonomies harder to grasp. Bird and Germain (1996), and Gordon (1998) developed taxonomies that were relatively comprehensive, but not tailored to the context of the construction industry. As a result, there is difficulty in classifying the incident information in some categories, especially the job-related factors, which differ from the construction context. Furthermore, some parts of the taxonomies were split into very detailed factors without sub-categorisations. Categorising incident investigation information into detailed factors allows more specific knowledge to be gained, but more often than not, such a categorisation approach causes data to be too sparse and hence unsuitable for data analysis. Whittington, Livingston and Lucas (1992) proposed a taxonomy based on the construction industry, but due to a difference in the underlying incident causation model and research objectives, their taxonomy could not be fully adopted.

There were also several works in the human error areas (Reason, 1990; Rasmussen, 1982) where the classification requires cognitive information that are frequently missing or inconclusive in construction accident investigation reports. Such classifications often require expertise and resources that are not readily obtainable in the construction industry. However, human error classifications that focus on behavioural aspects, for example Swain and Guttmann (1983), can be more easily adapted into the substandard act component of the MLCM.

Due to the fact that none of the taxonomies could be fully adopted into the MLCM framework, a compilation of the taxonomies from the literature review and the coding that were already in use in OSD was employed in the study. A draft taxonomy was first used to analyse forty accident cases. Following that, the taxonomy was evaluated and changes were made based on the evaluation. During the actual analysis, the taxonomy was constantly re-evaluated and minor changes were made as and when it was deemed necessary. The main categorisation of the taxonomy is summarised in Figure 30.3.

In the MLCM taxonomy presented in Figure 30.3, the type of work that the main participants of the incident were involved in is used as the situational variable. Inclusion of too many situational variables will limit the amount of data retrieved for statistical studies, thus presently only one key situational variable has been used.

The SMS components stated in SMS failures section are based on the SMS elements described in Singapore's code of practice for SMS for construction worksites (PSB, 1999). The alignment of the taxonomy to the code of practice was done to allow ease of codification of information. If necessary this section can always be replaced by the main elements of any organisation's SMS structure.

The source data for the study was based on the accident investigation reports that were produced by the OSD. The reports are in free-text format so that the incident variables have to be identified from these reports and classified according to the MLCM taxonomy. To minimise errors due to subjectivity during classification, the causal factors are identified only if clear statements describing them were stated in the reports, in this way subjective inference by the researchers could be kept to a minimum.

To ensure that the data for analysis were of sufficient quality, the study focused on accidents with at least one fatality. This would ensure a greater depth in the investigation with more objective description of the accident and the causal factors. After reviewing the fatal accident investigation reports between 1993 and 1999, 140 accident investigation reports were chosen for analysis.

	1. Type of Work	
Situational Variables	1.1 Architectural/Renovation/Finishing work 1.2 Building services work 1.3 Geotechnical work 1.4 Material/equipment handling/transportation	1.5 Plant/machinery/equipment maintenance/dismantling/installation 1.6 Structural work 1.7 Other types of work
Incident Sequence	*2. Types of Contact Event*	
	2.1 Fall of person 2.2 Struck by falling objects 2.3 Striking against or struck by objects 2.4 Caught in or between objects 2.5 Over-extertion or strenuous movement	2.6 Fire/explosion 2.7 Exposure/contact with extreme temp/pressure 2.8 Exposure/contact with electric current 2.9 Exposed to harmful substances/radiations 2.10 Other types of incidents
	3. Types of Breakdown Event	
	3.1 Collapse of object 3.2 Lost of balance 3.3 Object fall off surface	3.4 Loss control of plant/transport 3.5 Other types of breakdown event
Immediate Causes	*4. Types of Substandard Physical Conditions*	
	4.1 Substandard plant/machinery/equipment/tools 4.2 Substandard construction material 4.3 Substandard structures/parts of structure	4.4 Substandard work environment 4.5 Other substandard physical condition
	5. Types of Substandard Acts	
	5.1 Extraneous Acts 5.2 Improper equipment usage 5.3 Inappropriate response to emergency 5.4 Omission of basic safety measures	5.5 Spatial error 5.6 Improper work procedure 5.7 Other substandard acts

Figure 30.3 Main headings of the MLCM taxonomy

	6. Types of Personal Factors	
	6.1 Lack of knowledge/skill 6.2 Mental/psychological factors 6.3 Improper motivation	6.4 Physical/physiological factors 6.5 Other personal factors
SMS Failures	7. Types of SMS Failures	
	(A) Lack of measure (B) Inadequate measure (C) Inadequate execution	
	7.1 Safety policy 7.2 Safe work practices 7.3 Safety training 7.4 Group meetings 7.5 Incident investigation and analysis 7.6 In-house safety rules and regulations 7.7 Safety promotion	7.8 Evaluation, selection and control of sub-contractors 7.9 Safety inspections 7.10 Maintenance regime for all machinery and equipment 7.11 Hazard analysis 7.12 The control of movements and use of hazard substance and chemical 7.13 Emergency preparedness 7.14 Occupational health programme
Underlying Factors	8. Types of Job Factors	
	8.1 Factors related to designers 8.2 Factors related to operatives 8.3 Factors related to project management/corporate	8.4 Factors related to site management 8.5 Other job factors
	9. Types of Personal Factors	
	9.1 Lack of knowledge/skill 9.2 Mental/psychological factors 9.3 Improper motivation	9.4 Physical/physiological factors 9.5 Other personal factors

Figure 30.3 Main headings of the MLCM taxonomy (cont'd)

30.4 STATISTICAL RESULTS

The results of the study are shown in the eight histograms (Figures 30.4 to 30.13) below. These eight histograms show the distribution of the type of work (situational variable), contact event, breakdown event, substandard acts, substandard physical conditions, immediate personal factors, job factors based on job functions and job factors related to site management, according to the MLCM taxonomy.

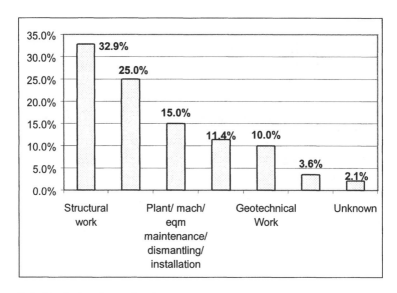

Figure 30.4 Distribution of type of work

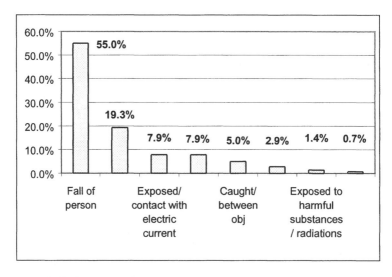

Figure 30.5 Distribution of type of contact event

Figure 30.6 shows a summary of the results depicting the main contributors in each component of the MLCM chain. In terms of type of work that resulted in fatal accidents, structural work and architectural/renovation/finishing work made up 58%.

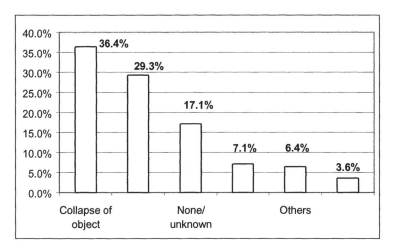

Figure 30.6 Distribution of type of breakdown event

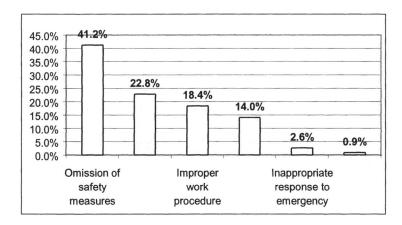

Figure 30.7 Distribution of type of substandard acts

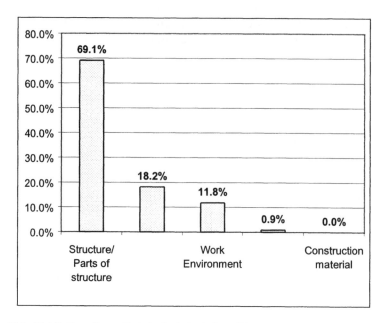

Figure 30.8 Distribution of type of substandard physical conditions

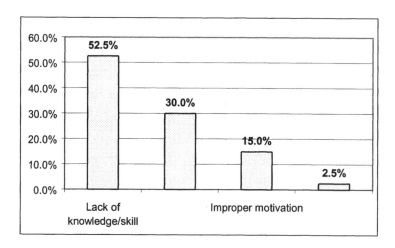

Figure 30.9 Distribution of types of immediate personal factors

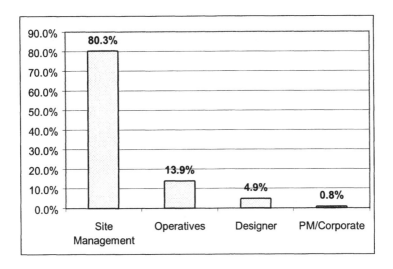

Figure 30.10 Distribution of type of job factors base on job function

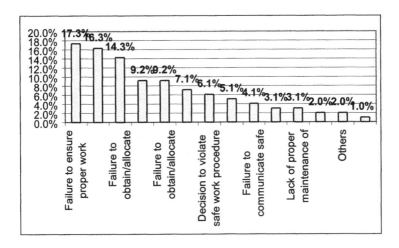

Figure 30.11 Distribution of type of job factors related to site management

With respect to incident sequence, the distribution of the type of contact event is in line with results presented in several other studies based in different countries (Hinze, Pedersen and Fredley, 1998; Jeong, 1998; Kartam and Bouz, 1998). Referring to Figure 30.12, fall of person is the main type of contact event in construction industry (55%). Struck by falling object is the next highest occurring type of incident, but at a lower relative frequency of 19%. As for breakdown event, collapse of object (36%) is the main type of breakdown event with the next highest occurrence being loss of balance (29%). Intuitively, the findings makes sense, as the high occurrence of loss of balance and collapse of objects naturally leads to a high occurrence of fall of persons and struck by falling objects.

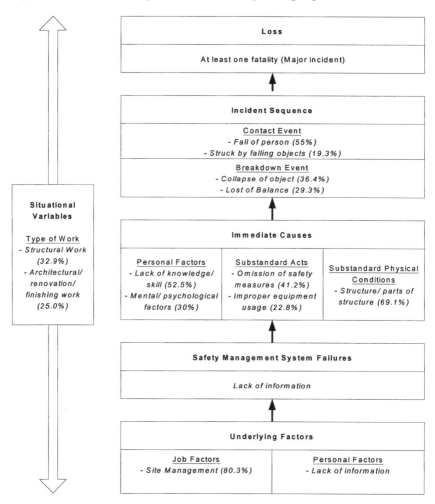

Figure 30.12   Summary of main findings

Findings on substandard acts reveal a high percentage of omission of basic safety measures (41%) such as the wearing of personal protective equipment (PPE) and checking behind the vehicle before reversing. The other main substandard act is improper equipment usage (22%); some common examples are workers using defective mobile scaffolds for work and using employee lifts to transport construction materials. With respect to substandard physical conditions, the main violation is in substandard structure/parts of structure category (69%). This usually refers to lack of safety structures like guardrails or barriers for open sides of buildings and shoring for trenches.

Relating to personal factors, the main underlying causes are lack of knowledge or skills (52%) and mental/psychological factors (30%). The recurring factors are usually related to the substandard acts of operatives and workers rather than other job categories. This would indicate that training and education of operatives can be a vital link in reducing substandard acts and physical conditions. However, deeper analysis of the factors would be needed to identify the appropriate strategies and these will vary in different countries and according to project type.

From the analysis it is realised that there is insufficient information on the types of SMS failures. In most cases the accident investigators had not explicitly defined which part of the SMS is related to the incident. This shows a lack of effort in focusing on the SMS during the investigations. This corroborates Rowlinson's view (Rowlinson, Mohamed and Lam, 2003) that these are systematic failures in SMS that are not addressed in the system maintenance process.

On underlying factors, the investigators only identified the job factors. No information on personal factors that influenced those contributory underlying job factors was available. Most of the job factors belong to the category of site management (80%). The high concentration of site management factors shows that site management plays an important role in construction SMS. A more detailed analysis as depicted in

Figure 30.11reveals that the top three factors with the highest occurrence are: failure to ensure proper work practices/monitor site work (17%), inadequate inspection (16%) and failure to obtain/allocate adequate/proper physical resources (14%). Again, this reflects the lack of understanding of the nature of SMS by management, as identified by Rowlinson, Mohamed and Lam (2003). The results also imply that there is some form of procedures and safety measures in place, but there is a lack of enforcement, implementation and maintenance of the system. This shows that when there is a failure in the safety management infrastructure, evidenced in a lack of close supervision on site and inadequate provision of physical resources to operatives (e.g. workers, technicians and plant operators), there is a higher potential that the SMS will fail, resulting in substandard acts by operatives and occurrence of substandard physical conditions.

It can be seen that the accident/incident investigation focuses primarily on the identification of incident sequence and immediate factors. The lack of identification of the specific SMS components that failed represents a missed opportunity to improve the SMS so that the recurrence of similar accidents can be presented. Moreover, underlying personal factors were not identified.

The analysis has shown that the MLCM and its taxonomy could be utilised to systematically codify incidents to reveal useful information and statistics that could be utilised in the risk assessment process for improving the design and planning of appropriate safety measures, monitoring of the effectiveness of safety measures and also allocation of safety resources. For instance, statistical results on immediate and underlying factors can be used as training material, inputs for training needs analysis, and to prioritise inspection, enforcement and promotional efforts. The continuous monitoring of the incident variables over time can also provide critical information on the effectiveness of safety measures to project management and government safety department. For example, after the implementation of a national campaign to emphasise the importance of safety inspection on sites, the number of job factors related to inadequate inspection should be reduced. If no reduction is observed it would imply that the campaign has not met its objectives.

30.5 APPLICATION OF MLCM-CODIFIED INCIDENT DATA IN RISK ASSESSMENT

In comparison to other components in the MLCM, information on situational variables and incident sequence are the most objective and are more readily available. The information can be used to improve the quality of risk assessment during safety planning.

Risk assessment consists of two main components, hazard identification and risk analysis. Hazard identification involves the explicit recognition of all possible breakdown events and contact events (incident sequence) that can happen in a set of situational variables, whereas risk analysis mainly involves the assessment of the risk imposed by each set of breakdown and contact events. After the risk levels of each possible breakdown event have been determined, more detailed studies can be conducted for high-risk breakdown events. The risk level also facilitates allocation of safety resources, so that more resources can be allocated to mitigate breakdown events with higher expected risk levels.

In order to ensure that the organisation learns from past incidents, incident investigation information can be incorporated into the risk assessment process. Figure 30.7 shows a risk assessment tree, which is a diagram made up of possible incident sequences that can occur in a situation described by a set of situational variables. The risk assessment tree in Figure 30.7 is based on the incident information gathered from the analysis of the 140 accident cases for work related to erection/dismantling of temporary structures. Altogether 34 cases were retrieved according to this work type. As mentioned earlier, other situational variables, like type of equipment/plant, and type of location can also be included to further describe the situation. However, this would form a more specific situation and less data would be retrieved for analysis.

Using the risk assessment tree of Figure 30.7, the breakdown and contact events that occurred in past incidents are identified. The risk assessment team can further augment the risk assessment tree with other possible incident sequences that had not occurred before. Incorporating past incident information ensure that these incidents are included in the analysis which could easily be omitted depending on the experience of the risk assessment team. Thus with a more extensive safety knowledge base, the risk assessment tree would be more comprehensive. Moreover, more complete sets of situational variables can help to construct more relevant and adapted trees.

The risk assessment tree derived in this way also facilitates risk analysis. As evident from. Figure 30.7, each breakdown and contact event branch has a number in brackets indicating the absolute number of occurrences of the event. For example, out of the 34 cases retrieved for work related to erection/dismantling of temporary structures, 9 cases involve loss of balance (breakdown event), and out of these 9 cases, 8 are related to fall of person and 1 drowning (contact events). Since all the cases used in the study involve fatalities, the numbers for consequences were excluded. Based on the number of occurrences, relative frequencies can be calculated. For example, the relative frequency of occurrence of loss of balance, given that the type of work in erection/dismantling of temporary structures, is 0.26 (9/34). The relative frequency can then be used to guide assignment of probability of occurrence and also assessment of risk levels for each breakdown event.

If other severity cases of incidents are included in the safety knowledge base, the expected risk level for each type of breakdown event can be determined as:

$$ERL(BE_m) = P(BE_m)[\sum P(CE_i) \times \sum (P(CSQ_j) \times CSQ_j)] \qquad (1)$$

where

$ERL(event)$ is the expected risk level of the event,

BE_m is the breakdown event m (for example, collapse of plant/machinery),

CE_i is the contact event i (each contact event under the breakdown event),

CSQ_j is the consequence j (catastrophic, major, minor or negligible), and

P($event$) is the probability of the event or consequence

To compute the expected risk level, the severity of consequences of incidents has to be quantified to reflect its impact on the organisation and its stakeholders. The discussion of the quantification of severity, is however, not within the scope of this chapter.

The calculation of expected risk level provides an objective and systematic way to prioritise risk control efforts based on the riskiness of breakdown events. As compared to the purely subjective assignment of risk, the MLCM-based risk assessment approach has obvious merits.

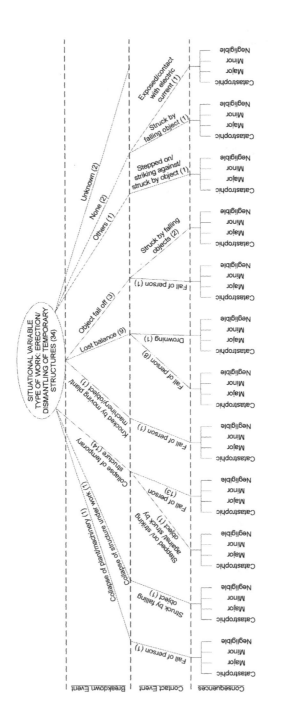

Figure 30.13 Risk assessment tree developed for erection/dismantling of temporary structures

30.6 CONCLUSIONS

In order to improve the safety performance of the construction industry, the industry needs to design and implement learning mechanisms that allows safety knowledge, such as incident investigation information, to be better utilised and be made more readily transferable. This chapter presents the MLCM as a framework to facilitate such learning mechanisms. Based on the MLCM, a taxonomy is developed to codify incident investigation information, so as to allow meaningful data analysis. The presented taxonomy is meant to be generic so as to cover all the possible situations and causal factors that can arise from construction incidents. With the MLCM as the framework, the taxonomy can be easily customised to suit more specific situations.

The MLCM taxonomy was applied to 140 fatal accident cases obtained from Singapore's MOM, OSD. The study identified several main characteristics and causal factors of fatal construction accidents. More importantly, the study demonstrated how the MLCM could be employed to facilitate in-depth statistical analysis. Some of the accident cases were also used to construct a risk assessment tree. The MLCM-based risk assessment tree can be utilised to ensure that past incident sequences are identified during safety planning, so as to allow appropriate measures to be designed to prevent recurrence of the incidents. Furthermore, the risk assessment tree also facilitates more objective determination of risk levels.

30.7 ACKNOWLEDGEMENTS

The study would not have been possible without the support of Ministry of Manpower, Occupational Safety Department and its numerous staff. The authors are also grateful for the comments by Dr John Anderson and the numerous exchanges of emails with him throughout the study.

30.8 REFERENCES

Brabrazon, P., Tipping, A. and Jones, J., 2000, Construction health and safety for the new Millennium, Technical Report. *Health and Safety Executive Contract Research Report, Report No 313/2000*, Entec UK Ltd., United Kingdom.
Bird, F.E. and Germain, G.L., 1996, *Practical Loss Control Leadership*, (Georgia: Det Norske Veritas).

BSI, 1996, Guide to occupational health and safety management systems. *BS 8800*, London.

Cattledge, G.H., Hendricks, S. and Stanevich, R., 1996, Fatal occupational falls in the US construction industry, 1980-1989. *Accident Analysis and Prevention*, 28, 5, pp. 647-654.

Chua, D.K.H. and Goh, Y.M., 2002, Application of case based reasoning in construction safety planning. In *Proceedings of ASCE and EG-ICE Joint Workshop on IT in Civil Engineering* (paper accepted for publication) Washington D.C.

Covey, S.R., 1989, *The 7 Habits of Highly Effective People* (London: Simon & Schuster UK Ltd).

Feyer, A.M. and Williamson, A.M., 1991, A classification system for the causes of occupational accidents for use in preventive strategies. *Scandinavian Journal of Work Environment and Health*, 17, pp. 302-311.

Gordon, R.P.E., 1998, The contribution of human factors to accident in the offshore oil industry. *Reliability Engineering and System Safety*, 61, pp. 95-108.

Henderson J., Whittington, C. and Wright, K., 2001, Accident investigation: The drivers, methods and outcome, Technical Report. *Health and Safety Executive Contract Research Report*, Human Reliability Associates, United Kingdom, Report No 344/2001.

Hinze, W.J., 1997, *Construction Safety* (New Jersey: Prentice Hall), USA.

Hinze, J., Pedersen, C. and Fredley, J., 1998, Identifying root causes of construction injuries. *Journal of Construction Engineering and Management*, 124, 1, pp. 67-71.

Jeong, B.Y., 1998, Occupational deaths and injuries in the construction industry. *Applied Ergonomics*, 29, 5, pp. 355-360.

Kartam, N.A. and Bouz, R.G., 1998, Fatalities and injuries in the Kuwaiti construction industry. *Accident Analysis and Prevention*, 30, 6, pp. 805-814.

MOM and MND (Ministry of Manpower and Ministry of National Development) 1999, *Construction 21*, MOM and MND, Singapore.

PSB (Singapore Productivity and Standards Board), 1999, Code of practice for Safety management system for construction worksites *CP79:1999*, Singapore.

Rasmussen, J., 1982, A taxonomy for describing human malfunction in industrial installations. *Journal of Occupational Accidents*, 4, pp. 311-333.

Reason, J., 1990, *Human Error* (New York: Cambridge University Press).

Rowlinson, S., Mohamed, S. and Lam, S.W., 2003, Hong Kong construction foremen's safety responsibilities: a case study of management oversight. *Engineering, Construction and Architectural Management*, Emerald, 10, 1.

Sawacha, E., Naoum, S. and Fong, D., 1999, Factors affecting safety performance on construction sites. *International Journal of Project Management*, 17, 5, pp. 309-315.

Swain, A. and Guttmann, H., 1983, *Handbook of Human Reliability Analysis with Emphasis on Nuclear Power Plant Applications* (NUREG/CR-1278), Washington (DC: Nuclear Regulatory Commission).

Tam, C.M. and Fung, I.W.H., 1998, Effectiveness of safety management strategies on safety performance in Hong Kong. *Construction Management and Economics*, 16, **1**, pp. 49-55.

Whittington, C., Livingston, A. and Lucas, D.A., 1992, Research into management, organizational and human factors in the construction industry. *Health and Safety Executive Contract Research Report No. 45/1992,* Sudbury (Suffolk: HSE Books).

CHAPTER 31

Construction Supervision in China

S. W. Poon, Y. S. Wang and Y. Zhang

INTRODUCTION

The number of construction projects undertaken in China have been increasing rapidly since adoption of the reforming policy in the 1980s. Despite the efforts of the Chinese Government in issuing the laws and regulations, the engineering quality and construction safety, however, remain as the two main problems in the construction industry. This chapter presents the regulations implemented in China to ensure construction quality and site safety. An investigation of construction accidents in China is also included in the latter part of the chapter.

31.1 CONSTRUCTION SUPERVISION

31.1.1 Construction law of the People's Republic of China

Supervision is an essential and important activity to ensure the quality of the works during construction. Chapter Four of the Construction Law outlines the scope of supervision of construction works in China.

The following are relevant clauses.

Clause 30. Implementation of supervision system for construction works.

Clause 31. The responsible department/client shall appoint the appropriate experienced construction supervision company to supervise the construction works by signing the appropriate contract.

Clause 32. The appointed construction supervision company shall supervise the contracting firm in assuring construction quality, construction planning and use of available capital in accordance with the laws, executive statutes, relevant technical standards, design documents and contract documents.

The supervising staff shall have the authority to demand the contracting firm to rectify if the construction works are not constructed in accordance with the design requirements, construction technical standards and contract specifications. The supervising staff shall report to the design institute for amendment and rectification if the design is found not complying with construction quality standard or construction quality requirements.

Clause 33. Before the supervising company assumes the supervision role, the client shall inform the contracting firm in writing the appointed supervision company about the extent and the authority of supervision works.

Clause 34. The supervising company shall, within their permitted scope of responsibilities, undertake the duties and work of construction supervision. The supervising company shall, in accordance with the appointment of the client, undertake the supervision duties objectively and impartially.

The supervising company and the contracting firm under supervision, together with their relevant suppliers shall not have any subordinate relationship or related interest.

The supervising company shall not transfer the supervision work to other organisation.

Clause 35. The supervising company shall inform the client the loss incurred if the supervision work is not carried out in accordance with the supervision contract, including absence in necessary checking or checking not complying with the regulations or specifications.

The supervising company together with the contracting firm shall be responsible for the compensation for the client's losses if both companies are found conspiring in making unlawful profits from the project concerned.

31.1.2   The construction supervision regulations

Each city in China will have their own regulations and specifications on construction supervision. The following is extracted from the Construction Supervision Regulations adopted in Shenzhen (Shenzhen Construction Safety Supervising Station, 1998).

The scope of supervision is divided in the following stages.

- Pre-design stage.
- Design stage.
- Pre-construction stage.
- Construction stage.
- Maintenance stage.

1) Pre-design stage
    - Consultation and decision for investment in the project.
    - Feasibility study of the project.
    - Organisation for design responsibility.

2) Design Stage
    Assist the client in preparing the brief and participate in selecting the design proposal.
    - Participate in tenderer's selection and contract signing for site

investigation; and supervise the investigation work.
- Remind the design institute in maximising the design functions.
- Assess the design in compliance with the planning design characteristics and the client's functional requirements.
- Assess the rationality of the technical and economic objectives of the design.
- Assess the design in satisfying the nation's framework requirements and design limitations.
- Analyse the feasibility and economics of materialisation of the design.

3) Pre-construction Stage
- On behalf of the client invite for tenders, organise tender examination and recommend acceptance of a tender.
- Audit the forecast budget for construction.
- Assist the client in signing relevant contracts.

4) Construction Stage
- Assist the client in applying for work commencement and assist both the client and the contractor in preparing the report on the commencement of the work.
- Endorse the subcontractors selected by the contractor.
- Organise a committee in checking the construction drawings.
- Assess the contractor's proposal on the design of construction organisational structure, method statement, construction process planning, construction quality assurance system and construction safety assurance system.
- Check and ensure the contractor is to carry out the contractual works in compliance with the nation's engineering and technology standards, and harmonise the relation between the client and the contractor.
- Check the materials, components and facilities in quality and quantity supplied by the contractor or client.
- Monitor the construction progress, quality and investment; ensure and check that the contractor has implemented safety assurance.
- Organise the checking of work packages and work to be covered up after completion; sign the documents for payment of the completed work.
- Sign the relevant certificates on site.
- Supervise and check the use of money received from pre-sale of the property development.
- Ensure the contractor in filing the documents and technical information.
- Organise the preliminary certification of construction works by the client and the contractor.

- Submit the application report for certifying the completion of works.
- Participate in certifying the construction works and audit the payments.

5) Maintenance Stage
   - Be responsible for checking the engineering works status.
   - Participate in assessing the quality.
   - Ensure re-visits by the contractor.
   - Supervise repairs until satisfying the quality requirements

31.1.3   Construction supervising engineer

All supervising engineers should have registered and possessed the Supervision Engineer Practice Certificate, and are appointed by the client to undertake the supervision duties. Unregistered personnel are prohibited to practise in the name of Supervising Engineer.

A practising Supervising Engineer must join a Supervising Engineer Company which is established with approval by the authority. When applying for the Practice Certificate, the following documents are required for consideration by the authority:

- Application Reports.
- Supervising Engineer Qualification Certificate.
- Evidence of two years' experience in supervision duties.
- Employment contract between the applicant and the Supervising Engineer Firm.
- Recommendation from two Supervising Engineers.
- Record of previous project supervision.
- Identity of citizenship.

31.2   CONSTRUCTION SAFETY

31.2.1   Safety supervision organisation

As reported by the Professional Construction Safety Committee, The China Association of Construction Industry (Qin, 1998), many cities in China have set up their construction safety organisation since 1988. The steps are carried out as follows:

1) Establish the safety management organisation by legislation
   The administrative departments have been set up to guide and supervise construction safety in all regions. In fact, the contracting firm is the

principal production unit in playing an important role in safe production. Coupled with appointment of sufficient safety supervisiong staff, safe production system should be set up and a promising administration network can be realised.

2) Set up the safety management system

The system should integrate the safe production, as part of activities of a contracting firm, which can prevent or reduce the construction accidents. Firstly, safety control should be linked to quality control of the firm. When firms wish to improve the quality or carry out the year-end evaluation, they are advised to consider the comments given by the safety administration department. Second, safety work records should be considered in project bidding. This not only raises the overall safety consciousness but also improves the quality of the construction industry at large. The fact that organisations with poor safety performance would be suspended from participating in future project bidding will be a strong signal of emphasising the importance of their commitment in reducing the accidents.

3) Strengthen the administrative function and provide services to construction firms

At present all levels of governments are undertaking the restructuring of their functions of work; and this requires the enforcement of macro supervision rather than micro supervision. Thus, there is a change from direct administration to indirect management. Seeing this, the government organisation should emphasis on supervising, giving assistance, delivering safety messages, organising safety management training for enterprises, providing assistance in establishing the safe production responsibility system, promoting the management of enterprises and upgrading the enterprise's capacity. Having taken into consideration all these factors involved, safety objectives can be achieved.

4) Reinforce the safety construction regulations, standards and norms and regularly organise competition for safe production among enterprises.

The government departments should work out the legislative matters in which safe construction body can be adhered to. Revision of regulations should also be made when necessary. At the same time, competition of safety objective achievement should be held among the firms according to the existing regulations and standards. In order to ensure the successful implementation of all these measures, the government should propagate safety laws, regulations and specifications.

5) Investigate the construction accidents thoroughly

The administration departments should work hand in hand with relevant departments according to the following principles: give a timely investigation, derive concrete and specific conclusion, and inflict punishment in accordance with the regulations. Punishment should be imposed in order to deter and to minimise the chances of accidents.

31.2.2   Construction safety regulations

Consisting of eight sections, the Construction Safety Regulations in Shenzhen (Shenzhen construction Supervising Engineers Association, 1998) have been effective since May 1998. The important areas are highlighted in the following sections.

Section 1 Principle

This regulation applies to new construction, extension, renovation and demolition works.
          The Authority Departments shall set up the Safe Construction Supervision Organisation and be responsible for the supervision work regarding to construction safety.

Section 2 Responsibilities of the Authority Departments and Safety Construction Supervising Organisation

The authority department is responsible for the guidance and supervision of the Safety Supervising Organisation, and for the assessment of the safety supervising and management staff and investigation of severe incidents.
          The Construction Safety Supervising Organisation is responsible for assessment prior to construction, routine supervision during construction and safety appraisal upon completion of the construction works.

Section 3 Clients and other related parties' responsibility

The client has to submit the following documents to the Construction Safety Supervising Station for registration.

- The permit for construction.
- The accepted tender.
- The construction contract.

The client is also responsible to pay the necessary fee for safety supervision. The supervising firms shall integrate the construction safety with construction quality, progress monitoring and investment control.

Section 4 Contractor's safety responsibilities

The Contractor shall establish their objectives and the measures with respect to construction safety, and improve the working condition and environment with plans and sequences.

Section 5 Construction site safety management

The main contractor should be responsible for the overall site safety and management, whereas the subcontractor is responsible for safety of the subcontract work.

With an aim to seeking the approval by the technical advisor of the contracting firm, specific safety design must be included in the following construction works.

- Foundation work.
- Underground construction.
- Installation and dismantling of the elevating scaffold.
- Installation and dismantling of vertical delivery machinery.
- Building demolition.
- Other dangerous activities.

Section 6 Investigation of severe incidents

The contractor shall inform the responsible Government department and safety management organisation immediately after the severe incident, and submit a written report within twenty-four hours.

The written report should include the items below:

- The organisation involved and the time and place of the incident.
- The brief description of the incident, casualties and the preliminary estimate of economic loss.
- The preliminary analysis and the reasons.
- The measures and the control implemented after the incident.
- The organisation submitting the report.

The investigation work shall complete within three months and in exceptional cases not more than six months and the investigation results should be announced publicly.

31.2.3   Standard of Construction Safety Inspection

Issued by the Ministry of Construction in 1999, the document 'Standard of Construction Safety Inspection' is adopted by Safety Supervising Engineers for checking and appraising the contracting firms' commitment to construction safety on sites.

In accordance with the document, the assessment of site activities are categorised into ten individual appraisal forms and one overall summary. The appraisal is centred on the activities which have caused frequent casualties. The ten activities are:

1) Safety management.
2) Civilised construction on site.
3) Scaffolding.
4) Excavation supports and formwork.
5) Personnel protective equipment such as safety helmet, safety belt and safety net, and protection at openings such as access, openings, stairs and lift shafts.
6) Use of electrical appliances.
7) Material hoist and external lift.
8) Tower crane.
9) Lifting activities and installation.
10) Construction plant and equipment.

Except activities numbered (5) and (10), there are general items and assuring items for appraisal purposes. The assuring items are crucial and important to fulfil.

For the individual activity appraisal, the full mark is 100. The overall score is the aggregation of marks under each individual item. For multiple activities under the same category, the marks for the category concerned will be the arithmetic mean for the activities. No negative marks will be applied. All deduction should not be more than the marks allocated to that particular item.

In the appraisal of assuring items of an activity, for any item which scores zero mark or the sub-total of assuring items is less than 40, no score will be given for this appraisal.

The overall score of the appraisal is calculated as below.
Score of a particular item

$$= \frac{\text{(Allocated marks to that item)} \times \text{(score for the item)}}{100}$$

The overall score is the summation of scores for the ten items.

If any of the listed activities is absent for appraisal, the overall score is calculated as follows:

Overall score = 100%×(scores calculated previously for activities appraised)/(summary of allocated marks to the appraised activities).

If more than one party takes part in the appraisal, their marks are combined in the following ratio.

Appointed Safety Officer = 0.6
Others   = 0.4

The safety consideration of a construction site is rated in three classes by the scores.

1) Excellent
    For assuring items satisfying the above requirements and the overall score is 80 or above.
2) Pass
    • Same as in (1) with overall score at 70 or above.
    • With one item without any positive score, but the overall score is 75 or above.
    • With no score in lifting activities and installation, or construction plant and equipment, but the overall score is 80 or above.
3) Fail
    • Overall score below 70.
    • With one item without positive score, but the overall score is below 75.
    • No score in lifting activities and installation on construction plant and equipment, and the overall score is less than 80.

31.3   ASSESSMENT OF SITE ACTIVITY

The individual marks for safety management with the overall summary are listed in the following section.

Safety management

Items (1)–(6) are assuring appraisal and items (7)–(10) are general items. Marks allocated to each item and deductions with respect to various inadequacies are shown.

1) Safe production responsibilities system (10 marks)
    • No establishment of the system - 10
    • No execution by various levels or departments - 4-6

- Absence of safety production target - 10
- No safety practice regulation - 10
- No safety officers which are required - 10
- Failure in auditing the management staff - 5

2) Management by objective (10 marks)
   - No safety management objectives - 10
   - No separation of safety responsibilities - 10
   - No audit of responsibility objectives - 8
   - Method appraised but not implemented or not properly implemented - 5

3) Design of construction organisation (10 marks)
   - No safety measures in the organization - 10
   - No approval of construction organization - 10
   - No individual safe construction design for items of work undertaken by professionals - 8
   - Safety measures not extensive - 2-4
   - Safety measures not directive - 6-8
   - Safety measures not implemented - 8

4) Submission of safety technology (10 marks)
   - No written safety technology - 10
   - Direction not clear or strong - 4-6
   - Not comprehensive - 4
   - Not properly signed - 2-4

5) Safety check (10 marks)
   - No periodic safety checking system - 5
   - No records of safety checking - 5
   - Checking incidents not leading to identification of who, when and what/how - 2-6
   - For major incidents, items of work listed are not finished on time - 5

6) Safety education (10 marks)
   - No safety education system - 10
   - No safety induction for newcomers - 10
   - No principal safety education content - 6-8
   - No safety education for workers undertaking the following trades - 10
   - Any one does not understand the safety practice in his/her trade - 2
   - No annual training for construction management staff as required - 5
   - No annual training assessment or failure in assessment of the appointed safety officer - 5

7) Safety activities before work commencement (10 marks)
   - No system establishment for safety activities before work commencement - 10
   - No record of such activities -2

8) Specialised activities requiring permit (10 marks)
   - Each person without training for performing specialized activity - 4
   - Each person without the permit - 2
9) Handling of construction accidents (10 marks)
   - No report which is required - 3-5
   - Accidents not investigated, analyzed and handled - 10
   - No filing system for accident records - 4
10) Safety signs (10 marks)
   - No site safety sign location plans - 5
   - No erection of signs as required - 5

Total = 100

31.4 SUMMARY: CONSTRUCTION SAFETY APPRAISAL

The summary of the appraisal will include the following information.

Name of enterprise:
Name of site:
Area:
Type of structure:
Overall marks (100 maximum):
Individual categories:
1) Safety management (10)
2) Civilised construction site (20)
3) Scaffolding (10)
4) Excavation supports and formwork (10)
5) Personnel protective equipment and protection at openings (10)
6) Use of electrical appliances (10)
7) Material hoist and external lift (10)
8) Tower crane (10)
9) Lifting activities and installation (5)
10) Construction plant and equipment (5)

Remarks:
Checking Party: _____   Person Responsible: _____
Checking Categories: _____   Project Manager: _____
Date: _____

(Source: Ministry of Construction, 1999)

31.4.1   A study of construction site accidents and a major accident

31.4.1.1 Site accidents

A total of 307 cases of construction accident in Southern China had been collected. The aim was to analyse the main causes of accidents, the places of high occurrence, the type of work undertaken, etc. Suggestions were also made with an attempt to improving safety management on site.

  The information was collected by sending questionnaires, visiting managers on sites and asking for records from relevant government departments. Altogether 136 companies were investigated, and 307 safety accident reports involving 209 injuries and 98 deaths were analysed. The results are highlighted as follows.

31.4.1.2 Age distribution of persons involved in accident

The age of labourers and workers on construction sites are mostly in the range of 20-50 years old. The age distribution of persons involved in accidents as shown in Figure 31.1 is quite the same as that of the workers on site. This indicates that there is no significant relationship between age of the worker and occurrence of an accident. It can be interpreted that the probability of an accident occurrence for workers is almost equal irrespective of their age. Therefore, safety education should be introduced to all workers, young or old, on site.

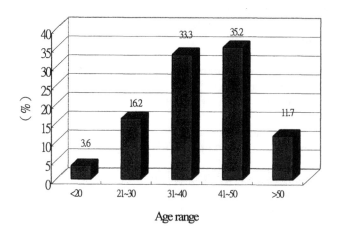

Figure 31.1 The age distribution of persons involved in accident

31.4.1.3 Types of accident

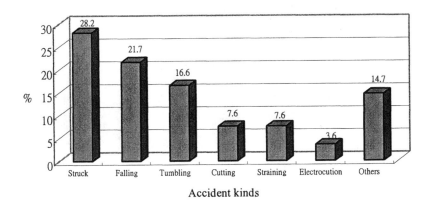

Figure 31.2 Types of accident

Frequency of different types of accident is shown in Figure 31.2. Half of the accidents involved workers being struck by objects or falling from height. The other frequent accidents are tumbling, cutting, straining and electrocuting in descending order of frequency. These are the main threats to workers on construction sites. Precautions and education are necessary to prevent these accidents.

31.4.1.4 Locations of accident

Most of the accidents occurred at hoists, openings such as holes and wells, scaffoldings, cranes and ladders (see Figure 31.3). Signs should be displayed at these places to warn workers of possible dangers. Besides, detailed instructions of works to be carried out near these areas are required in order to reduce the occurrence of accidents.

31.4.1.5 Types of activity

The data displayed in Figure 31.4 suggests that frequent accidents are found in activities in connection with the scaffolding work, earthwork, piling, building framework construction, concreting work and mechanical plant maintenance. Special training to workers involved in these works is recommended to ensure that all workers are qualified in their work as well as in taking necessary precautions.

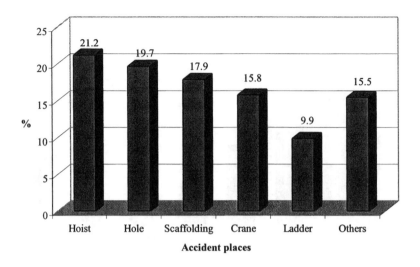

Figure 31.3 Locations of accident

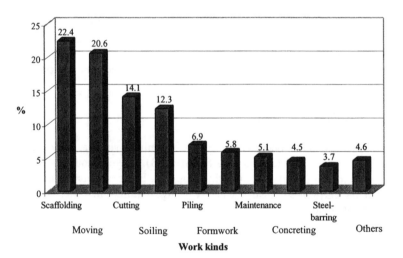

Figure 31.4 Activities

31.4.1.6 Recommendations

Based on the discussion above, the following are recommendations for improving the safety conditions on construction sites.

- The government departments should pay more attention to safety improvement on construction sites. It is necessary to organise activities not just the technical training but also seminars on safety for all levels and all ages of construction professionals.
- Safety management of the industry needs to be strengthened. While legislation of legal regulations would take time to be effective, a sound safety culture must be promoted in the construction industry. As an incentive, the rewards and punishment of contracting firms should link with their safety records.
- Self-regulation should be emphasized in taking precautions on safety. All parties should understand that safety is one of the most important prerequisites in pursuing profit.
- Provide personal protective equipment such as safety helmets, belts, boots, gloves, goggles, ear-plugs, etc. and enhance the sharing of knowledge of advanced technology and experience on safety.
- Develop and implement accident insurance systems for workers on sites. This is the proper system to provide compensation to workers in case of an accident. The premium paid by the contracting firms would be affected by their safety performance and can be an incentive to improve safety conditions on site.

31.4.1.7 Concluding remarks

The survey gives an understanding of the safety issues on construction sites in Southern China. The occurrence of an accident has no specific relation to the age of the workers. Thus, education and training should be provided to workers irrespective of their age. The accidents are usually involved with being struck by objects, falling from height, tumbling, cutting, straining and electrocuting. The dangerous places on site are found near to hoists, holes and wells, scaffoldings, cranes and ladders, etc. The activities in accidents are always involved with scaffolding work, earthwork, piling, building framework construction, concreting work and mechanical plant maintenance.

In order to reduce and prevent the accidents, it is recommended that measures including strengthening the role of the government departments, executing self-discipline in the industry, educating the workers, promoting safety awareness, encouraging self-regulation, providing personal protective equipment, and implementing mandatory accident insurance systems in the construction industry are necessary.

31.4.2 A major accident

A severe accident occurred at a bridge construction site during concrete pouring in December 1996, near Ru Yuan Town, Guangdong, China. It was a box arch bridge of reinforced concrete construction in a highway engineering project. The beams

forming the arch were twisted and collapsed totally to the bottom of the valley.
Thirty-four workers died and twenty-seven severely injured in the accident.

It was one of the most severe construction failures in recent years in China.
The original drawings and related information of the bridge were kept confidential
by the Authority. Even visiting the site within one month after the failure
occurrence was strictly forbidden. The authors made great efforts to visit the site
twice. The first visit was made one week and the second one was three months
after the accident. Information was collected by visiting the site and interviewing
the workers. The possible causes for the failure were identified as poor
management, lack of safety control and instability of falsework.

31.4.3   Bridge construction

This bridge was one of the two main arches in a joint-venture project which was
scheduled to be completed by March 1997. The bridge was 12m in width and
163m long. The top part of the 110m centre span would be 74m above the valley.
The height of the arch was about 17m which means the ratio of height to span was
1/6.5. Details of the bridge are shown in Figure 31.5.

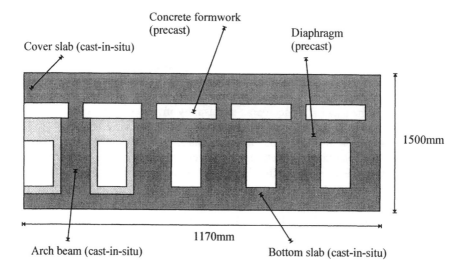

Figure 31.5 Elevation and plan of the arch bridge

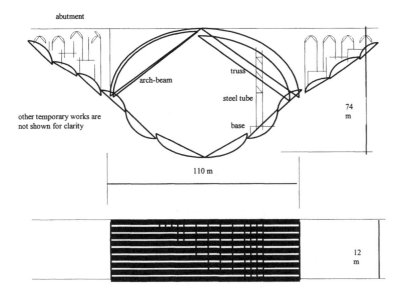

abutment

truss

arch-beam

steel tube

base

other temporary works are
not shown for clarity

74
m

110 m

12
m

Figure 31.6 Cross-section of the box arch

The bridge deck comprised nine arch beams to form eight boxes with the bottom slab, cover slab and diaphragms (Figure 31.6). The reinforced concrete arch beams were cast *in situ*. The precast concrete diaphragms spaced at a defined distance were used to increase the stability and stiffness of the arch. Concrete was mixed by three mixers near by. A steel truss tower had been erected near each abutment to support steel ropes which were attached with two trolleys and suspending hooks for transportation of materials across the valley.

The procedures of the arch bridge construction are shown in Figure 31.7. The failure happened during concrete pouring of the top part of the arch. The concrete abutments of both sides had been completed earlier and remained intact after the collapse.

Falsework erection was one of the key activities in the arch bridge construction. Timber and lattice frames to form the arch were supported by columns made up of steel tubes and small trusses. According to the design requirements concrete should be poured in a continuous process. Workers were divided in groups for the non-stop concreting work. A couple of days before the accident, displacement of the forms had been noticed. The problem was solved by raising and restoring the forms at their positions with the suspending hooks. However, this operation might have buried the root of the tragedy.

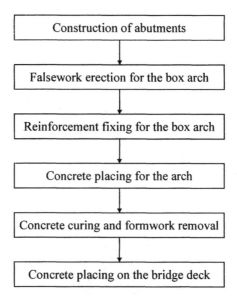

Figure 31.7 Flow diagram of the bridge construction

After several days' hard work, there was an area of less than 10m long at the top and near the centre of the arch yet to be concreted. On 20 December 1996, over one hundred workers and technicians were scattered on the top of the bridge for concrete pouring. At about 9:10 a.m., a labourer standing at the west edge of the newly placed concrete heard a strange noise under his feet. Then, the west part of the deck suddenly wisted and crashed with the arch to the bottom of the valley.

31.4.4   Causes of the failure

Falsework failure occurs very often at the time of concrete pouring due to its biggest load during construction. The design of the arch bridge was adequate according to the official results released. The following are the main causes which attributed to the collapse during construction.

1) Falsework failure
    *   The falsework consisted of a variety of components which were made of different materials and shapes. Any displacement, loosening, breaking in any part of the system would lead to the redistribution of stress and falsework failure.
    *   Settlement of temporary foundations

It was observed that the natural surface of the valley was of highly weathered rocks. A few places were selected as the bases for falsework erection. The use of the 700mm × 700mm concrete blocks as temporary foundations appeared to be over-simplified.

• Joint failure of the falsework
  The falsework was supported by trusses (600mm × 600mm) which were made of angle steel (50mm × 50mm) and U steel (80mm × 40mm) and steel tubes (600mm diameter). Steel tubes were placed on the base without adequate connection. The trusses were linked by the tubes and the connection between them was by welding four pieces of steel bars. Small steel tubes were also used as the supporting falsework.

• Failure of the falsework
  Erection based on the used and insufficient materials was popular for contractors so as to reduce the cost in construction. As a result, use of inadequate materials may lead to the failure of falsework.

• Instability of the falsework
  The ratios of length to section size of the supporting columns were large and this could easily affect the stability of the falsework. Horizontal forces during construction could trigger the collapse of the falsework.

2) Poor management on site
  It was reported that there had been a lack of emphasises of quality assurance and safety control on site. The contractor had ignored the warning by the displacement of forms which happened a few days ago before the collapse. As it was close to the Chinese new year, the contractor and the workers wanted to complete the work quickly so that they could go home before the festival. Taking chances and cutting corners could lead to failures.

According to the official announcement on 1 November 1997 by China's Central TV Station, the failure was due to inadequate design of the falsework.

31.5   CONCLUSION

This chapter has reviewed the supervision of works for quality and safety on sites in China. China has developed rules and regulations regarding the standards of quality of construction works and safety on sites. However, many poor quality and safety cases have been reported from time to time. Thus, the effectiveness of implementation of the supervision systems adopted in China requires further investigation.

31.6   REFERENCES

Ministry of Construction, China, 1999, *Standard of Construction Safety Inspection*, JGJ59-99, Mar, 49 pp.

Poon, S.W., Zhang, Yan and Wang, Y.S., 1998, Collapse of an arch bridge during construction.   In *Proceedings of the Sixth East-Pacific Conference on Structural Engineering and Construction*, Jan, Taiwan, pp. 1017-1022.

Qin, C.F., 1998, Construction safety management in China. In *Proceedings of the First Conference of Safety and Health Organizations in the Asia-Pacific Region Construction Industry*, June, Tokyo, Japan.

Shenzhen Construction Safety Supervising Station, 1998, *Construction Safety Regulations in Shenzhen*, Apr., 15 pp.

Shenzhen Construction Supervising Engineers Association, 1998, *Construction Supervision Regulations in Shenzhen*, Oct., 161 pp.

The National Sector Standards of the Peoples' Republic of China, 1999, *Standards of Construction Safety Inspection*, JGJ59-99, 16 pp.

Wang, Y.S., Zhang, Y., Poon, S.W. and Huang, H.Y., 2002, A study of construction site accident statistics.   In *Proceedings of Triennial Conference CIB W099: Implementation of Safety and Health on Construction Sites*, May, Hong Kong, pp. 223-227.

Index